Brownian Motion and Classical Potential Theory

This is a volume in
PROBABILITY AND MATHEMATICAL STATISTICS
A Series of Monographs and Textbooks

Editors: Z. W. Birnbaum and E. Lukacs

A complete list of titles in this series appears at the end of this volume.

Brownian Motion and Classical Potential Theory

SIDNEY C. PORT and CHARLES J. STONE

Department of Mathematics
University of California, Los Angeles
Los Angeles, California

ACADEMIC PRESS New York San Francisco London 1978

A Subsidiary of Harcourt Brace Jovanovich, Publishers

ACADEMIC PRESS, INC.
111 Fifth Avenue, New York, New York 10003

United Kingdom Edition published by
ACADEMIC PRESS, INC. (LONDON) LTD.
24/28 Oval Road, London NW1 7DX

Library of Congress Cataloging in Publication Data

Port, Sidney C
Brownian motion and classical potential theory.

(Probability and mathematical statistics)
Bibliography: p.
1. Brownian movements. 2. Potential, Theory of.
3. Markov processes. I. Stone, Charles Joel,
Date joint author. II. Title.
QC184.P67 519.2'82 78-6772
ISBN 0-12-561850-6

AMS (MOS) 1970 Subject Classifications: 60J45, 60J65

PRINTED IN THE UNITED STATES OF AMERICA

Contents

Preface

The important and beautiful connection between Brownian motion and classical potential theory was first investigated by Kakutani [1], Kac [1], and Doob [1]. Hunt [2] then showed that potential theory could be developed by probabilistic methods for a large class of transient Markov processes. There is now a large literature on potential theory for transient Markov processes, including books by Dynkin [1], Itô and McKean [2], Meyer [1], and Blumenthal and Getoor [1]. Recurrent potential theory was first developed for random walks by Spitzer and for Markov chains by Kemeny and Snell. Further work in this area has been done by a number of people, including the present authors Port and Stone [1] and Stone [1] (see books by Spitzer [1], Kemeny, Snell, and Knapp [1], and Revuz [1]).

In spite of the large amount of written material on probabilistic potential theory, the connection between Brownian motion and classical potential theory remains largely inaccessible to nonexperts. The purpose of this book is to give an account of this connection, accessible to readers familiar with real variable theory and probability theory as usually developed at the graduate level (a knowledge of Chung [1] should suffice). No prior knowledge of either Brownian motion or classical potential theory is assumed.

In the first three chapters of the book the emphasis is on developing properties of Brownian motion with results from potential theory mainly being introduced as tools when needed. In the last three chapters the emphasis is on using Brownian motion to obtain the main theorems of classical potential theory. Harmonic functions and the Dirichlet problem are studied in Chapter 4, and superharmonic functions are studied in Chapter 5. The transient potential theory of Green potentials, including Newtonian potentials as a special case, and the recurrent potential theory of logarithmic potentials are thoroughly developed in Chapter 6. This probabilistic approach to classical potential theory is self-contained and often leads to simpler and more intuitive proofs than does the purely analytic approach. But the reader who is not already familiar with the analytic approach would get a much better overall perspective of potential theory by reading portions of Kellogg [1], Brelot [8], Helms [1], and Landkof [1] along with the material in the last three chapters of this book.

In order to keep the level of the book as elementary as possible hitting times are considered only for F_σ sets (i.e., increasing limits of compact sets), and a corresponding restriction is made in developing the potential theory. The reader who wishes to see how probabilistic techniques can be combined with the capacitibility theorem of Choquet to handle arbitrary Borel sets should refer to the book by Blumenthal and Getoor.

We wish to thank Thomas Liggett and Ronald Getoor for reading portions of the manuscript and for their helpful suggestions. Richard Durrett carefully read the entire manuscript. His many detailed comments led to substantial improvement in the exposition. Finally, it is our pleasure to thank Elaine Barth for her excellent typing, and especially for her patience and sympathy while making numerous alterations to the "final" draft. A portion of the research related to writing this book was supported by NSF Grant No. MPF 72-04591.

Glossary of Notation

Set Theory

A^c	complement of A; $A^c = \{x : x \notin A\}$
$A \setminus B$	$A \cap B^c$
$A_n \uparrow A$	$A_1 \subset A_2 \subset \cdots$ and $\bigcup_n A_n = A$
$A_n \downarrow A$	$A_1 \supset A_2 \supset \cdots$ and $\bigcap_n A_n = A$
I_A	indicator function of A; $I_A(x) = 1$ if $x \in A$ and $I_A(x) = 0$ if $x \notin A$
$f \circ g$	composition of f and g; $f \circ g(x) = f(g(x))$
	used to denote a function of the indicated variable; e.g., $\log(y + e^{\,})$ denotes the function f defined by $f(x) = \log(y + e^x)$.

Real Line

$\mathbf{R}$	real line
$x \vee y$	maximum of x and y
$x \wedge y$	minimum of x and y
x^+	positive part of x; $x^+ = x \vee 0$
x^-	negative part of x; $x^- = (-x) \vee 0$
$\|x\|$	absolute value of x

Euclidean Space

$\mathbf{R}^d$	d-dimensional Euclidean space
$x \cdot y$	inner product of x and y; $x \cdot y = \Sigma_1^d x_i y_i$
$\\|x\\|$	norm of x; $\\|x\\| = (\Sigma_1^d x_i^2)^{1/2} = (x \cdot x)^{1/2}$

$c + bA$	$\{c + ba : a \in A\}$ for $c \in \mathbb{R}^d$ and $b \in \mathbb{R}$
$d(x, A)$	distance from x to A; $d(x, A) = \inf[\|a - x\| : a \in A]$
∂A	boundary of A; $\partial A = \{x : d(x, A) = d(x, A^c) = 0\}$
$\bar{A}$	closure of A; $\bar{A} = A \cup \partial A$
$\mathring{A}$	interior of A; $\mathring{A} = A \setminus \partial A$
$\|A\|$	Lebesgue measure of A
$\mathscr{B}$	collection of F_σ sets; $B \in \mathscr{B}$ if and only if B is an increasing limit of compact sets
$B_r(c)$	closed ball with center c and radius r; $B_r(c) = \{x : \|x - c\| \leq r\}$
B_r	$B_r(0)$
$\mathring{B}_r(c)$	open ball with center c and radius r; $\mathring{B}_r(c) = \{x : \|x - c\| < r\}$
$\mathring{B}_r$	$\mathring{B}_r(0)$
$S_r(c)$	sphere with center c and radius r; $S_r(c) = \{x : \|x - c\| = r\}$
S_r	$S_r(0)$
$\sigma_r(c)$	uniform probability distribution on $S_r(c)$
σ_r	$\sigma_r(0)$
δ_c	probability distribution concentrated at c

Brownian Motion and Potential Theory

CHAPTER 1

Ω	probability space; Ω = collection of continuous functions from $[0, \infty)$ to $\mathbb{R}^d$
ω	element of Ω
$X(t)$	position of Brownian motion path at time t; $X(t) = X(t, \omega) = \omega(t)$
θ_t	shift transformation on Ω; $\theta_t\omega(s) = \omega(s + t)$ and $X(s, \theta_t\omega) = X(s + t, \omega)$
$\sigma(Y_\alpha, \alpha \in A)$	smallest σ-field on Ω for which the functions Y_α, $\alpha \in A$, are measurable
$\mathfrak{F}_\infty$	$\sigma(X(t), t \geq 0)$
$\mathfrak{F}_t$	$\sigma(X(s), 0 \leq s \leq t)$
$\mathfrak{F}_{t+}$	$\bigcap_{s>t} \mathfrak{F}_s$
τ	stopping time; $0 \leq \tau \leq \infty$ and $\{\tau \leq t\} \in \mathfrak{F}_{t+}$ for all $t > 0$
τ_B	hitting time of B; $\tau_B = \inf[t > 0 : X(t) \in B]$
$\mathfrak{F}_{\tau+}$	σ-field associated with stopping time τ; $\mathfrak{F}_{\tau+}$ = collection of sets A such that $A \cap \{\tau \leq t\} \in \mathfrak{F}_{t+}$ for all $t > 0$
$p(t, \cdot)$	normal density on $\mathbb{R}^d$ given by $p(t, x) = (2\pi t)^{-d/2} \exp(-\|x\|^2/2t)$
$p(t, x, y)$	$p(t, y-x)$
P_x	probability measure on Ω corresponding to Brownian motion starting at x; $P_x(X(t) \in A) = \int_A p(t, x, y)\, dy$
E_xY	expectation of Y relative to P_x; $E_x Y = \int Y(\omega)P_x(d\omega)$
$E_x(Y; A)$	$\int_A Y(\omega)P_x(d\omega)$

CHAPTER 2

B^r	points which are regular for B; $B^r = \{x : P_x(\tau_B = 0) = 1\}$
$p^t(x, A)$	$P_x(X(t) \in A) = \int_A p(t, x, y)\, dy$
L_B	last exit time from B; $L_B = \sup[t > 0 : X(t) \in B]$ if $\tau_B < \infty$ and $L_B = 0$ if $\tau_B = \infty$
$h_B(x, \cdot)$	hitting distribution of B for Brownian motion starting at x; $h_B(x, A) = P_x(\tau_B < \infty, X(\tau_B) \in A)$
$r_B(t, x, y)$	$E_x(p(t-\tau_B, X(\tau_B), y); \tau_B < t)$
$q_B(t, x, y)$	$p(t, x, y) - r_B(t, x, y)$
$q_B^t(x, A)$	$P_x(\tau_B > t, X(t) \in A) = \int_A q_B(t, x, y)\, dy$

T_D	exit time of D; $T_D = \tau_{D^c} = \inf[t > 0; X(t) \notin D]$
$Q_D(t, x, y)$	$q_{D^c}(t, x, y)$ restricted to $x, y \in D$
$Q^t_D(x, A)$	$P_x(T_D > t, X(t) \in A) = q^t_{B^c}(x, A) = \int_A Q_D(t, x, y)\, dy$ restricted to $x \in D$ and $A \subset D$
g^λ	λ-potential kernel; $g^\lambda(x) = \int_0^\infty e^{-\lambda t} p(t, x)\, dt$
$g^\lambda(x, y)$	$g^\lambda(y-x) = \int_0^\infty e^{-\lambda t} p(t, x, y)\, dt$
$g^\lambda \mu$	λ-potential of μ; $g^\lambda \mu(x) = \int g^\lambda(x, y)\mu(dy)$
$g^\lambda_B(x, y)$	$\int_0^\infty e^{-\lambda t} q_B(t, x, y)\, dt$
$h^\lambda_B(x, A)$	$E_x(e^{-\lambda \tau_B}; \tau_B < \infty$ and $X(\tau_B) \in A)$
μ^λ_B	λ-equilibrium measure of B; $\mu^\lambda_B(A) = \int \lambda\, dy\, h^\lambda_B(y, A)$
$C^\lambda(B)$	λ-capacity of B; $C^\lambda(B) = \mu^\lambda_B(\mathbf{R}^d)$

CHAPTER 3

$g_B(x, y)$	$\int_0^\infty q_B(t, x, y)\, dt \quad (d \geq 3)$
g	Newtonian potential kernel; $g(x) = \int_0^\infty p(t, x)\, dt = (2\pi^{d/2})^{-1}\Gamma(d/2-1)\|x\|^{2-d}$
$g(x, y)$	$g(y-x) = \int_0^\infty p(t, x, y)\, dt$
$g\mu$	Newtonian potential of μ; $g\mu(x) = \int g(x, y)\mu(dy)$
μ_B	equilibrium measure of B; μ_B is the unique measure concentrated on B^r such that $g\mu_B = 1$ on B^r
$C(B)$	capacity of B; $C(B) = \mu_B(\mathbf{R}^d) \quad (d \leq 2)$
$k^\lambda(x)$	$(d = 2)\ g^\lambda(x) - g^\lambda(u)$, where $\|u\| = 1$ $(d = 1)\ g^\lambda(x) - g^\lambda(0)$
$k^\lambda(x, y)$	$k^\lambda(y-x)$
k	$(d = 2)$ logarithmic potential kernel; $k(x) = \pi^{-1}\log(1/\|x\|)$ $(d = 1)$ linear potential kernel; $k(x) = -\|x\|$
$k(x, y)$	$k(y-x)$
$k\mu$	logarithmic $(d = 2)$ or linear $(d = 1)$ potential of μ; $k\mu(x) = \int k(x, y)\mu(dy)$
$W^\lambda_B(x)$	$(d = 2)\ g^\lambda(u)(1 - E_x(e^{-\lambda\tau_B}))$, where $\|u\| = 1$ $(d = 1)\ g^\lambda(0)(1 - E_x(e^{-\lambda\tau_B}))$
$W_B(x)$	$\lim_{\lambda\to 0} W^\lambda_B(x)$
μ_B	equilibrium measure of B; μ_B is the unique probability measure concentrated on B^r such that $k\mu_B$ is constant on B^r
$R(B)$	Robin constant of B; $k\mu_B = R(B)$ on B^r

CHAPTER 4

Δf	Laplacian of f; $\Delta f(x) = \Sigma_1^d\, \partial^2 f(x)/\partial x_j^2$
$\tilde{\Delta} f$	generalized Laplacian of f; $\tilde{\Delta} f(x) = \lim_{r\to 0} r^{-1} 2d[\int f(y)\sigma_r(x, dy) - f(x)]$
x^*	inversion (relative to $S_\rho(b)$); $x^* = b + \dfrac{\rho}{\|x-b\|^2}(x-b)$
D^*	$\{x^* : x \in D\}$
f^*	Kelvin transformation of f (relative to $S_\rho(b)$); $f^*(x^*) = \dfrac{\rho^{d-2}}{\|x^*-b\|^{d-2}} f(x)$
$H_D(x, \cdot)$	exit distribution (harmonic measure) of D for Brownian motion starting at $x \in D$; $H_D(x, A) = h_{D^c}(x, A) = P_x(T_D < \infty, X(T_D) \in A)$
G_D	Green function of D; $G_D(x, y) = g_{D^c}(x, y)$ for $x, y \in D$

CHAPTER 5

$\underline{f}$	lower regularization of f on D; $\underline{f}(x) = \lim_{\epsilon \to 0} \inf[f(y) : y \in D \text{ and } 0 \le \|y-x\| \le \epsilon]$ for $x \in \bar{D}$ and $\underline{f}(\infty) = \lim_{r \to \infty} \inf[f(y) : y \in D \text{ and } \|y\| \ge r]$ if D is unbounded

CHAPTER 6

$\mu\vert_B$	restriction of measure μ to B; $\mu\vert_B(A) = \mu(A \cap B)$
$\mu^+, \mu^-, \vert\mu\vert$	positive part, negative part and total variation of signed measure μ; $\mu = \mu^+ - \mu^-$ and $\vert\mu\vert = \mu^+ + \mu^-$

Green Potentials

$G_D\mu$	Green potential of measure μ on D; $G_D\mu = \int G_D(\cdot, y)\mu(dy)$
$\mathscr{M}_D$	collection of measures μ on D such that $G_D\mu$ is superharmonic on D
$\mathscr{M}_D(B)$	collection of measures $\mu \in \mathscr{M}_D$ which are concentrated on B
$h_{B,D}(x, \cdot)$	hitting distribution of B for Brownian motion on D starting at $x \in D$; $h_{B,D}(x, A) = h_{B \cup D^c}(x, A) = P_x(\tau_B < T_D, X(\tau_B) \in A)$ for $A \subset D$
$\mu h_{B,D}$	measure on D given by $\mu h_{B,D}(A) = \int \mu(dy) h_{B,D}(y, A)$
$I_D(\mu)$	energy of measure μ on D; $I_D(\mu) = \int G_D\mu \, d\mu$
$\mathscr{E}_D$	collection of signed Radon measures μ on D such that $\vert\mu\vert$ has finite energy
$\mathscr{E}_D(B)$	collection of elements $\mu \in \mathscr{E}_D$ such that $\vert\mu\vert$ is concentrated on B
$\mathscr{E}_D^+, \mathscr{E}_D^+(B)$	collection of (nonnegative) measures in $\mathscr{E}_D$, $\mathscr{E}_D(B)$, respectively
(μ, ν)	mutual energy of $\mu, \nu \in \mathscr{E}_D$; $(\mu, \nu) = \int G_D\mu \, d\nu$
$\\|\mu\\|$	$\sqrt{(\mu, \mu)}$
$\mu_{B,D}$	equilibrium measure of equilibrium set B; $\mu_{B,D}$ is the unique measure $\mu \in \mathscr{M}_D(B^r)$ such that $G_D\mu = 1$ on $B^r \cap D$
$C_D(B)$	capacity of B; $C_D(B) = \mu_{B,D}(D)$ if B is an equilibrium set and $C_D(B) = \infty$ otherwise
$\underline{C}_D(B)$	inner capacity of B; $\underline{C}_D(B) = \sup[C_D(A) : A$ is a compact subset of $B]$
$\bar{C}_D(B)$	outer capacity of B; $\bar{C}_D(B) = \inf[C_D(U) : U$ is an open subset of D containing $B]$
$L_{B,D}$	last exit time from B relative to D; $L_{B,D} = \sup[t < T_D : X(t) \in B]$ on $\{\tau_B < T_D\}$ and $L_{B,D} = 0$ on $\{\tau_B \ge T_D\}$

Logarithmic Potentials

$\mathscr{M}$	collection of finite measures on $\mathbf{R}^2$ having compact support
$\mathscr{M}(B)$	collection of measures $\mu \in \mathscr{M}$ which are concentrated on B
μh_B	measure given by $\mu h_B(A) = \int \mu(dy)\, h_B(y, A)$
$I(\mu)$	energy of $\mu \in \mathscr{M}$; $I(\mu) = \int k\mu \, d\mu$
$\mathscr{E}$	collection of signed measures μ on $\mathbf{R}^2$ such that $\vert\mu\vert \in \mathscr{M}$ and $\vert\mu\vert$ has finite energy
$\mathscr{E}(B)$	collection of elements $\mu \in \mathscr{E}$ such that $\vert\mu\vert$ is concentrated on B
$\mathscr{E}^+, \mathscr{E}^+(B)$	collection of (nonnegative) measures in $\mathscr{E}$, $\mathscr{E}(B)$, respectively
(μ, ν)	inner product on $\mathscr{E}(C)$; $(\mu, \nu) = \int k\mu \, d\nu - \mu(\mathbf{R}^2)\nu(\mathbf{R}^2)R(C_1)$ where C_1 is a fixed compact set containing the nonpolar compact set C in its interior
$\\|\mu\\|$	$\sqrt{(\mu, \mu)}$

Chapter 1

Brownian Motion as a Strong Markov Process

In this chapter the existence of Brownian motion as a particular stochastic process having continuous paths in d-dimensional Euclidean space is verified. The strong Markov property is shown to hold for this process.

1. Definition of Brownian Motion

For $t > 0$ let $p(t, \cdot)$ denote the normal density on d-dimensional Euclidean space $\mathbb{R}^d$ defined by

$$p(t, y) = (2\pi t)^{-d/2} \exp(-\|y\|^2/2t), \qquad y \in \mathbb{R}^d,$$

and set $p(t, x, y) = p(t, y - x)$ for $x, y \in \mathbb{R}^d$. It follows from the uniqueness theorem for characteristic functions that if U and V are independent $\mathbb{R}^d$-valued random variables having normal densities $p(s, \cdot)$ and $p(t, \cdot)$, respectively, where $s, t > 0$, then $U + V$ has the normal density $p(s + t, \cdot)$. Consequently the densities $p(t, x, y)$ satisfy the semigroup property

$$(1) \qquad p(s + t, x, y) = \int p(s, x, z)p(t, z, y)\, dz, \qquad s, t > 0 \quad \text{and} \quad x, y \in \mathbb{R}^d.$$

Let Ω denote the collection of all continuous functions (paths) from $[0, \infty)$ to $\mathbb{R}^d$. Set $X(t) = X(t, \omega) = \omega(t)$ for $t \geq 0$ and $\omega \in \Omega$. Let $\mathfrak{F}_\infty = \sigma(X(t), t \geq 0)$ denote the smallest σ-field on Ω such that $X(t)$ is $\mathfrak{F}_\infty$-measurable for all

$t \geq 0$. An *event* is a set in $\mathfrak{F}_\infty$. A *random variable* is an $\mathfrak{F}_\infty$-measurable function. Let P be a probability distribution on Ω. Then P determines the joint distribution of $X(t_1), \ldots, X(t_n)$ for $0 \leq t_1 < \cdots < t_n$.

Let $x \in \mathbb{R}^d$ and let P_x be a probability distribution on Ω. The stochastic process $X(t)$, $t \geq 0$, is called *d-dimensional Brownian motion starting at x* if the random variables $X(t_1), \ldots, X(t_n)$ have joint density

$$(2) \quad p(t_1, x, x_1) p(t_2 - t_1, x_1, x_2) \cdot \cdots \cdot p(t_n - t_{n-1}, x_{n-1}, x_n), \qquad x_1, \ldots, x_n \in \mathbb{R}^d,$$

for $n \geq 1$ and $0 < t_1 < \cdots < t_n$. Since $X(t) = X(t, \omega) = \omega(t)$ for $t \geq 0$, Brownian motion automatically has continuous paths, from which it follows easily that $P_x(X(0) = x) = 1$. It is left to the reader to show that if $X(t)$, $t \geq 0$, is Brownian motion starting at x, then $X(t) - X(s)$ has density $p(t - s, \cdot)$ for $0 \leq s < t$ and $X(t_1), X(t_2) - X(t_1), \ldots, X(t_n) - X(t_{n-1})$ are independent for $n \geq 1$ and $0 < t_1 < \cdots < t_n$.

Brownian motion gets its name from the corresponding physical process reported on in 1828 by Robert Brown, an English botanist; he observed that pollen grains suspended in water perform a continuous swarming motion. Einstein in 1905, unaware of the accumulating experimental data, derived Brownian motion from statistical mechanical considerations (see Einstein [1] and Nelson [1] for interesting historical discussions). Slightly earlier in 1900 Bachelier [1] used Brownian motion to model the fluctuations of prices of such securities as futures and options and to study various methods of hedging. Brownian motion was first given the following mathematical formulation by Wiener [1].

Theorem 1.1. *For each $x \in \mathbb{R}^d$ there is a unique probability distribution P_x corresponding to Brownian motion starting at x.*

The proof is carried out through a number of lemmas.

Lemma 1.2. *Let $Y_1, \ldots, Y_n$ be independent $\mathbb{R}^d$-valued random variables on a probability space such that each Y_j has the same distribution as $-Y_j$. Set $S_j = Y_1 + \cdots + Y_j$ for $1 \leq j \leq n$. Then*

$$P\left(\max_{1 \leq j \leq n} \|S_j\| > a\right) \leq 2P(\|S_n\| > a), \qquad a \geq 0.$$

Proof. Choose $a \geq 0$ and let A_j denote the event $\{\|S_j\| > a$ and $\|S_i\| \leq a$ for $1 \leq i < j\}$. Then $A_1, \ldots, A_n$ are disjoint and $\bigcup_1^n A_j = \{\max_{1 \leq j \leq n} \|S_j\| > a\}$. Let $x \cdot y$ denote the usual inner product of $x, y \in \mathbb{R}^d$. If $\|S_j\| > a$ and $(S_n - S_j) \cdot S_j \geq 0$, then

$$\|S_n\|^2 = \|S_j\|^2 + \|S_n - S_j\|^2 + 2(S_n - S_j) \cdot S_j > a^2.$$

Thus

$$A_j \cap \{(S_n - S_j) \cdot S_j \geq 0\} \subset A_j \cap \{\|S_n\| > a\}.$$

It follows from the assumptions of independence and symmetry that

$$P(A_j \cap \{(S_n - S_j) \cdot S_j > 0\}) = P(A_j \cap \{(S_n - S_j) \cdot S_j < 0\}).$$

Consequently

$$P(A_j) \leq 2P(A_j \cap \{(S_n - S_j) \cdot S_j \geq 0\}) \leq 2P(A_j \cap \{\|S_n\| > a\}).$$

Therefore

$$\begin{aligned} P\left(\max_{1 \leq j \leq n} \|S_j\| > a\right) &= \sum_{j=1}^{n} P(A_j) \\ &\leq 2 \sum_{j=1}^{n} P(A_j \cap \{\|S_n\| > a\}) \\ &\leq 2P(\|S_n\| > a), \end{aligned}$$

as desired.

Choose $x \in \mathbb{R}^d$ and set $T = \{2^{-m}k : k, m = 0, 1, 2, \ldots\}$. It follows from the semigroup property (1) and the Kolmogorov extension theorem (stated, for example, in Chung [1]) that there is a probability space $(\Omega', \mathfrak{F}, P)$ and $\mathbb{R}^d$-valued random variables $Y(t)$, $t \in T$, defined on that space having joint densities given by (2) and with $Y(0) = x$. The next three lemmas will be used to show that $Y(t)$, $t \in T$, depends continuously on t except on a set of points in Ω' having probability zero.

Lemma 1.3. $\lim_n 2^n P(\|Y(2^{-n}) - x\| > \delta) = 0, \quad \delta > 0.$

Proof. Observe that for a positive number $t \in T$

$$\begin{aligned} P(\|Y(t) - x\| > \delta) &= \int_{\|y\| > \delta} p(t, y)\, dy \\ &= t^{-d/2} \int_{\|y\| > \delta} p(1, t^{-1/2} y)\, dy \\ &= \int_{\|y\| > t^{-1/2}\delta} p(1, y)\, dy \\ &\leq \delta^{-2} t \int_{\|y\| > t^{-1/2}\delta} \|y\|^2 p(1, y)\, dy. \end{aligned}$$

Thus

$$2^n P(\|Y(2^{-n}) - x\| > \delta) \leq \delta^{-2} \int_{\|y\| > \delta 2^{n/2}} \|y\|^2 p(1, y)\, dy.$$

Since the right-hand side of this inequality approaches zero as $n \to \infty$, the lemma is proven.

Set $U_n = \sup[\|Y(t) - x\| : t \in T \text{ and } t \le 2^{-n}]$.

Lemma 1.4. $\lim_n 2^n P(U_n > \delta) = 0, \quad \delta > 0.$

Proof. For $m \ge n$ set

$$U_{nm} = \max_{1 \le j \le 2^{m-n}} \|Y(2^{-m}j) - x\|.$$

By Lemma 1.2

$$P(U_n > \delta) = \lim_m P(U_{nm} > \delta) \le 2P(\|Y(2^{-n}) - Y(0)\| > \delta).$$

The desired result now follows from Lemma 1.3.

Let N be a positive integer and set $V_n = \sup[\|Y(t) - Y(s)\| : (s, t) \in T \cap [0, N] \text{ and } |t - s| \le 2^{-n}]$.

Lemma 1.5. $P(\lim_n V_n = 0) = 1.$

Proof. Set $V_{nm} = \sup[\|Y(t) - Y(2^{-n}m)\| : t \in T \cap [2^{-n}m, 2^{-n}(m+1)]]$. It is easily seen that $V_n \le 3 \max_{0 \le m \le 2^n N} V_{nm}$. Since each V_{nm} is distributed as U_n, it follows from Lemma 1.4 that for $\delta > 0$

$$\begin{aligned}\overline{\lim_n}\, P(V_n > 3\delta) &\le \overline{\lim_n}\, P\left(\max_{0 \le m \le 2^n N} V_{nm} > \delta\right) \\ &\le \lim_n 2^n N P(U_n > \delta) = 0.\end{aligned}$$

Thus V_n converges to zero in probability as $n \to \infty$. Since V_n is nondecreasing in n, the lemma holds as desired.

The proof of Theorem 1.1 is now easily completed. It follows from Lemma 1.5 that for each positive integer N, $Y(\cdot)$ is uniformly continuous on $T \cap [0, N]$ once an appropriate subset of Ω' having probability zero is deleted from Ω'. Thus $Y(t)$, $t \in T$, extends uniquely to $Y(t)$, $t \ge 0$, which is continuous in t for $\omega' \in \Omega'$. By the continuity of $p(t, y)$ in t the joint distributions of $Y(t)$, $t \ge 0$, are given by (2). Consider the map M from Ω' to Ω defined by $(M\omega')(t) = Y(t, \omega')$, $t \ge 0$. Let P_x be the distribution on $(\Omega, \mathfrak{F}_\infty)$ induced by M and $(\Omega', \mathfrak{F}, P)$. Then P_x corresponds to Brownian motion starting at x. The uniqueness of P_x follows from the fact that a probability measure on a field of sets has at most one extension to the probability measure on the induced σ-field. This completes the proof of the theorem.

From now on P_x corresponds to Brownian motion starting at x. Let $Y(t)$, $t \geq 0$, be $\mathbb{R}^d$-valued random variables on $(\Omega, \mathfrak{F}_\infty, P_x)$ which depend continuously on t and which induce the distribution P_y on $(\Omega, \mathfrak{F}_\infty)$ for some $y \in \mathbb{R}^d$. Then $Y(t)$, $t \geq 0$, can be considered as Brownian motion starting at y.

Many quantities associated with Brownian motion are most readily calculated by appealing to one of the following three invariance properties. These properties follow easily from the change of variable formula for multiple integrals, the details being left to the reader.

Let $X(t)$, $t \geq 0$, be Brownian motion starting at x.

Translation Invariance. If $a \in \mathbb{R}^d$, then $a + X(t)$, $t \geq 0$, is Brownian motion starting at $a + x$.

Scale Invariance. If $c > 0$, then $c^{-1/2}X(ct)$, $t \geq 0$, is Brownian motion starting at $c^{-1/2}x$.

Orthogonal Invariance. If Q is an orthogonal linear transformation on $\mathbb{R}^d$, then $QX(t)$, $t \geq 0$, is Brownian motion starting at Qx (in particular $-X(t)$, $t \geq 0$, is Brownian motion starting at $-x$).

2. Monotone Class Theorem

Let m be a nonnegative integer and let $\mathcal{G}_1$ denote the Borel subsets of $\mathbb{R}^m$ ($\mathbb{R}^0 = \{0\}$). Given a nonempty subset T of $[0, \infty)$, set $\mathcal{G}_2 = \sigma(X(t), t \in T)$. Let $\mathcal{G}$ denote the σ-field $\mathcal{G}_1 \times \mathcal{G}_2$ on $\mathbb{R}^m \times \Omega$. The following *monotone class theorem* will be used to verify that Brownian motion is a strong Markov process.

Theorem 2.1. *Let $\mathcal{V}$ be a vector space of $\mathcal{G}$-measurable functions on $\mathbb{R}^m \times \Omega$. Suppose that the following two conditions are satisfied:*

(i) *$f_0(x) \cdot f_1(X(t_1, \omega)) \cdot \cdots \cdot f_r(X(t_r, \omega))$ defines a function in $\mathcal{V}$ whenever f_0 is a bounded continuous function on $\mathbb{R}^m$, $t_1, \ldots, t_r \in T$, and $f_1, \ldots, f_r$ are bounded continuous functions on $\mathbb{R}^d$; and*

(ii) *$f_n \in \mathcal{V}$, $f_n \geq 0$, $f_n \uparrow f$, and f bounded imply that $f \in \mathcal{V}$. Then $\mathcal{V}$ contains all bounded $\mathcal{G}$-measurable functions on $\mathbb{R}^m \times \Omega$.*

Proof. Let $\mathcal{J}_2$ denote the collection of all "open intervals" in $\mathbb{R}^d$, i.e., sets of the form $\{(x_1, \ldots, x_d) : a_1 < x_1 < b_1, \ldots, a_d < x_d < b_d\}$. Let $\mathcal{J}_1$ denote the collection of open intervals in $\mathbb{R}^m$. Then $\mathcal{J}_1$ and $\mathcal{J}_2$ are closed under finite intersections and generate the Borel sets in $\mathbb{R}^m$ and $\mathbb{R}^d$, respectively.

Let $\mathscr{M}$ denote the collection of all subsets A of $\mathbb{R}^m \times \Omega$ such that $I_A \in \mathscr{V}$. Then $\mathscr{M}$ contains the collection $\mathscr{L}$ of sets L such that

$$I_L(x, \omega) = I_{J_0}(x) \cdot I_{J_1}(X(t_1, \omega)) \cdot \cdots \cdot I_{J_r}(X(t_r, \omega))$$

for $J_0 \in \mathscr{J}_1, J_1, \ldots, J_r \in \mathscr{J}_2$, and $t_1, \ldots, t_r \in T$. Note that $\mathscr{L}$ is closed under finite intersections and generates $\mathscr{G}$.

Consider subcollections $\mathscr{S}$ of $\mathscr{G}$ which satisfy the following three properties:

(iii) $\mathbb{R}^m \times \Omega \in \mathscr{S}$;
(iv) $A, B \in \mathscr{S}$ and $A \subset B$ imply that $B \backslash A \in \mathscr{S}$; and
(v) $A_n \in \mathscr{S}$ and $A_n \uparrow A$ imply that $A \in \mathscr{S}$.

Observe that $\mathscr{M}$ satisfies (iii)–(v). Let $\mathscr{S}_0$ denote the smallest subcollection of $\mathscr{G}$ containing $\mathscr{L}$ and satisfying these three properties. It will next be shown that $\mathscr{S}_0 = \mathscr{G}$. To this end set

$$\mathscr{S}_1 = \{A \in \mathscr{S}_0 : A \cap L \in \mathscr{S}_0 \quad \text{for all} \quad L \in \mathscr{L}\}.$$

Then $\mathscr{S}_1$ contains $\mathscr{L}$ and is easily seen to satisfy (iii)–(v). Thus $\mathscr{S}_1 = \mathscr{S}_0$. In other words $A \cap L \in \mathscr{S}_0$ for all $A \in \mathscr{S}_0$ and $L \in \mathscr{L}$. Set

$$\mathscr{S}_2 = \{A \in \mathscr{S}_0 : A \cap C \in \mathscr{S}_0 \quad \text{for all} \quad C \in \mathscr{S}_0\}.$$

Then $\mathscr{S}_2$ contains $\mathscr{L}$ by what was just shown. It is easily seen that $\mathscr{S}_2$ satisfies (iii)–(v) and hence that $\mathscr{S}_2 = \mathscr{S}_0$. Consequently $\mathscr{S}_0$ is closed under finite intersections. It now follows from (iii)–(v) that $\mathscr{S}_0$ is a σ-field. Since it contains $\mathscr{S}$ it must be true that $\mathscr{S}_0 = \mathscr{G}$ as was to be shown.

Since $\mathscr{M}$ satisfies (iii)–(v), $\mathscr{M} = \mathscr{G}$. Thus as $\mathscr{V}$ is a vector space it contains all $\mathscr{G}$-measurable simple functions. Hence by (ii) it contains all nonnegative bounded $\mathscr{G}$-measurable functions. Since it is a vector space it contains all bounded $\mathscr{G}$-measurable functions. This completes the proof of the theorem.

3. Markov Property

For $x \in \mathbb{R}^d$, P_x determines an expectation operator E_x acting on random variables Y according to the formula $E_x Y = \int Y(\omega) P_x(d\omega)$ (with the usual restriction on Y to prevent the integral from being of the indeterminate form $\infty - \infty$). The next result is required for the proof of the theorem which follows it.

Proposition 3.1. *Let f be a bounded Borel function on $(\mathbb{R}^d)^n$. Then $E_x f(X(t_1), \ldots, X(t_n))$ is jointly continuous in $x, t_1, \ldots, t_n$ for $x \in \mathbb{R}^d$ and $0 < t_1 < \cdots < t_n$.*

Proof. This result follows easily from the formula for the joint density of $X(t_1), \ldots, X(t_n)$ given by (1.2).

Set $\mathfrak{F}_t = \sigma(X(s), 0 \le s \le t)$ for $t \ge 0$ and note that $\mathfrak{F}_t = \sigma(X(s), 0 < s \le t)$ for $t > 0$. Also set $\mathfrak{F}_{t+} = \bigcap_{s>t} \mathfrak{F}_s$ for $t \ge 0$. Clearly $\mathfrak{F}_t \subset \mathfrak{F}_{t+}$. Actually $\mathfrak{F}_t$ is a proper subset of $\mathfrak{F}_{t+}$. For let U be a nonempty open subset of $\mathbb{R}^d$ and let A denote the event that $\inf[s > t : X(s) \in U] = t$. Then $A \in \mathfrak{F}_{t+}$ but $A \notin \mathfrak{F}_t$.

For $t \ge 0$ let θ_t be the *shift transformation* from Ω to itself defined by $(\theta_t\omega)(s) = \omega(s + t)$, $s \ge 0$. It is easily seen that θ_t is measurable for all $t \ge 0$. Moreover θ_0 is the identity transformation on Ω, $\theta_{s+t} = \theta_s \circ \theta_t$ for $s, t \ge 0$, and $X(s, \theta_t\omega) = X(s + t, \omega)$ for $\omega \in \Omega$ and $s, t \ge 0$.

Let Y be a bounded random variable. Then $E.Y$ is a Borel function on $\mathbb{R}^d$ by Proposition 3.1 and the monotone class theorem. Consequently $E_{X(t)}Y$ is $\mathfrak{F}_t$-measurable. The next result states that Brownian motion possesses the *Markov property*, i.e., that the conditional expectation of a function of future values of Brownian motion given past and present values depends only on the present value.

Theorem 3.2. *Let $x \in \mathbb{R}^d$ and $t \ge 0$ and let Y be a bounded random variable. Then $E_x(Y \circ \theta_t \mid \mathfrak{F}_{t+}) = E_{X(t)}Y$ a.s. (P_x).*

Proof. Let Z be a bounded $\mathfrak{F}_{t+}$-measurable random variable. Let $0 < t_1 < \cdots < t_n$, and let f be a bounded continuous function on $(\mathbb{R}^d)^n$. Choose $h \in (0, t_1)$. Let $0 < s_1 < \cdots < s_m \le t + h$ and let f_1 be a bounded continuous function on $(\mathbb{R}^d)^m$. Then by (1.2)

$$\begin{aligned} &E_x(f_1(X(s_1), \ldots, X(s_m))f(X(t + t_1), \ldots, X(t + t_n))) \\ &\quad = E_x(f_1(X(s_1), \ldots, X(s_m))E_{X(t+h)}f(X(t_1 - h), \ldots, X(t_n - h))). \end{aligned}$$

Thus by the monotone class theorem

$$E_x(Zf(X(t + t_1), \ldots, X(t + t_n))) = E_x(ZE_{X(t+h)}f(X(t_1 - h), \ldots, X(t_n - h))).$$

By Proposition 3.1 it is permissible to let $h \to 0$ and conclude that

$$E_x(Zf(X(t_1, \theta_t\omega), \ldots, X(t_n, \theta_t\omega))) = E_x(ZE_{X(t)}f(X(t_1), \ldots, X(t_n))).$$

Therefore $E_x(Z(Y \circ \theta_t)) = E_x(ZE_{X(t)}Y)$ by another application of the monotone class theorem. This is equivalent to the conclusion of the theorem.

Proposition 3.3. *Let $x \in \mathbb{R}^d$ and $t \ge 0$ and let Y be a bounded random variable. Then $E_x(Y \mid \mathfrak{F}_{t+}) = E_x(Y \mid \mathfrak{F}_t)$ a.s. P_x.*

Proof. By the monotone class theorem it suffices to verify the result for random variables Y of the form $Z(W \circ \theta_t)$, where Z and W are bounded random variables and Z is $\mathfrak{F}_t$-measurable. But in this case the result follows from Theorem 3.2 since $E_{X(t)}W$ is $\mathfrak{F}_t$-measurable and $\mathfrak{F}_t \subset \mathfrak{F}_{t+}$.

Proposition 3.4. *Let $x \in \mathbb{R}^d$ and $t \geq 0$ and let $A \in \mathfrak{F}_{t+}$. Then there is an $A_1 \in \mathfrak{F}_t$ such that $P_x(A \backslash A_1) = P_x(A_1 \backslash A) = 0$.*

Proof. Observe first that $I_A = E_x(I_A \mid \mathfrak{F}_{t+}) = E_x(I_A \mid \mathfrak{F}_t)$ a.s. (P_x). Set $A_1 = \{E_x(I_A \mid \mathfrak{F}_t) = 1\}$. Then $A_1 \in \mathfrak{F}_t$ and $I_{A_1} = I_A$ a.s. (P_x), which is equivalent to the conclusion of the proposition.

The next result is known as *Blumenthal's zero-one law.*

Proposition 3.5. *Let $A \in \mathfrak{F}_{0+}$. Then for each $x \in \mathbb{R}^d$, $P_x(A)$ equals zero or one.*

Proof. By the previous result it can be assumed that $A \in \mathfrak{F}_0$, so that $A = \{X(0) \in B\}$ for some Borel subset B of $\mathbb{R}^d$. Then $P_x(A) = P_x(X(0) \in B) = I_B(x)$, which equals zero or one.

For an illustration of Blumenthal's zero-one law, let $d \geq 2$, let U be a nonempty bounded open set such that ∂U is connected, let $x \in \partial U$ and let A denote the event that $\inf[s > 0: X(s) \in U] = 0$. Then $A \in \mathfrak{F}_{0+}$ and hence $P_x(A)$ equals zero or one. Actually $P_x(A) = 1$ if $d = 2$ (see Theorem 2.7.2 below); if $d \geq 3$ it is possible to make $P_x(A)$ equal zero by letting U come to a sharp point at x (see Proposition 3.3.7).

4. Stopping Times

In this section stopping times are defined and studied. The hitting time of a closed (or more generally F_σ) set is shown to be a stopping time.

Let τ be a nonnegative random variable which may assume the value ∞ with positive probability. This random variable is called a *stopping time* if $\{\tau \leq t\} \in \mathfrak{F}_{t+}$ for all $t \geq 0$. Obviously a nonnegative constant is a stopping time and if τ is a stopping time, so is $\tau + t$ for $t > 0$.

Proposition 4.1. *Let τ be a nonnegative random variable which may assume the value ∞. Then τ is a stopping time if and only if $\{\tau < t\} \in \mathfrak{F}_t$ for all $t > 0$.*

Proof. If τ is a stopping time, then

$$\{\tau < t\} = \bigcup_{n \geq 1} \left\{\tau \leq t - \frac{1}{n}\right\} \in \bigcup_{n \geq 1} \mathfrak{F}_{(t-1/n)+} \subset \mathfrak{F}_t.$$

Suppose conversely that $\{\tau < t\} \in \mathfrak{F}_t$ for all $t > 0$. Then

$$\{\tau \le t\} = \bigcap_{n \ge 1} \left\{\tau < t + \frac{s-t}{n}\right\} \in \mathfrak{F}_s, \qquad s > t \ge 0.$$

Thus $\{\tau \le t\} \in \bigcap_{s>t} \mathfrak{F}_t = \mathfrak{F}_{t+}$ for $t \ge 0$ and hence τ is a stopping time.

Proposition 4.2. *Let τ_n be stopping times. Then $\inf_n \tau_n$ and $\sup_n \tau_n$ are stopping times.*

Proof. Since $\{\inf_n \tau_n < t\} = \bigcup_n \{\tau_n < t\}$ and $\{\sup_n \tau_n \le t\} = \bigcap_n \{\tau_n \le t\}$, this result follows from Proposition 4.1 and the definition of a stopping time.

Corresponding to a stopping time τ is the σ-field $\mathfrak{F}_{\tau+}$ defined to be the collection of events A such that $A \cap \{\tau \le t\} \in \mathfrak{F}_{t+}$ for all $t \ge 0$ (that $\mathfrak{F}_{\tau+}$ is closed under complementation follows from the identity $A^c \cap \{\tau \le t\} = \{\tau \le t\} \backslash (A \cap \{\tau \le t\})$). It is easily seen that τ is $\mathfrak{F}_{\tau+}$-measurable. Intuitively a random variable is $\mathfrak{F}_{\tau+}$-measurable if it depends on the process only up to time $\tau+$.

Proposition 4.3. *Let σ and τ be stopping times. If $A \in \mathfrak{F}_{\sigma+}$, then $A \cap \{\sigma \le \tau\} \in \mathfrak{F}_{\tau+}$. If $\sigma \le \tau$, then $\mathfrak{F}_{\sigma+} \subset \mathfrak{F}_{\tau+}$. Finally $\{\sigma < \tau\}$, $\{\sigma = \tau\}$, and $\{\sigma > \tau\}$ are in $\mathfrak{F}_{\sigma+}$ and $\mathfrak{F}_{\tau+}$.*

Proof. Let $A \in \mathfrak{F}_{\sigma+}$ and choose $t \ge 0$. Then

$$A \cap \{\sigma \le \tau\} \cap \{\tau \le t\} = (A \cap \{\sigma \le t\}) \cap \{\tau \le t\} \cap \{\sigma \wedge t \le \tau \wedge t\} \in \mathfrak{F}_{t+}$$

since $\sigma \wedge t$ and $\tau \wedge t$ are $\mathfrak{F}_{t+}$-measurable. Thus the first conclusion is valid.

Suppose $\sigma \le \tau$ and choose $A \in \mathfrak{F}_{\sigma+}$. Then $A = A \cap \{\sigma \le \tau\} \in \mathfrak{F}_{\tau+}$ by the first conclusion, so the second conclusion is valid.

It follows from the first conclusion that $\{\sigma \le \tau\} \in \mathfrak{F}_{\tau+}$ and hence that $\{\sigma > \tau\} \in \mathfrak{F}_{\tau+}$. Now τ is $\mathfrak{F}_{\tau+}$-measurable and, by Proposition 4.2 and the second conclusion, $\sigma \wedge \tau$ is $\mathfrak{F}_{\tau+}$-measurable. Thus $\{\sigma < \tau\} = \{\sigma \wedge \tau < \tau\} \in \mathfrak{F}_{\tau+}$ and hence $\{\sigma = \tau\} \in \mathfrak{F}_{\tau+}$. The proof of $\mathfrak{F}_{\sigma+}$-measurability follows by interchanging the roles of σ and τ.

Let $\mathfrak{F}$ be a sub σ-field of $\mathfrak{F}_\infty$ and let A be an $\mathfrak{F}$-measurable subset of Ω. A function on A is said to be *$\mathfrak{F}$-measurable on A* if it is measurable with respect to the σ-field $\{B : B \subset A \text{ and } B \in \mathfrak{F}\}$ on A. A random variable on an event A is an $\mathfrak{F}_\infty$-measurable function on A. Given $x \in \mathbb{R}^d$ and a random variable Y defined on an event A, set $E_x(Y; A) = \int_A Y(\omega) P_x(d\omega)$. If $P_x(A) = 1$, $E_x(Y; A)$ may safely be written as $E_x Y$.

Proposition 4.4. *Let τ be a stopping time, let $t \geq 0$, and let Y be an $\mathfrak{F}_t$-measurable random variable. Then $Y \circ \theta_\tau$ is an $\mathfrak{F}_{(t+\tau)+}$-measurable random variable on $\{\tau < \infty\}$.*

Proof. By the monotone class theorem it suffices to show that if f is a bounded measurable function on $\mathbb{R}^d$, then $f(X(t + \tau))$ is $\mathfrak{F}_{(t+\tau)+}$-measurable on $\{\tau < \infty)$. Since $t + \tau$ is a stopping time and $\{t + \tau < \infty\} = \{\tau < \infty\}$, it suffices to show that $f(X(\tau))$ is $\mathfrak{F}_{\tau+}$-measurable on $\{\tau < \infty\}$. To this end it is enough to show that $X(\tau)$ is $\mathfrak{F}_{\tau+}$-measurable on $\{\tau < \infty\}$. Thus it suffices to show that $X(\tau)$ is $\mathfrak{F}_{t+}$-measurable on $\{\tau \leq t\}$ for $t > 0$.

Let τ_n be the stopping time defined by $\tau_n = \infty$ if $\tau = \infty$ and

$$\tau_n = j2^{-n}t \qquad \text{if} \qquad (j-1)2^{-n}t < \tau \leq j2^{-n}t, \qquad j = 0, 1, 2, \ldots .$$

It is easily seen that $X(\tau_n)$ is $\mathfrak{F}_{t+}$-measurable on $\{\tau \leq t\}$ and that $\tau_n \downarrow \tau$ as $n \to \infty$. By continuity of paths $X(\tau) = \lim_n X(\tau_n)$ is $\mathfrak{F}_{t+}$-measurable on $\{\tau \leq t\}$ as desired.

The next result is a special case of Proposition 4.4.

Proposition 4.5. *Let τ be a stopping time and let f be a Borel function on $\mathbb{R}^d$. Then $X(\tau)$ and $f(X(\tau))$ are $\mathfrak{F}_{\tau+}$-measurable on $\{\tau < \infty\}$.*

If σ and τ are stopping times, then $\tau + \sigma \circ \theta_\tau$ is defined to equal ∞ on $\{\tau = \infty\}$.

Proposition 4.6. *If σ and τ are stopping times, then $\tau + \sigma \circ \theta_\tau$ is a stopping time.*

Proof. Let Q^+ denote the positive rationals. Then for $t > 0$

$$\{\tau + \sigma \circ \theta_\tau < t\} = \bigcup_{r \in Q^+} \{\tau < t - r, \sigma \circ \theta_\tau < r\}.$$

Now

$$\{\tau < \infty, \sigma \circ \theta_\tau < r\} = \{\tau < \infty, I_{\{\sigma < r\}} \circ \theta_\tau > 0\} \in \mathfrak{F}_{(r+\tau)+}$$

by Propositions 4.1 and 4.4. Thus $\{\tau < t - r,\ \sigma \circ \theta_\tau < r\} \in \mathfrak{F}_t$ and hence $\{\tau + \sigma \circ \theta_\tau < t\} \in \mathfrak{F}_t$. Consequently $\tau + \sigma \circ \tau$ is a stopping time.

The *hitting time* τ_B of a subset B of $\mathbb{R}^d$ is defined as $\tau_B = \inf[t > 0 : X(t) \in B]$ ($\tau_B = \infty$ if $X(t) \notin B$ for all $t > 0$). Since the infimum is taken over all $t > 0$ it is possible that $\tau_B > 0$ even if $X(0) \in B$. Observe that for $t \geq 0$, $t + \tau_B \circ \theta_t$ is the first hitting time of B after time t; that is

$$t + \tau_B \circ \theta_t = \inf[s > t : X(s) \in B]$$

if $X(s) \in B$ for some $s > t$ and $t + \tau_B \circ \theta_t = \infty$ otherwise. Clearly τ_B satisfies the *terminal time property*: $\tau_B = t + \tau_B \circ \theta_t$ on $\{\tau_B > t\}$ for $t \geq 0$.

A subset B of $\mathbb{R}^d$ is called an F_σ set if B is a countable union of closed sets or equivalently if B is an increasing limit of compact sets. Let $\mathscr{B}$ denote the collection of all F_σ subsets of $\mathbb{R}^d$. Then $\mathscr{B}$ contains all open and closed subsets of $\mathbb{R}^d$ and $\mathscr{B}$ is closed under finite intersections and finite or countable unions. Note that $\mathscr{B}$ does not contain all Borel sets. For example, it follows from the Baire category theorem that the set of irrational points in $[0, 1]$ is not an F_σ set. The results that are proven in this book for sets in $\mathscr{B}$ can invariably be extended to Borel sets by using the capacitability theorem of Choquet (see the end of Section 6.5).

It will now be shown that hitting times of F_σ sets are stopping times. This result would not be true if in the definition of a stopping time it were required that $\{\tau \leq t\} \in \mathfrak{F}_t$ for $t \geq 0$.

Theorem 4.7. *Let $B \in \mathscr{B}$. Then τ_B is a stopping time. For each $x \in \mathbb{R}^d$, $P_x(\tau_B = 0)$ equals zero or one and $P_x(\tau_B = t) = 0$ for $t > 0$.*

Proof. Suppose first that B is closed. Set $B_n = \{x \in \mathbb{R}^d : d(x, B) < 1/n\}$, where $d(x, B) = \inf[\|y - x\| : y \in B]$ is the distance from x to B. Then B_n is open and $B_n \downarrow B$ as $n \to \infty$. Let Q^+ denote the positive rationals. For $r > 0$ set $\tau_B^r = \inf[t \geq r : X(t) \in B]$. Since B is closed, it follows that for $0 < r < t$

$$\begin{aligned}
\{\tau_B^r \leq t\} &= \{X(s) \in B \text{ for some } s \in [r, t]\} \\
&= \bigcap_n \{X(s) \in B_n \text{ for some } s \in [r, t]\} \\
&= \bigcap_n \{X(s) \in B_n \text{ for some } s \in [r, t] \cap Q^+\} \in \mathfrak{F}_t.
\end{aligned}$$

Now $\{\tau_B \leq t\} = \bigcup_{m \geq 1} \{\tau_B^{1/m} \leq t\}$, so $\{\tau_B \leq t\} \in \mathfrak{F}_t$ and hence τ_B is a stopping time. If $B = \bigcup_m B_m$ where each B_m is closed, then $\{\tau_B < t\} = \bigcup_m \{\tau_{B_m} < t\} \in \mathfrak{F}_t$ so τ_B is a stopping time by Proposition 4.1.

Let $x \in \mathbb{R}^d$. Since τ_B is a stopping time, $\{\tau_B = 0\} \in \mathfrak{F}_{0+}$. Thus $P_x(\tau_B = 0)$ equals zero or one by Blumenthal's zero-one law.

Let n be a positive integer. Then

$$\int_{n \leq \|y\| < n+1} P_y(\tau_B \leq t)\, dy < \infty$$

for every $t > 0$ and hence

$$\int_{n \le \|y\| < n+1} P_y(\tau_B = s)\, dy > 0$$

for only a finite or countably infinite number of values of s. Therefore $\int P_y(\tau_B = s)\, dy > 0$ for only a finite or countably infinite number of values of s.

Let $t > 0$. By the previous paragraph there is an $s \in (0, t)$ such that $\int P_y(\tau_B = s)\, dy = 0$. By the terminal time property of τ_B

$$\{\tau_B = t\} = \{\tau_B > t - s, t - s + \tau_B \circ \theta_{t-s} = t\} \subset \{\tau_B \circ \theta_{t-s} = s\}.$$

Thus by the Markov property

$$P_x(\tau_B = t) \le E_x(P_{X(t-s)}(\tau_B = s)) = \int p(t - s, x, y) P_y(\tau_B = s)\, dy = 0.$$

This completes the proof of the theorem.

5. Strong Markov Property

Let τ be a stopping time and let Y be a bounded random variable. Also let $x \in \mathbb{R}^d$ and $t \ge 0$. The Markov property states that $E_x(Y \circ \theta_t | \mathfrak{F}_{t+}) = E_{X(t)} Y$ a.s. (P_x). The *strong Markov property* states that this formula remains valid if t is replaced by τ; that is

(1) $$E_x(Y \circ \theta_\tau | \mathfrak{F}_{\tau+}) = E_{X(\tau)} Y \text{ a.s. } (P_x) \qquad \text{on } \{\tau < \infty\}.$$

A generalization of (1) is required. Let F be a bounded measurable function on $[0, \infty) \times \Omega$. Let the quantity $E_x F(s, \omega)$ evaluated at $x = X(\tau)$ and $s = \tau$ be denoted by $E_{X(\tau)} F(s, \omega)|_{s=\tau}$. It will be shown that

(2) $$E_x(F(\tau, \theta_\tau \omega) | \mathfrak{F}_{\tau+}) = E_{X(\tau)} F(s, \omega)|_{s=\tau} \text{ a.s. } (P_x) \qquad \text{on } \{\tau < \infty\}.$$

Formula (2) is also referred to as the strong Markov property. Observe that (1) follows from (2) with $F(s, \omega) = Y(\omega)$.

For an illustration of the use of (2), let $t > 0$, $x \in \mathbb{R}^d$ and $B \in \mathscr{B}$ and let A be a Borel subset of $\mathbb{R}^d$. Then

(3) $$P_x(\tau_B \le t, X(t) \in A) = E_x\left(\int_A p(t - \tau_B, X(\tau_B), y)\, dy; \tau_B < t \right).$$

For set $F(s, \omega) = 1$ if $s < t$ and $X(t - s) \in A$ and $F(s, \omega) = 0$ otherwise. Then

$$\begin{aligned} G(x, s) = E_x F(s, \omega) &= \int_A p(t - s, x, y)\, dy \qquad &\text{if } s < t, \\ &= 0 \qquad &\text{if } s \ge t. \end{aligned}$$

Now $P_x(\tau_B = t) = 0$ by Theorem 4.7, so by (2)

$$\begin{aligned} P_x(\tau_B \le t, X(t) \in A) &= E_x(I_A(X(t - \tau_B, \theta_{\tau_B}\omega)); \tau_B < t) \\ &= E_x(F(\tau_B, \theta_{\tau_B}\omega); \tau_B < t) \\ &= E_x(E_x(F(\tau_B, \theta_{\tau_B}\omega) \mid \mathfrak{F}_{\tau_B+}); \tau_B < t) \\ &= E_x(G(X(\tau_B), \tau_B); \tau_B < t) \\ &= E_x\left(\int_A p(t - \tau_B, X(\tau_B), y)\, dy; \tau_B < t\right) \end{aligned}$$

and hence (3) holds as desired.

Hunt [1] proved that Brownian motion possesess the strong Markov property. His result was applicable to processes with stationary independent increments. Extensions to more general Markov processes were obtained independently by Blumenthal [1] and Dynkin and Yushkevich [1].

Theorem 5.1. *Let τ be a stopping time and let F be a bounded measurable function on $[0, \infty) \times \Omega$. Then (2) holds.*

Proof. Let $r \ge 1$ and $0 < t_1 < \cdots < t_r$ and let $f_1, \ldots, f_r$ be bounded continuous functions on $\mathbb{R}^d$. Let τ_n be the stopping time defined by $\tau_n = \infty$ if $\tau = \infty$ and

$$\tau_n = 2^{-n}j \qquad \text{if} \quad (j-1)2^{-n} < \tau \le j2^{-n}, \qquad j = 0, 1, 2, \ldots.$$

Then $\{\tau_n = \infty\} = \{\tau = \infty\}$ and $\tau_n \downarrow \tau$ as $n \to \infty$. Also $\{\tau_n = 2^{-n}j\} \in \mathfrak{F}_{2^{-n}j+}$. Thus by the Markov property

$$\begin{aligned} E_x\left(\prod_{i=1}^r f_i(X(t_i + \tau_n)); \tau < \infty\right) &= \sum_{j=0}^\infty E_x\left(\prod_{i=1}^r f_i(X(t_i + \tau_n)); \tau_n = 2^{-n}j\right) \\ &= \sum_{j=0}^\infty E_x\left(E_{X(\tau_n)} \prod_{i=1}^r f_i(X(t_i)); \tau_n = 2^{-n}j\right) \\ &= E_x\left(E_{X(\tau_n)} \prod_{i=1}^r f_i(X(t_i)); \tau < \infty\right). \end{aligned}$$

Consequently by Proposition 3.1

$$(4) \qquad E_x\left(\prod_{i=1}^r f_i(X(t_i + \tau)); \tau < \infty\right) = E_x\left(E_{X(\tau)} \prod_{i=1}^r f_i(X(t_i)); \tau < \infty\right).$$

Let $A \subset \{\tau < \infty\}$ be in $\mathfrak{F}_{\tau+}$ and let σ be the stopping time defined by $\sigma = \tau$ on A and $\sigma = \infty$ on A^c. By (4) applied to σ

$$(5) \qquad E_x\left(\prod_{i=1}^r f_i(X(t_i + \tau); A\right) = E_x\left(E_{X(\tau)} \prod_{i=1}^r f_i(X(t_i)); A\right).$$

Let Z be a bounded $\mathfrak{F}_{\tau+}$-measurable random variable and let f be a bounded continuous function on $\mathbb{R}$. Then by (5)

$$E_x\left(Zf(\tau)\prod_{i=1}^{r} f_i(X(t_i, \theta_\tau\omega)); \tau < \infty\right) = E_x\left(Zf(\tau)E_{X(\tau)}\prod_{i=1}^{r} f_i(X(t_i)); \tau < \infty\right).$$

It now follows from the monotone class theorem that

$$E_x(ZF(\tau, \theta_\tau\omega); \tau < \infty) = E_x(ZE_{X(\tau)}F(s, \omega)\Big|_{s=\tau}; \tau < \infty),$$

and hence (2) holds. This completes the proof of the theorem.

Observe that $E_x X(t) = x$ and $E_x\|X(t) - x\|^2 = dt$. The strong Markov property will now be used to show that these formulas remain valid if t is replaced by a stopping time τ.

Proposition 5.2. *Let τ be a stopping time and let $x \in \mathbb{R}^d$ be such that $E_x\tau < \infty$. Then $E_x X(\tau) = x$ and $E_x\|X(\tau) - x\|^2 = dE_x\tau$.*

Proof. Let $0 \le s < t < \infty$. It will first be shown that

(6) $$E_x\|X(t \wedge \tau) - X(s \wedge \tau)\|^2 = d\int_s^t P_x(\tau \ge u)\,du.$$

Since $s \vee (t \wedge \tau)$ is a stopping time, in proving (6) it can be assumed that $P_x(s \le \tau \le t) = 1$. Note that

$$\begin{aligned} d(t - s) &= E_x\|X(t) - X(s)\|^2 \\ &= E_x\|X(\tau) - X(s) + X(t) - X(\tau)\|^2 \\ &= E_x\|X(\tau) - X(s)\|^2 + E_x\|X(t) - X(\tau)\|^2 \\ &\quad + 2E_x((X(\tau) - X(s)) \cdot (X(t) - X(\tau))). \end{aligned}$$

Now $P_x(\tau < \infty) = 1$, so by the strong Markov property

$$\begin{aligned} E_x\|X(t) - X(\tau)\|^2 &= E_x\|X(\tau + t - \tau) - X(\tau)\|^2 \\ &= E_x\left(E_{X(\tau)}\|X(t - u) - X(0)\|^2\Big|_{u=\tau}\right) \\ &= d(t - E_x\tau). \end{aligned}$$

Since $X(\tau) - X(s)$ is $\mathfrak{F}_{\tau+}$-measurable, it follows from the strong Markov property that

$$E_x((X(\tau) - X(s)) \cdot (X(t) - X(\tau))) = 0.$$

Thus $E_x\|X(\tau) - X(s)\|^2 = d(t - s - (t - E_x\tau)) = d(E_x\tau - s)$. Now

$$E_x\tau = \int_0^\infty P_x(\tau \geq u)\,du = s + \int_s^t P_x(\tau \leq u)\,du,$$

so (6) holds as desired. Since $E_x\tau < \infty$, it follows from (6) that

$$\lim_{s,t\to\infty} E_x\|X(t \wedge \tau) - X(s \wedge \tau)\|^2 = 0. \tag{7}$$

By the strong Markov property $E_x(X(t) - X(t \wedge \tau)) = 0$. Now $E_x(X(t) - x) = 0$, so

$$E_x X(t \wedge \tau) = x. \tag{8}$$

Since $\lim_{t\to\infty} X(t \wedge \tau) = X(\tau)$ a.s. (P_x) it follows from (7) that $\lim_{t\to\infty} E_x\|X(t \wedge \tau) - X(\tau)\|^2 = 0$. Thus by (8), $E_x X(\tau) = \lim_{t\to\infty} E_x X(t \wedge \tau) = x$, which verifies the first conclusion of the proposition. It follows from (6) and (7) that

$$E_x\|X(\tau) - x\|^2 = \lim_{t\to\infty} E_x\|X(t \wedge \tau) - x\|^2 = \lim_{t\to\infty} d\int_0^t P_x(\tau \geq u)\,du$$
$$= d\int_0^\infty P_x(\tau \geq u)\,du = dE_x\,\tau.$$

This completes the proof of the proposition.

Chapter 2

Hitting Times

Throughout the remainder of this book subsets of $\mathbb{R}^d$ are understood to be Borel sets unless the contrary is noted. Similarly a function on $\mathbb{R}^d$ or on a subset of $\mathbb{R}^d$ is understood to be a Borel function on its domain. A real-valued function is allowed to assume the values ∞ and $-\infty$ except as otherwise noted. Also "continuous" will be taken to mean finite valued and continuous, with "continuous in the extended sense" being used when infinite values are permitted.

1. Auxiliary Analytical Results

In this preliminary section some analytical definitions and results are gathered together.

For $x \in \mathbb{R}^d$ and $r > 0$ set $B_r(x) = \{y \in \mathbb{R}^d : \|y - x\| \leq r\}$, $\mathring{B}_r(x) = \{y \in \mathbb{R}^d : \|y - x\| < r\}$, and $S_r(x) = \{y \in \mathbb{R}^d : \|y - x\| = r\}$. Let $\sigma_r(x) = \sigma_r(x, \cdot)$ denote the uniform probability distribution on $S_r(x)$, i.e., surface area on $S_r(x)$ normalized to have total measure one. The abbreviated notations $B_r = B_r(0)$, $\mathring{B}_r = \mathring{B}_r(0)$, $S_r = S_r(0)$, and $\sigma_r = \sigma_r(0)$ will also be used, where 0 is the origin of $\mathbb{R}^d$.

Proposition 1.1. *Let f be a function on $\mathbb{R}^d$ such that $\int f(y)\,dy$ is well defined. Then*

$$\int f(y)\,dy = \frac{2\pi^{d/2}}{\Gamma(d/2)} \int_0^\infty \left(\int f(y) \sigma_r(dy) \right) r^{d-1}\,dr.$$

Proof. Let s_d denote the surface area of the unit sphere S_1 in $\mathbb{R}^d$. Then S_r has surface area $r^{d-1}s_d$ for $r > 0$. Thus

$$\int f(y)\,dy = s_d \int_0^\infty \left(\int f(y)\sigma_r(dy)\right) r^{d-1}\,dr.$$

It follows by letting f be the standard normal density $(2\pi)^{-d/2}\exp(-\|\cdot\|^2/2)$ on $\mathbb{R}^d$ that

$$1 = s_d(2\pi)^{-d/2}\int_0^\infty \exp(-r^2/2)r^{d-1}\,dr = s_d\Gamma(d/2)/2\pi^{d/2}$$

and hence that $s_d = 2\pi^{d/2}/\Gamma(d/2)$, which completes the proof of the proposition.

The probability distribution σ_r is preserved under orthogonal linear transformations on $\mathbb{R}^d$. By the next result, which will be used to prove Proposition 2.15, this property uniquely characterizes σ_r.

Proposition 1.2. *Let $r > 0$ and let μ be a probability distribution on S_r which is preserved under orthogonal linear transformations on $\mathbb{R}^d$. Then $\mu = \sigma_r$.*

Proof. Let φ denote the characteristic function of μ defined by $\varphi(x) = \int e^{ix\cdot y}\mu(dy)$ where $i = \sqrt{-1}$. Since μ is preserved under orthogonal linear transformations on $\mathbb{R}^d$, φ depends on x only through $\|x\|$. Thus there is a function ψ on $\mathbb{R}$ such that $\varphi(x) = \psi(\|x\|)$ for $x \in \mathbb{R}^d$. Similarly the characteristic function φ_1 of σ_1 is of the form $\varphi_1(x) = \psi_1(\|x\|)$ for $x \in \mathbb{R}^d$, where ψ_1 is a function on $\mathbb{R}$. Observe that the characteristic function φ_r of σ_r satisfies

$$\varphi_r(x) = \int e^{ix\cdot y}\sigma_r(dy) = \int e^{irx\cdot y}\sigma_1(dy) = \psi_1(r\|x\|), \qquad x \in \mathbb{R}^d.$$

Let $s > 0$. Then

$$\psi(s) = \int \varphi(x)\sigma_s(dx) = \int \sigma_s(dx)\int e^{ix\cdot y}\mu(dy) = \int \mu(dy)\psi_1(s\|y\|) = \psi_1(rs).$$

Consequently $\varphi(x) = \psi(\|x\|) = \psi_1(r\|x\|) = \varphi_r(x)$ for $x \in \mathbb{R}^d$, so $\mu = \sigma_r$ by the uniqueness theorem for characteristic functions.

Let μ be a measure on an open subset D of $\mathbb{R}^d$. It is called a *Radon* measure if $\mu(C) < \infty$ for all compact subsets C of D. A property is said to hold *almost everywhere with respect to* μ (a.e. (μ)) on B if there is a subset A of D such that $\mu(A) = 0$ and the property holds on $(B\backslash A) \cap D$. If μ is Lebesgue measure "a.e. (μ)" will be abbreviated to "a.e." and if $B = D$, the phrase "on B" will be omitted. The measure μ is said to be *concentrated* on B if $\mu(D\backslash B) = 0$. It is said to be *supported* on B if B is relatively closed in D and μ is concentrated on B. The *support* of μ is the set of points $x \in D$ such that $\mu(\mathring{B}_r(x)) > 0$ for all $r > 0$. It is easily seen that the support of μ is a relatively closed set upon which μ is concentrated. For $x \in \mathbb{R}^d$ let δ_x be the probability measure on $\mathbb{R}^d$ which is concentrated at x.

Let μ_n, $n \geq 1$, and μ be Radon measures on D. Then μ_n is said to *converge vaguely* to μ if $\lim_n \int \varphi \, d\mu_n = \int \varphi \, d\mu$ for every continuous function φ on D having compact support in D (i.e., vanishing outside a compact subset of D). If μ_n converges vaguely to μ and $\varphi : D \to [0, \infty]$ is continuous in the extended sense, then $\lim_n \int \varphi \, d\mu_n \geq \int \varphi \, d\mu$. The measures μ_n are said to *converge completely* to μ if $\mu(D) < \infty$ and $\lim_n \int \varphi \, d\mu_n = \int \varphi \, d\mu$ for every bounded continuous function φ on D. Suppose μ_n converges vaguely to μ and $\mu(D) < \infty$. Then μ_n converges completely to μ if and only if $\mu_n(D) \to \mu(D)$ or, alternatively, if and only if for every $\varepsilon > 0$ there is a compact subset C of D such that $\overline{\lim}_n \, \mu_n(D \backslash C) \leq \varepsilon$.

Let μ_n, $n \geq 1$, be Radon measures on D. If $\mu_n(C)$ is bounded in n for every compact subset C of D, there is a strictly increasing sequence $\{n_j\}$ of positive integers and a Radon measure μ on D such that μ_{n_j} converges vaguely to μ.

Let $L(x, \cdot)$ be a measure on C for each $x \in A$ and suppose that $L(x, B)$ is a Borel function of x for each Borel subset B of C. Consider the operator L defined by $Lf(x) = \int L(x, dy) f(y)$, $x \in A$, where f is a Borel function on C. Clearly Lf is well defined (and a Borel function on the Borel set A) if $f \geq 0$ on C. If $f = f^+ - f^-$ is the decomposition of f into its positive and negative parts, Lf is well defined as long as, for each $x \in A$, $Lf^+(x)$ and $Lf^-(x)$ are not both infinite. Also Lf makes sense if the domain of f is a Borel subset C_1 of C, provided that $L(x, \cdot)$ is concentrated on C_1 for each $x \in A$. If $L(x, \cdot) = \delta_x$ for each $x \in A$, then $Lf = f$ on A, in which case L is called the *identity operator*. A nonnegative kernel $L(x, y)$, $x \in A$ and $y \in C$, yields the measures $L(x, \cdot)$, $x \in A$, defined by $L(x, B) = \int_B L(x, y) \, dy$. In this case $Lf(x) = \int L(x, y) f(y) \, dy$.

Let $L(x, \cdot)$, $x \in A$, and $M(z, \cdot)$, $z \in C$, be of the above form. Then LM is defined by $LM(x, B) = \int L(x, dz) M(z, B)$. Clearly $LMf(x) = \int L(x, dz) Mf(z)$ holds at least when $f \geq 0$. A collection L^t, $t \geq 0$, of operators of the above form is said to satisfy the *semigroup property* if $L^{s+t} = L^s L^t$ for all $s, t \geq 0$.

Let p^t, $t \geq 0$, be the operators defined by $p^t f(x) = E_x f(X(t))$ for $x \in \mathbb{R}^d$. Then p^0 is the identity operator and $p^t f(x) = \int p(t, x, y) f(y) dy$ for $t > 0$ *and* $x \in \mathbb{R}^d$. It follows from (1.1.1) that the operators $p^t, t \geq 0$, satisfy the semigroup property. It is also convenient to set $p^t(x, y) = p(t, x, y)$ for $t > 0$ and $x, y \in \mathbb{R}^d$.

Proposition 1.3. *Let f be a bounded function on $\mathbb{R}^d$. Then*

$$\lim_{t \to \infty} (p^t f(x) - p^t f(y)) = 0$$

uniformly for x and y in compact sets.

Proof. Choose M such that $|f| \le M$ and let $x, y \in \mathbb{R}^d$. Then for $t > 0$

$$\begin{aligned}|p^t f(x) - p^t f(y)| &= \left|\int (p(t,x,z) - p(t,y,z)) f(z)\,dz\right| \\ &\le M \int |p(t,x,z) - p(t,y,z)|\,dz \\ &= M \int |p(1,z) - p(1, z + t^{-1/2}(x-y))|\,dz,\end{aligned}$$

which approaches zero as $t \to \infty$ uniformly for x and y in compact sets since $p(1,\cdot)$ is continuous and integrable.

Proposition 1.4. *Let f be a bounded function on $\mathbb{R}^d$ such that $p^t f = f$ for some $t > 0$. Then f is constant on $\mathbb{R}^d$.*

Proof. Let $t > 0$ be such that $p^t f = f$. Then $p^{nt} f = f$ for all positive integers n by the semigroup property. Thus by the previous proposition

$$f(x) - f(y) = \lim_n (p^{nt} f(x) - p^{nt} f(y)) = 0 \qquad \text{for } x, y \in \mathbb{R}^d,$$

so f is constant on $\mathbb{R}^d$ as desired.

Let D be an open subset of $\mathbb{R}^d$ and let f be a function defined on D. Then f is said to be *lower semicontinuous* on D if $\underline{\lim}_{y \to x} f(y) \ge f(x) > -\infty$ for $x \in D$. It is easily seen that f is lower semicontinuous on D if and only if $f > -\infty$ on D and $\{x \in D: f(x) > c\}$ is open for each $c \in \mathbb{R}$. If f is lower semicontinuous on D and B is a compact subset of D, there is an $x \in B$ such that $f(x) = \min_{y \in B} f(y)$. In particular a lower semicontinuous function on D is bounded below on compact subsets of D. If f_1 and f_2 are lower semicontinuous on D, so are $f_1 \wedge f_2$ and $c_1 f_1 + c_2 f_2$ for $c_1, c_2 \ge 0$. An increasing limit of lower semicontinuous functions on D is lower semicontinuous on D. The function f is said to be *upper semicontinuous* on D if $\overline{\lim}_{y \to x} f(y) \le f(x) < \infty$ for $x \in D$ or equivalently if $-f$ is lower semicontinuous on D. Observe that f is continuous on D if and only if it is both lower semicontinuous and upper semicontinuous on D.

Let D be an open subset of $\mathbb{R}^d$ and let $L(x, y)$, $x \in D$ and $y \in D$, be a kernel which is symmetric in x and y. Given a measure μ on D, the *L-potential* $L\mu$ of μ is defined by

$$L\mu(x) = \int L(x,y)\mu(dy) = \int \mu(dy)\, L(y,x).$$

If μ is of the form $\mu(dx) = f(x)\,dx$, then $L\mu = \int L(\cdot, y) f(y)\,dy = Lf$. Given a second measure ν on D, set $\int L\mu\, d\nu = \int L\mu(x)\nu(dx)$ if this integral is well defined. By Fubini's theorem and the symmetry of $L(x, y)$ in x and y, $\int L\mu \, dv = \int Lv \, d\mu$ provided that $\int L^- \mu \, dv < \infty$. It follows from Fatou's lemma that if L is nonnegative and continuous in the extended sense on $D \times D$, then $L\mu$ is lower semicontinuous on D.

2. Elementary Properties of Hitting Times

In this section a number of properties related to the hitting time τ_B of a set $B \in \mathscr{B}$ are obtained.

Proposition 2.1. *Let $B \in \mathscr{B}$ and $t > 0$. Then $P_{\cdot}(\tau_B \le t)$ is lower semicontinuous on $\mathbb{R}^d$.*

Proof. Let $0 < s < t$. Then $P_x(X(u) \in B \text{ for some } u \in (s, t)) = \int p(s, x, y) P_y(\tau_B \le t - s) \, dy$, which is continuous in x and increases to $P_x(\tau_B < t) = P_x(\tau_B \le t)$ as $s \downarrow 0$. Thus $P_{\cdot}(\tau_B \le t)$ is lower semicontinuous on $\mathbb{R}^d$ as desired.

Proposition 2.2. *Let $B \in \mathscr{B}$. Then $\lim_{t \to \infty} P_{\cdot}(t \le \tau_B < \infty) = 0$ uniformly on compacts.*

Proof. Let C be a compact subset of $\mathbb{R}^d$. Choose $\varepsilon > 0$ and let A be a compact subset of $\mathbb{R}^d$ such that $P_x(X(1) \notin A) \le \varepsilon$ for $x \in C$. Now $p(1, \cdot) \le 1$, so for $t \ge 1$ and $x \in C$

$$\begin{aligned} P_x(t \le \tau_B < \infty) &\le P_x(t - 1 \le \tau_B \circ \theta_1 < \infty) \\ &\le P_x(X(1) \notin A) + \int_A p(1, x, y) P_y(t - 1 \le \tau_B < \infty) \, dy \\ &\le \varepsilon + \int_A P_y(t - 1 \le \tau_B < \infty) \, dy \end{aligned}$$

and hence $\lim_{t \to \infty} \sup_{x \in C} P_x(t \le \tau_B < \infty) \le \varepsilon$. Since ε can be made arbitrarily small, the conclusion of the proposition is valid.

Let $B \in \mathscr{B}$. It is said to be *polar* if $P_x(\tau_B < \infty) = 0$ for all $x \in \mathbb{R}^d$ and *nonpolar* otherwise. If B has positive measure it is clearly nonpolar, since $P_x(\tau_B \le t) \ge P_x(X(t) \in B) > 0$ for all $t > 0$ and $x \in \mathbb{R}^d$. Clearly if B is polar and $B_1 \in \mathscr{B}$ is a subset of B, then B_1 is polar.

Proposition 2.3. *If $B_n \in \mathscr{B}$ is polar for $n \ge 1$, then $\bigcup_n B_n$ is polar.*

Proof. Set $B = \bigcup_n B_n$. Since $\{\tau_B < \infty\} = \bigcup_n \{\tau_{B_n} < \infty\}$, $P_x(\tau_B < \infty) \le \sum_n P_x(\tau_{B_n} < \infty) = 0$ for all $x \in \mathbb{R}^d$ and hence B is polar.

Proposition 2.4. *If $B \in \mathscr{B}$ is nonpolar there is a compact nonpolar subset of B.*

Proof. Let B_n be compact sets such that $B = \bigcup_n B_n$. By Proposition 2.3, one of the sets B_n must be nonpolar.

Let $B \in \mathscr{B}$. For $t \geq 0$, $\{\tau_B \circ \theta_t < \infty\} = \{X(s) \in B \text{ for some } s > t\}$ denotes the event that Brownian motion visits B after time t. Consequently $P_x(\tau_B \circ \theta_t < \infty) \uparrow P_x(\tau_B < \infty)$ as $t \downarrow 0$. By the Markov property

(1) $$P_x(\tau_B \circ \theta_t < \infty) = \int p(t, x, y) P_y(\tau_B < \infty)\, dy, \qquad x \in \mathbb{R}^d \quad \text{and} \quad t > 0.$$

A simpler proof of the next result will be given in the beginning of Section 6.

Proposition 2.5. *If $d \geq 2$ every one-point set is polar.*

Proof. It suffices to show that $\{0\}$ is polar. Since $d \geq 2$ it is easily seen that $\int_0^1 p(t, 0)\, dt = \infty$. Thus by Fatou's lemma

(2) $$\lim_{y \to 0} \int_0^1 p(t, y)\, dt = \infty.$$

Choose $\varepsilon > 0$ and $M > \varepsilon$. Since $p(t, x, y) \leq p(\varepsilon, 0)$ for $t \geq \varepsilon$ and $x, y \in \mathbb{R}^d$, there is an $N > 0$ such that

$$\int_\varepsilon^{M+1} p(t, x, y)\, dt \leq N, \qquad t \geq \varepsilon \quad \text{and} \quad x, y \in \mathbb{R}^d.$$

Thus for $r > 0$

$$E_{\cdot} \int_\varepsilon^{M+1} I_{B_r}(X(t))\, dt \leq N|B_r|,$$

where $|B_r|$ is the Lebesgue measure of the ball B_r. Set $\tau_0 = \tau_{\{0\}}$. By the strong Markov property

$$E_{\cdot} \int_\varepsilon^{M+1} I_{B_r}(X(t))\, dt \geq P_{\cdot}(\varepsilon + \tau_0 \circ \theta_\varepsilon \leq M) \int_{B_r} \left(\int_0^1 p(t, y)\, dt \right) dy,$$

so

(3) $$P_{\cdot}(\varepsilon + \tau_0 \circ \theta_\varepsilon \leq M) \int_{B_r} \left(\int_0^1 p(t, y)\, dt \right) dy \leq N|B_r|.$$

It follows from (2) and (3) by letting $r \to 0$ that $P_{\cdot}(\varepsilon + \tau_0 \circ \theta_\varepsilon \leq M) = 0$. Since ε can be made arbitrarily small and M can be made arbitrarily large, $P_{\cdot}(\tau_0 < \infty) = 0$ and hence $\{0\}$ is polar as desired.

It follows from Propositions 2.3 and 2.5 that if $d \geq 2$ and B is a finite or countable set, then B is polar.

Proposition 2.6. *If $B \in \mathscr{B}$ is nonpolar and $t > 0$, then $P_{.}(\tau_B \leq t)$ is bounded away from zero on compacts.*

Proof. Suppose $P_x(\tau_B \leq t) = 0$ for all $x \in \mathbb{R}^d$. Then

$$P_x(nt < \tau_B \leq (n+1)t) = E_x(P_{X(nt)}(\tau_B \leq t); \tau_B > nt) = 0$$

for all $x \in \mathbb{R}^d$ and positive numbers n. Consequently B is polar.

Suppose now that B is nonpolar and let $t > 0$. Then $P_{x_0}(\tau_B \leq t) > 0$ for some $x_0 \in \mathbb{R}^d$ and hence

$$\int p(\delta, x_0, y) P_y(\tau_B \leq t - \delta)\, dy = P_{x_0}(\tau_B \circ \theta_\delta \leq t - \delta) > 0$$

for some $\delta \in (0, t)$. Thus $P_y(\tau_B \leq t - \delta) > 0$ for y in a set having positive Lebesgue measure, so

$$P_x(\tau_B \leq t) \geq \int p(\delta, x, y) P_y(\tau_B \leq t - \delta)\, dy > 0$$

for all $x \in \mathbb{R}^d$. Let C be a compact subset of $\mathbb{R}^d$. Since $P_{.}(\tau_B \leq t)$ is lower semicontinuous on $\mathbb{R}^d$ by Proposition 2.1, $\inf_{x \in C} P_x(\tau_B \leq t) = P_{x_0}(\tau_B \leq t) > 0$ for some $x_0 \in C$. This completes the proof of the proposition.

Let $B \in \mathscr{B}$. Set

$$g_B(x, A) = E_x\left(\int_0^{\tau_B} I_A(X(t))\, dt\right), \qquad x \in \mathbb{R}^d \quad \text{and} \quad A \subset \mathbb{R}^d.$$

Observe that $g_B(x, A)$ is the expected amount of time that Brownian motion starting at x spends in A before hitting B.

Proposition 2.7. *Let $B \in \mathscr{B}$ and let A be bounded. If $d \geq 3$ or if B is nonpolar, then $\sup_x g_B(x, A) < \infty$.*

Proof. Suppose first that $d \geq 3$. Then

$$g_B(x, A) \leq \int_0^\infty \left(\int_A p(t, x, y)\, dy\right) dt = \int_A g(y - x)\, dy,$$

where

$$g(y) = \int_0^\infty p(t, y)\, dt = (2\pi)^{-d/2} \int_0^\infty t^{-d/2} e^{-\|y\|^2/2t}\, dt \tag{4}$$

$$= \frac{\Gamma(d/2 - 1)}{2\pi^{d/2}} \|y\|^{2-d}, \qquad y \in \mathbb{R}^d.$$

Now $\lim_{\|y\|\to\infty} g(y) = 0$ and g is integrable on compacts by Proposition 1.1. The desired result follows easily from these observations.

Suppose next that B is nonpolar. Let f be a continuous function on $\mathbb{R}^d$ having compact support C and such that $I_A \le f \le 1$. For $r > 0$ set $T_r = \inf[t > 0: \int_0^t f(X(u))\,du \ge r]$. It is easily seen that $T_r \ge r$ and that $X(T_r) \in C$ on $\{T_r < \infty\}$. Let $s > 0$. Then

$$T_{r+s} = \inf\left[t > 0: \int_0^t f(X(u))\,du \ge r + s\right]$$

$$= T_r + \inf\left[t > 0: \int_{T_r}^{t+T_r} f(X(u))\,du \ge s\right]$$

$$= T_r + \inf\left[t > 0: \int_0^t f(X(u, \theta_{T_r}\omega))\,du \ge s\right]$$

and hence $T_{r+s} = T_r + T_s \circ \theta_{T_r}$ on $\{T_r < \infty\}$. Observe that $\{T_r \le t\} = \{\int_0^t f(X(u))\,du \ge r\}$. It follows by using Riemann sums to approximate $\int_0^t f(X(u))\,du$ that this integral is $\mathfrak{F}_t$-measurable. Consequently T_r is a stopping time. By Proposition 2.6 there is a $\delta > 0$ such that $P_y(T_1 < \tau_B) \le P_y(\tau_B > 1) \le 1 - \delta$ for $y \in C$. Choose $x \in \mathbb{R}^d$. Then

$$P_x(T_{r+1} < \tau_B) = E_x(P_{X(T_r)}(T_1 < \tau_B); T_r < \tau_B) \le (1 - \delta)P_x(T_r < \tau_B)$$

and hence $P_x(T_n < \tau_B) \le (1 - \delta)^{n-1}$ for every positive integer n. Now for any nonnegative random variable $Y, E_x Y = \int_0^\infty P_x(Y \ge y)\,dy \le \sum_0^\infty P_x(Y > n)$. Consequently

$$g_B(x, A) \le E_x\left(\int_0^{\tau_B} f(X(t))\,dt\right) \le \sum_{n=0}^{\infty} P_x\left(\int_0^{\tau_B} f(X(t))\,dt > n\right)$$

$$\le \sum_{n=0}^{\infty} P_x(T_n < \tau_B) \le 1 + \frac{1}{\delta} < \infty$$

and hence $\sup_x g_B(x, A) < \infty$ as desired.

Proposition 2.8. *Let $B \in \mathscr{B}$ be such that B^c is bounded. Then $P_.(\tau_B < \infty) = 1$ on $\mathbb{R}^d$ and $\sup_x E_x(\tau_B) < \infty$.*

Proof. Clearly B is nonpolar since it has positive Lebesgue measure. Since $X(t) \notin B$ for $0 < t < \tau_B$, $E_x(\tau_B) = g_B(x, B^c)$. Thus $\sup_x E_x(\tau_B) < \infty$ by Proposition 2.7 and hence $P_x(\tau_B < \infty) = 1$ for all $x \in \mathbb{R}^d$.

Let $B \in \mathscr{B}$. The *last exit time* L_B from B is defined by $L_B = \sup[t > 0: X(t) \in B]$ if $\tau_B < \infty$ and $L_B = 0$ if $\tau_B = \infty$. Note that $\{L_B > t\} = \{\tau_B \circ \theta_t < \infty\}$

for $t \geq 0$. Thus L_B is $\mathfrak{F}_\infty$-measurable and hence a random variable. The set B is said to be *recurrent* if $P_.(L_B = \infty) = 1$ on $\mathbb{R}^d$ and *transient* if $P_.(L_B = \infty) = 0$ on $\mathbb{R}^d$. It follows from Proposition 2.8 and the next result that if B^c is bounded, then B is recurrent.

Proposition 2.9. *Let $B \in \mathscr{B}$. Then B is recurrent if $P_.(\tau_B < \infty) = 1$ on $\mathbb{R}^d$ and transient otherwise. If B is transient, then $\inf_{x \in B^c} P_x(\tau_B < \infty) = 0$ and*

$$\lim_{t\to\infty} P_.(L_B > t) = \lim_{t\to\infty} P_.(\tau_B \circ \theta_t < \infty) = 0 \tag{5}$$

uniformly on compacts.

Proof. If B is recurrent, then $P_.(\tau_B < \infty) = P_.(L_B > 0) = 1$ on $\mathbb{R}^d$. If $P_.(\tau_B < \infty) = 1$ on $\mathbb{R}^d$, then $P_.(L_B > t) = P_.(\tau_B \circ \theta_t < \infty) = 1$ on $\mathbb{R}^d$ for $t > 0$ by (1), so $P_.(L_B = \infty) = 1$ on $\mathbb{R}^d$ and hence B is recurrent. This proves the first result.

Suppose B is transient. Choose $y \in \mathbb{R}^d$ such that $P_y(\tau_B = \infty) = 1 - P_y(\tau_B < \infty) > 0$. Then for $t > 0$

$$P_y(t < \tau_B < \infty) = E_y(P_{X(t)}(\tau_B < \infty); \tau_B > t) \geq P_y(\tau_B > t) \inf_{x \in B^c} P_x(\tau_B < \infty).$$

Thus $0 = P_y(\tau_B = \infty) \inf_{x \in B^c} P_x(\tau_B < \infty)$, so $\inf_{x \in B^c} P_x(\tau_B < \infty) = 0$. Choose $\varepsilon > 0$. Let $y \in \mathbb{R}^d$ now be such that $P_y(\tau_B < \infty) \leq \varepsilon$. Then $P_y(\tau_B \circ \theta_t < \infty) \leq \varepsilon$ for all $t > 0$. By (1) and Proposition 1.3

$$\lim_{t\to\infty} (P_x(\tau_B \circ \theta_t < \infty) - P_y(\tau_B \circ \theta_t < \infty)) = 0$$

uniformly for x in compacts. Thus for any compact set C,

$$\limsup_{t\to\infty} \sup_{x \in C} P_x(\tau_B \circ \theta_t < \infty) \leq \varepsilon.$$

Consequently (5) holds, which completes the proof of the proposition.

Proposition 2.10. *Suppose $d \leq 2$ and $B \in \mathscr{B}$ is nonpolar. Then B is recurrent.*

Proof. Set $\varphi_B = P_.(\tau_B < \infty)$ and let $t > 0$. Then for $u > 0$

$$\int_0^u p^s(\varphi_B - p^t\varphi_B)\, ds = \int_0^t p^s\varphi_B\, ds - \int_u^{u+t} p^s\varphi_B\, ds \leq t.$$

Since $\varphi_B \geq p^t\varphi_B$ it follows by letting $u \to \infty$ that $\int_0^\infty p^s(\varphi_B - p^t\varphi_B)\, ds < \infty$. Now $\int_0^\infty p(s, x, y)\, ds = \infty$ for $x, y \in \mathbb{R}^d$ since $d \leq 2$, so that $\varphi_B = p^t\varphi_B$ a.e. Consequently for $s > 0$, $p^s\varphi_B = \varphi_B = p^t\varphi_B$ a.e. Since $p^s\varphi_B$ and $p^t\varphi_B$ are continuous on $\mathbb{R}^d$, $p^s\varphi_B = p^t\varphi_B$ on $\mathbb{R}^d$. Suppose B is transient. Then

$\lim_{s\to\infty} p^s\varphi_B = 0$ by (1) and Proposition 2.9. Consequently $P_.(\tau_B \circ \theta_t < \infty) = p^t\varphi_B = 0$ on $\mathbb{R}^d$. Therefore $P_.(\tau_B < \infty) = \lim_{t\to 0} P_.(\tau_B \circ \theta_t < \infty) = 0$ on $\mathbb{R}^d$ and hence B is polar. This completes the proof of the proposition.

It follows from Proposition 2.10 that if $d \leq 2$ and $B \in \mathscr{B}$, then $P_.(\tau_B < \infty)$ is either identically zero or identically one. The situation is quite different for $d \geq 3$.

Proposition 2.11. *Suppose $d \geq 3$ and $B \in \mathscr{B}$ is bounded. Then B is transient and* $\lim_{\|x\|\to\infty} P_x(\tau_B < \infty) = 0$.

Proof. Let A be a compact subset of $\mathbb{R}^d$ having positive Lebesgue measure. Then

$$E_x\left(\int_0^\infty I_A(X(t))\,dt\right) = \int_{A-x} g(y)\,dy,$$

where g is given by (4). Thus $E_x(\int_0^\infty I_A(X(t))\,dt)$ is bounded away from zero on compacts and approaches zero as $\|x\| \to \infty$. Observe that

$$\begin{aligned} E_x\left(\int_0^\infty I_A(X(t))\,dt\right) &\geq E_x\left(\int_{\tau_B}^\infty I_A(X(t))\,dt\right) \\ &= E_x\left(E_{X(\tau_B)}\int_0^\infty I_A(X(t))\,dt;\tau_B < \infty\right) \\ &\geq P_x(\tau_B < \infty)\inf_{y\in\bar{B}} E_y\left(\int_0^\infty I_A(X(t))\,dt\right). \end{aligned}$$

Thus $\lim_{\|x\|\to\infty} P_x(\tau_B < \infty) = 0$ and hence B is transient.

Brownian motion on $\mathbb{R}^d$ is said to be a *recurrent process* if

$$P_.(\lim_{t\to\infty} \|X(t)\| = \infty) = 0$$

on $\mathbb{R}^d$ and a *transient process* if $P_.(\lim_{t\to\infty} \|X(t)\| = \infty) = 1$ on $\mathbb{R}^d$.

Proposition 2.12. *Brownian motion on $\mathbb{R}^d$ is recurrent if $d \leq 2$ and transient if $d \geq 3$.*

Proof. Suppose first that $d \leq 2$. Then $B_1(0)$ is nonpolar and hence recurrent by Proposition 2.10. Thus

$$P_.\left(\lim_{t\to\infty} \|X(t)\| = \infty\right) \leq P_.(L_{B_1(0)} < \infty) = 0 \qquad \text{on } \mathbb{R}^d$$

and hence the Brownian motion process is recurrent. Suppose next that $d \geq 3$. By Proposition 2.11

$$P_{\cdot}\left(\underline{\lim}_{t \to \infty} \|X(t)\| \geq r\right) \geq P_{\cdot}(L_{B_r(0)} < \infty) = 1 \qquad \text{on } \mathbb{R}^d, \qquad r > 0,$$

so $P_{\cdot}(\lim_{t \to \infty} \|X(t)\| = \infty) = 1$ on $\mathbb{R}^d$ and hence the Brownian process is transient.

A subset C of $\mathbb{R}^d$ is called a *cone with vertex* $b \in \mathbb{R}^d$ if there is an $\alpha > 0$ and a unit vector $u \in \mathbb{R}^d$ such that $C = \{y \in \mathbb{R}^d : |(y - b) \cdot u| \geq \alpha \|y - b\|\}$. The sufficient condition in the next result is called the *cone condition for recurrence.*

Proposition 2.13. *Let $B \in \mathscr{B}$ satisfy the condition that there is a cone C and an $r \geq 0$ such that $\{y \in C : \|y\| \geq r\} \subset B$. Then B is recurrent.*

Proof. Let b be the vertex of the cone. By translation and scale invariance there is a constant $\gamma > 0$ such that $P_b(X(t) \in C) = \gamma$ for all $t > 0$. Since $\lim_{t \to \infty} P_b(\|X(t)\| < r) = 0$,

$$\begin{aligned} \lim_{t \to \infty} P_b(\tau_B \circ \theta_t < \infty) &\geq \underline{\lim}_{t \to \infty} P_b(X(t) \in B) \\ &\geq \lim_{t \to \infty} P_b(X(t) \in C \text{ and } X(t) \geq r) \\ &= \gamma > 0, \end{aligned}$$

so that B is recurrent by Proposition 2.9.

Some further applications of invariance will now be given.

Proposition 2.14. *Let $B \in \mathscr{B}$, $x \in \mathbb{R}^d$, $t \geq 0$, $r > 0$, and $A \subset \mathbb{R}^d$. Then*

$$P_{rx}(\tau_{rB} \leq r^2 t, X(\tau_{rB}) \in rA) = P_x(\tau_B \leq t, X(\tau_B) \in A)$$

and

$$P_{rx}(\tau_{rB} < \infty, X(\tau_{rB}) \in rA) = P_x(\tau_B < \infty, X(\tau_B) \in A).$$

Proof. Set

$$\sigma_B = \inf[t > 0 : r^{-1} X(r^2 t) \in B] = r^{-2} \tau_{rB}.$$

By scale invariance

$$P_{rx}(\tau_{rB} \leq r^2 t, X(\tau_{rB}) \in rA) = P_{rx}(\sigma_B \leq t, r^{-1} X(r^2 \sigma_B) \in A) = P_x(\tau_B \leq t, X(\tau_B) \in A),$$

so the first result of the proposition is valid. The second result follows from the first result.

Proposition 2.15. *Let $r > 0$, $t \geq 0$, and $A \subset S_r$. Then*

$$P_0(\tau_{S_r} \leq t, X(\tau_{S_r}) \in A) = P_0(\tau_{S_r} \leq t)\sigma_r(A).$$

Proof. It follows from orthogonal invariance that if Q is an orthogonal linear transformation on $\mathbb{R}^d$, then

$$P_0(\tau_{S_r} \leq t, X(\tau_{S_r}) \in QA) = P_0(\tau_{S_r} \leq t, X(\tau_{S_r}) \in A).$$

The desired conslusion now follows from Proposition 1.2.

The proof of Theorem 4.2.19 requires another geometric condition for recurrence. A subset D of $\mathbb{R}^d$ is said to be *starshaped about the origin* if $rD \subset D$ for all $r \in (0, 1)$ and *starshaped* if $D - b$ is starshaped about the origin for some $b \in \mathbb{R}^d$.

Proposition 2.16. *Suppose $B \in \mathscr{B}$ is nonpolar and B^c is starshaped. Then B is recurrent.*

Proof. Set $D = B^c$. By translation invariance it can be assumed that D is starshaped about the origin. Since B is nonpolar it follows from Proposition 2.6 that there is a $\delta > 0$ such that $P_x(\tau_B < \infty) \geq \delta$ for $\|x\| \leq 1$. Suppose now that $\|x\| > 1$ and set $r = 1/\|x\|$. Then $rB = rD^c = (rD)^c \supset D^c = B$ and hence by Proposition 2.14

$$P_x(\tau_B < \infty) = P_{rx}(\tau_{rB} < \infty) \geq P_{rx}(\tau_B < \infty) \geq \delta.$$

Thus $P_x(\tau_B < \infty) \geq \delta > 0$ for all $x \in \mathbb{R}^d$, so B is recurrent by Proposition 2.9.

The next four results pertain to linear Brownian motion.

Proposition 2.17. *Let $d = 1$ and $x, y \in \mathbb{R}$. Then $P_x(\tau_{\{x\}} = 0) = 1$ and, for $y \neq x$, $P_x(\tau_{\{y\}} \in \cdot)$ is a probability distribution on $(0, \infty)$ having density $|y - x|(2\pi t^3)^{-1/2} \exp(-(y - x)^2/2t)$, $t > 0$.*

Proof. Observe that $P_x(\tau_{[x,\infty)} \leq t) \geq P_x(X(t) > x) = \frac{1}{2}$ for $t > 0$. Thus $P_x(\tau_{[x,\infty)} = 0) \geq \frac{1}{2}$ and hence $P_x(\tau_{[x,\infty)} = 0) = 1$ by Blumenthal's zero–one law. Similarly $P_x(\tau_{(-\infty,x]} = 0) = 1$, so $P_x(\tau_{\{x\}} = 0) = 1$ by continuity of paths. Suppose now that $x < y$ and $t > 0$. Since $P_y(X(s) > y) = \frac{1}{2}$ for all $s > 0$ and

$P_x(\tau_{\{y\}} = t) = 0$, it follows from the strong Markov property that

$$P_x(X(t) > y) = P_x(\tau_{\{y\}} \le t, X(t) > y) = P_x(\tau_{\{y\}} \le t)/2.$$

Consequently by the change of variables $s = (y - x)^2 t/z^2$

$$\begin{aligned} P_x(\tau_{\{y\}} \le t) &= 2P_x(X(t) > y) \\ &= 2\int_{|y-x|}^{\infty} (2\pi t)^{-1/2} \exp(-z^2/2t)\, dz \\ &= \int_0^t |y - x|(2\pi s^3)^{-1/2} \exp(-(y - x)^2/2s)\, ds \end{aligned}$$

as desired. A similar proof works if $y < x$.

Proposition 2.18. *If $d = 1$ and $B \in \mathscr{B}$ is nonempty, then B is nonpolar.*

Proof. Choose $x \in B$. Then $P_x(\tau_B = 0) \ge P_x(\tau_{\{x\}} = 0) = 1$ and hence B is nonpolar.

Proposition 2.19. *If $d = 1$ and $B \in \mathscr{B}$, then $P_.(\tau_B = \tau_{\bar{B}}) = 1$ on $\mathbb{R}$.*

Proof. Let $y \in \bar{B}$. Now $\lim_{z \to y} P_y(\tau_{\{z\}} \le t) = 1$ for all $t > 0$ by Proposition 2.17 and hence $P_y(\tau_B = 0) = 1$. Thus by the strong Markov property

$$P_x(\tau_B > \tau_{\bar{B}}) \le E_x(P_{X(\tau_{\bar{B}})}(\tau_B > 0); \tau_B < \infty) = 0$$

for all $x \in \mathbb{R}$.

Proposition 2.20. *Let $d = 1$, let $-\infty < a < x < b < \infty$, and set $B = R \setminus (a, b)$. Then $P_x(\tau_B = b) = 1 - P_x(\tau_B = a) = (x - a)/(b - a)$ and $E_x(\tau_B) = (x - a)(b - x)$.*

Proof. By Proposition 2.8, $P_x(\tau_B < \infty) = 1$ and $E_x(\tau_B) < \infty$. By continuity of paths, $P_x(X(\tau_B) = a \text{ or } b) = 1$. By Proposition 1.5.2 $E_x(X(\tau_B)) = x$ and $E_x(X(\tau_B) - x)^2 = E_x(\tau_B)$. The desired conclusions follow easily from these observations.

Let $B \in \mathscr{B}$. The *hitting distribution* $h_B(x, \cdot)$ of B for Brownian motion starting at x is defined by $h_B(x, A) = P_x(\tau_B < \infty, X(\tau_B) \in A)$. Note that $h_B(x, \cdot)$ is supported on $\bar{B}$ and has total measure $P_x(\tau_B < \infty)$. Thus $h_B(x, \cdot)$ is a probability measure for each $x \in \mathbb{R}^d$ if and only if B is recurrent. For $f \ge 0$, $h_B f(x) = E_x(f(X(\tau_B)); \tau_B < \infty)$.

Proposition 2.21. *Let $r > 0$. Then $h_{S_r}(0, \cdot) = \sigma_r$ and $E_0(\tau_{S_r}) = r^2/d$.*

Proof. By Proposition 2.8, $P_0(\tau_{S_r} < \infty) = 1$ and $E_0(\tau_{S_r}) < \infty$. The first result now follows from Proposition 2.15. By Proposition 1.5.2, $E_0(\tau_{S_r}) = E_0\|X(\tau_{S_r})\|^2/d = r^2/d$, so the second result is also valid.

A *hyperplane* in $\mathbb{R}^d$ is a set of the form $\{x \in \mathbb{R}^d : x \cdot b = a\}$, where $b \in \mathbb{R}^d\backslash\{0\}$ and $a \in \mathbb{R}$.

Proposition 2.22. *Let $d \geq 2$, let P be a hyperplane in $\mathbb{R}^d$ and let $\sum_P$ denote surface area measure on P. Then for $x \in \mathbb{R}^d\backslash P$*

$$h_P(x, dy) = \frac{\Gamma(d/2)d(x, P)}{\pi^{d/2}\|y - x\|^d} \textstyle\sum_P (dy),$$

Proof. By translation and orthogonal invariance it can be assumed that $P = \{(y_1, \ldots, y_d) : y_d = 0\}$ and that $x = (0, \ldots, 0, x_d)$ where $x_d > 0$. Let $X(t)$, $t \geq 0$, be Brownian motion starting at x and let $Z(t)$ be the first $d - 1$ coordinates of $X(t)$. Then $Z(t)$, $t \geq 0$, is $(d - 1)$-dimensional Brownian motion starting at the origin.

Let f be bounded on $\mathbb{R}^d$ and write $f(z_1, \ldots, z_{d-1}, 0)$ as $f(z, 0)$ for $z = (z_1, \ldots, z_{d-1}) \in \mathbb{R}^{d-1}$. Now $h_P f(x) = E_x(f(X(\tau_P))) = E_x(f(Z(\tau_P), 0))$. Since τ_P depends only on the last coordinate of $X(t)$, $t \geq 0$, it is independent of $Z(t)$, $t \geq 0$. Note that for $t > 0$

$$E_x(f(Z(t), 0)) = \int_{\mathbb{R}^{d-1}} f(z, 0)(2\pi t)^{-(d-1)/2} \exp(-\|z\|^2/2t)\, dz.$$

By Proposition 2.17, τ_p has the density $x_d(2\pi t^3)^{-1/2} \exp(-x_d^2/\mathrm{dt})$, $t > 0$. Therefore

$$\begin{aligned} h_P f(x) &= \int_0^\infty x_d(2\pi t^3)^{-1/2} \exp(-x_d^2/2t) \\ &\quad \times \left(\int_{\mathbb{R}^{d-1}} f(z, 0)(2\pi t)^{-(d-1)/2} \exp(-\|z\|^2/2t)\, dz\right) dt \\ &= \int_{\mathbb{R}^{d-1}} f(z, 0) \frac{\Gamma(d/2)x_d}{\pi^{d/2}(x_d^2 + \|z\|^2)^{d/2}}\, dz \\ &= \int_P f(y) \frac{\Gamma(d/2)x_d}{\pi^{d/2}\|y - x\|^d} \textstyle\sum_P (dy), \end{aligned}$$

which is equivalent to the desired conclusion.

Let $x \in \mathbb{R}\backslash\{0\}$. For planar Brownian motion starting at $(0, x) \in \mathbb{R}^2$ the hitting distribution of the first coordinate axis has the Cauchy density $|x|/\pi(x^2 + y^2)$, $y \in \mathbb{R}$. This result follows immediately from the previous proposition.

It has the following interesting application, whose derivation is left as an exercise for the reader. Consider planar Brownian motion. Let b and c be two points on the boundary of a convex open set D and let A be a boundary arc from b to c. Choose $x \in D$ and let α be the angle in radians determined by A and bxc. Then $h_{D^c}(x, A) \leq \alpha/\pi$.

3. Regular Points

Let $B \in \mathscr{B}$. Since τ_B is a stopping time $\{\tau_B = 0\} \in \mathfrak{F}_{0+}$. Thus by Blumenthal's zero–one law for each $x \in \mathbb{R}^d$, $P_x(\tau_B = 0)$ equals zero or one. The point x is said to be *regular* for B if $P_x(\tau_B = 0) = 1$ and *irregular* for B if $P_x(\tau_B = 0) = 0$. If Brownian motion starts at an irregular point of B, then with probability one there is a $t = t(\omega) > 0$ such that $X(s) \notin B$ for $0 < s < t$. If it starts at a regular point of B, then with probability one there is no such t. Regularity is a local property. For suppose $A \in \mathscr{B}$ and there is an open neighborhood U of x such that $A \cap U = B \cap U$. Then $\{\tau_A = 0\} = \{\tau_B = 0\}$ and hence $P_x(\tau_A = 0) = P_x(\tau_B = 0)$. Thus x is regular for A if and only if it is regular for B.

The set of points $x \in \mathbb{R}^d$ which are regular for B is denoted by B^r. By continuity of paths $\mathring{B} \subset B^r \subset \bar{B}$. If B is polar then B^r is clearly empty. The converse to this result is nontrivial and will be verified in Section 6. In this section some elementary properties related to regularity are obtained.

Proposition 3.1. *Let $x \in \mathbb{R}^d$. Then x is regular for $\{x\}$ if and only if $d = 1$.*

Proof. This result follows from Propositions 2.5 and 2.17.

Proposition 3.2. *If $d = 1$ and $B \in \mathscr{B}$, then $B^r = \bar{B}$.*

Proof. This result follows from Propositions 2.17 and 2.19.

The sufficient condition in the next result is called the *cone condition for regularity*.

Proposition 3.3. *Let $B \in \mathscr{B}$ and $x \in \mathbb{R}^d$. If there is a cone C having vertex x such that $C \cap B_r(x) \subset B$ for some $r > 0$, then x is regular for B.*

Proof. By translation invariance it can be assumed that $x = 0$. Set $C_r = C \cap B_r$. By scale invariance, for $t > 0$

$$P_0(\tau_B \leq t) \geq P_0(X(t) \in B) \geq P_0(X(t) \in C_r) = P_0(X(1) \in C_{t^{-1/2}r}),$$

which approaches $P_0(X(1) \in C) > 0$ as $t \to 0$. Thus $P_0(\tau_B = 0) > 0$ and hence 0 is regular for B as desired.

Proposition 3.4. *All points on a sphere are regular for the sphere.*

Proof. Choose $c \in \mathbb{R}^d$, $r > 0$, and $x \in S_r(c)$. By Proposition 3.3, x is regular for both $\{y : \|y - c\| \leq r\}$ and $\{y : \|y - c\| \geq r\}$. Thus by continuity of paths x is regular for $S_r(c)$. (This argument extends in an obvious manner to other smooth surfaces.

Proposition 3.5. *Let $B \in \mathscr{B}$, $b \in B^r$, and $t > 0$ and let U be an open neighborhood of b. Then*

$$\lim_{x \to b} P_x(\tau_B \leq t, X(\tau_B) \in U) = 1.$$

Proof. Choose $\varepsilon > 0$. By translation invariance and continuity of paths there is an $s \in (0, t)$ such that

$$\varliminf_{x \to b} P_x(X(u) \in U \text{ for } 0 \leq u \leq s) = \varliminf_{x \to b} P_b(X(u) \in U + b - x$$
$$\text{for} \quad 0 \leq u \leq s) \geq 1 - \varepsilon.$$

Now $P_{\cdot}(\tau_B \leq s)$ is lower semicontinuous on $\mathbb{R}^d$ by Proposition 2.1 and $P_b(\tau_B = 0) = 1$, so $\lim_{x \to b} P_x(\tau_B > s) = 0$. Consequently

$$\varliminf_{x \to b} P_x(\tau_B \leq t, X(\tau_B) \in U) \geq \varliminf_{x \to b} P_x(\tau_B \leq s, X(\tau_B) \in U) \geq 1 - \varepsilon$$

and hence the desired conclusion holds.

Proposition 3.6. *Let $B \in \mathscr{B}$ and $b \in \partial B$. If $b \in B^r$, then $h_B(x, \cdot)$ converges completely to δ_b as $x \to b$. Conversely if $h_B(x, \cdot)$ converges completely to δ_b as $x \to b$ within B^c, then $b \in B^r$.*

Proof. It follows immediately from the previous proposition that if $b \in B^r$, then $h_B(x, \cdot)$ converges completely to δ_b as $x \to b$. In proving the converse result it can be assumed that $d \geq 2$ since $B^r = \bar{B}$ if $d = 1$ by Proposition 3.2.

Suppose then that $d \geq 2$ and that $h_B(x, \cdot)$ converges completely to δ_b as $x \to b$ within B^c. It will first be shown that $h_B(b, \{b\}) = 1$. Let U be an open neighborhood of b and choose $\varepsilon > 0$. Since $h_B(x, U) \to 1$ as $x \to b$ within B^c, there is an open neighborhood V of b such that $h_B(x, U) \geq 1 - \varepsilon$ for $x \in B^c \cap V$. By continuity of paths there is a $t > 0$ such that $P_b(X(s) \in U \cap V$

for $0 \le s \le t) \ge 1 - \varepsilon$. Now

$$\begin{aligned} h_B(b, U) &= P_b(\tau_B \le t, X(\tau_B) \in U) + E_b(h_B(X(t), U); \tau_B > t) \\ &\ge (1 - \varepsilon)[P_b(\tau_B \le t, X(\tau_B) \in U) + P_b(\tau_B > t, X(t) \in V)] \\ &\ge (1 - \varepsilon)P_b(X(s) \in U \cap V \text{ for } 0 \le s \le t) \\ &\ge (1 - \varepsilon)^2. \end{aligned}$$

Since ε can be made arbitrarily small, $h_B(b, U) = 1$. Thus $h_B(b, \{b\}) = 1$.

By the last result $P_b(\tau_B < \infty$ and $X(\tau_B) = b) = 1$. Since $\{b\}$ is polar by Proposition 2.5, $P_b(\tau_B = 0) = 1$ and hence $b \in B^r$ as desired.

The next result states in particular that the set of points in B which are irregular for B has Lebesgue measure zero. In Section 6 it will be shown that this set is actually polar.

Proposition 3.7. *Let $B \in \mathscr{B}$. Then $(B^r)^c \in \mathscr{B}$, $B \backslash B^r \in \mathscr{B}$, and $B \backslash B^r$ has Lebesgue measure zero.*

Proof. For n a positive integer $P_{\cdot}(\tau_B \le 1/n)$ is a lower semicontinuous function and hence $\{x : P_x(\tau_B \le 1/n) > 1 - 1/n\}$ is open. Now

$$B^r = \{x : P_x(\tau_B = 0) = 1\} = \bigcap_n \{x : P_x(\tau_B \le 1/n) > 1 - 1/n\}.$$

Thus $(B^r)^c \in \mathscr{B}$ and hence $B \backslash B^r = B \cap (B^r)^c \in \mathscr{B}$.

Let A be a compact subset of $B \backslash B^r$. Then for $x \in A$

$$\overline{\lim_{t \to 0}}\, P_x(X(t) \in A) \le \lim_{t \to 0} P_x(\tau_B \le t) = 0.$$

Define the bounded function f on $\mathbb{R}^d$ by $f(y) = \int I_A(y + x) I_A(x)\, dx$. It is easily seen that $E_0 f(X(t)) = \int_A P_x(X(t) \in A)\, dx$. Now f is continuous (approximate $I_A(y + x)$ by $g(y + x)$ where g is continuous and has compact support) and hence

$$\int I_A(x)\, dx = f(0) = \lim_{t \to 0} E_0 f(X(t)) = \lim_{t \to 0} \int_A P_x(X(t) \in A)\, dx = 0,$$

so A has Lebesgue measure zero. Consequently $B \backslash B^r$ has Lebesgue measure zero.

The dependence of τ_B on B will now be studied. Observe first that if $A \subset B$, then $\tau_A \ge \tau_B$.

Proposition 3.8. *If $B_n \uparrow B$, then $\tau_{B_n} \downarrow \tau_B$. If each B_n is closed and $B_n \downarrow B$, then $\tau_{B_n} \uparrow \tau_B$ on $\{X(0) \in B^c\}$ and $P_{\cdot}(\tau_{B_n} \uparrow \tau_B) = 1$* on $B^c \cup B^r$.

Proof. By hypothesis $\tau = \lim_n \tau_{B_n}$ exists. Suppose first that $B_n \uparrow B$. Then $\tau \geq \tau_B$ so if $\tau_B = \infty$, then $\tau = \tau_B$. Assume that $\tau_B < \infty$ and let $\varepsilon > 0$. There is a $t \in (0, \tau_B + \varepsilon)$ such that $X(t) \in B$. For n sufficiently large $X(t) \in B_n$ and hence $\tau_{B_n} \leq t < \tau_B + \varepsilon$. Consequently $\tau \leq \tau_B$ and hence $\lim_n \tau_{B_n} = \tau = \tau_B$.
Suppose now that each B_n is closed and that $B_n \downarrow B$. Then B is closed. Clearly $\tau \leq \tau_B$ so if $\tau = \infty$, then $\tau = \tau_B$. Assume that $\tau < \infty$. Then $X(\tau_{B_n}) \to X(\tau)$ by continuity of paths and hence $X(\tau) \in \bigcap_n B_n = B$. If $X(0) \in B^c$, then $\tau > 0$ so $\tau_B \leq \tau$ and hence $\tau_B = \tau$. If $x \in B^r$, then $P_x(\tau_B = 0) = P_x(\tau_{B_n} = 0$ for all $n) = 1$ and hence $P_x(\tau = \tau_B) = 1$. Thus

$$P_x\left(\lim_n \tau_{B_n} = \tau = \tau_B\right) = 1$$

for $x \in B^c \cup B^r$.

4. Transition Operators for the Killed Process

Let $B \in \mathscr{B}$. Set $q_B^t(x, A) = P_x(\tau_B > t, X(t) \in A)$ for $t \geq 0$, $x \in \mathbb{R}^d$, and $A \subset \mathbb{R}^d$. Then $q_B^t f = E_{\cdot}(f(X(t)); \tau_B > t)$ for $t \geq 0$ and $f \geq 0$ on $\mathbb{R}^d$. If B is polar, $q_B^t = p^t$ for $t \geq 0$. As shown in the beginning of the proof of Theorem 4.3 below, the operators q_B^t, $t \geq 0$, satisfy the semigroup property. These operators are called the *transition operators for Brownian motion killed when it hits B*. In this section densities $q_B(t, x, \cdot)$ for the distributions $q_B^t(x, \cdot)$, $t > 0$, will be constructed and symmetry, continuity, and positivity properties of these densities will be obtained.
Let $t > 0$, $x \in \mathbb{R}^d$, and $A \subset \mathbb{R}^d$. Then

$$P_x(X(t) \in A) = P_x(\tau_B > t, X(t) \in A) + P_x(\tau_B \leq t, X(t) \in A)$$

and hence

$$p^t(x, A) = q_B^t(x, A) + P_x(\tau_B \leq t, X(t) \in A). \tag{1}$$

By (1) and (1.5.3)

$$\begin{aligned} q_B^t(x, A) &= p^t(x, A) - E_x\left(\int_A p(t - \tau_B, X(\tau_B), y)\, dy; \tau_B < t\right) \\ &= \int_A [p(t, x, y) - E_x(p(t - \tau_B, X(\tau_B), y); \tau_B < t)]\, dy. \end{aligned}$$

Set

$$r_B(t, x, y) = E_x(p(t - \tau_B, X(\tau_B), y); \tau_B < t), \qquad t > 0 \quad \text{and} \quad x, y \in \mathbb{R}^d, \tag{2}$$

and let $q_B^t(x, y) = q_B(t, x, y)$ be defined by

(3) $$q_B(t, x, y) = p(t, x, y) - r_B(t, x, y) \qquad \text{for } t > 0 \quad \text{and} \quad x, y \in \mathbb{R}^d.$$

Then $q_B^t(x, A) = \int_A q_B(t, x, y)\, dy$.

Let $t > 0$ and $x \in \mathbb{R}^d$. Set $A = \{y \in \mathbb{R}^d : q_B(t, x, y) < 0\}$. Then $\int_A q_B(t, x, y)\, dy = q_B^t(x, A) \geq 0$ and therefore $q_B(t, x, \cdot) \geq 0$ a.e. on $\mathbb{R}^d$. It follows from (2) and Fatou's lemma that $r_B(t, x, \cdot)$ is lower semicontinuous on $\mathbb{R}^d$. Thus $q_B(t, x, \cdot)$ is upper semicontinuous on $\mathbb{R}^d$ and hence $q_B(t, x, \cdot) \geq 0$ everywhere on $\mathbb{R}^d$. Consequently $q_B(t, x, \cdot)$ is a density for the measure $q_B^t(x, \cdot)$. Clearly $r_B(t, x, \cdot) \geq 0$, so $q_B(t, x, \cdot) \leq p(t, x, \cdot)$ on $\mathbb{R}^d$. If B is polar, then $r_B(t, x, \cdot) = 0$ and hence $q_B(t, x, \cdot) = p(t, x, \cdot)$ on $\mathbb{R}^d$.

The next result implies in particular that if $x \notin \bar{B}$, then $\lim_{t \to 0} q_B(t, x, x)/p(t, x, x) = 1$.

Proposition 4.1. *Let $B \in \mathscr{B}$ and let $a \in (\bar{B})^c$. There is an $r > 0$ such that $\lim_{t \to 0} q_B(t, x, y)/p(t, x, y) = 1$ uniformly for $x, y \in B_r(a)$.*

Proof. Choose $r > 0$ such that $3r < d(a, B)$. Choose $x, y \in B_r(a)$ and set $\alpha = d(y, B) > d(a, B) - r$. Then $\|x - y\| \leq 2r < d(a, B) - r < \alpha$. It is easily seen that $(2\pi u)^{-d/2} \exp(-\alpha^2/2u)$ is increasing in u for $0 \leq u \leq \alpha^2/d$. Consequently

$$p(t - s, z, y) \leq (2\pi(t - s))^{-d/2} \exp[-\alpha^2/2(t - s)] \leq (2\pi t)^{-d/2} \exp(-\alpha^2/2t)$$

for $0 \leq s < t \leq \alpha^2/d$ and $z \in B$. Thus by (2)

$$r_B(t, x, y) \leq (2\pi t)^{-d/2} \exp(-\alpha^2/2t) \qquad \text{for} \quad 0 < t \leq \alpha^2/d$$

and hence

$$r_B(t, x, y)/p(t, x, y) \leq \exp[-(\alpha^2 - \|y - x\|^2/2t] \leq \exp[-(\alpha^2 - 4r^2)/2t]$$

for $0 < t \leq \alpha^2/d$. Therefore $\lim_{t \to 0} r_B(t, x, y)/p(t, x, y) = 0$ uniformly for $x, y \in B_r(a)$. The desired conclusion now follows from (3).

An important step in proving Theorem 4.3 below is to go from "almost everywhere" to "everywhere." The next result provides the technical tool for carrying out this step.

Proposition 4.2. *Let $B \in \mathscr{B}$, $t > 0$, and $x, y \in \mathbb{R}^d$. Then*

$$\int q_B(t - \varepsilon, x, z) p(\varepsilon, z, y)\, dz \downarrow q_B(t, x, y) \qquad \text{as } \varepsilon \downarrow 0$$

and

$$\int p(\varepsilon, x, z) q_B(t - \varepsilon, z, y)\, dz \downarrow q_B(t, x, y) \qquad \textit{as } \varepsilon \downarrow 0.$$

Proof. The first result is equivalent to

$$(4) \qquad \int r_B(t - \varepsilon, x, z) p(\varepsilon, z, y)\, dz \uparrow r_B(t, x, y) \qquad \text{as } \varepsilon \downarrow 0.$$

Now (4) follows immediately from the formula

$$\int r_B(t - \varepsilon, x, z) p(\varepsilon, z, y)\, dz = E_x(p(t - \tau_B, X(\tau_B), y); \tau_B < t - \varepsilon),$$

which is a consequence of (2) and the semigroup property of $p(t, x, y)$.

The second result is equivalent to

$$(5) \qquad \int p(\varepsilon, x, z) r_B(t - \varepsilon, z, y)\, dz \uparrow r_B(t, x, y) \qquad \text{as } \varepsilon \downarrow 0.$$

To prove (5) let $\varepsilon \in (0, t)$ and note that by (2)

$$r_B(t, x, y) = E_x(p(t - \tau_B, X(\tau_B), y); \tau_B < \varepsilon) + E_x(p(t - \tau_B, X(\tau_B), y); \varepsilon < \tau_B < t).$$

Apply the semigroup property of $p(t, x, y)$ to the first term on the right-hand side of this equation and the Markov property to the second term to obtain

$$(6) \qquad r_B(t, x, y) = \int r_B(\varepsilon, x, z) p(t - \varepsilon, z, y)\, dz + \int q_B(\varepsilon, x, z) r_B(t - \varepsilon, z, y)\, dz.$$

Since $p(t - \varepsilon, z, y) \geq r_B(t - \varepsilon, z, y)$ and $q_B(\varepsilon, x, z) + r_B(\varepsilon, x, z) = p(\varepsilon, x, z)$, it follows from (6) that

$$(7) \qquad r_B(t, x, y) \geq \int p(\varepsilon, x, z) r_B(t - \varepsilon, z, y)\, dz.$$

It follows easily from (7) and the semigroup property of $p(t, x, y)$ that the left-hand side of (5) increases as ε decreases for $0 < \varepsilon < t$. Thus to complete the proof of (5) it is enough to show that

$$(8) \qquad \underline{\lim}_{\varepsilon \to 0} \int p(\varepsilon, x, z) r_B(t - \varepsilon, z, y)\, dz \geq r_B(t, x, y).$$

Suppose $x \in B^r$. It follows from (2) and Proposition 3.5 that

$$\underline{\lim}_{\substack{s \to t \\ z \to x}} r_B(s, z, y) \geq p(t, x, y) = r_B(t, x, y)$$

and hence that (8) holds. Suppose now that $x \notin B^r$. Then

$$\overline{\lim}_{\varepsilon \to 0} \int r_B(\varepsilon, x, z) p(t - \varepsilon, z, y)\, dz \leq p(t, 0) \lim_{\varepsilon \to 0} \int r_B(\varepsilon, x, z)\, dz$$

$$= p(t, 0) \lim_{\varepsilon \to 0} P_x(\tau_B < \varepsilon) = p(t, 0) P_x(\tau_B = 0) = 0.$$

Thus the first term in (6) approaches zero as $\varepsilon \to 0$. It now follows from (6) that

$$r_B(t, x, y) = \lim_{\varepsilon \to 0} \int q_B(\varepsilon, x, z) r_B(t - \varepsilon, z, y)\, dz$$
$$\leq \varliminf_{\varepsilon \to 0} \int p(\varepsilon, x, z) r_B(t - \varepsilon, z, y)\, dz$$

and hence that (8) again holds. This completes the proof of the proposition.

A number of important properties of q_B are summarized in the next result, which is due to Hunt [1].

Theorem 4.3. *Let $B \in \mathscr{B}$. Consider $q_B = q_B(t, x, y)$ as (t, x, y) ranges over $(0, \infty) \times \mathbb{R}^d \times \mathbb{R}^d$. Then q_B satisfies the semigroup property*

$$q_B(s + t, x, y) = \int q_B(s, x, z) q_B(t, z, y)\, dz. \tag{9}$$

Also $q_B(t, x, y)$ is symmetric in x and y and vanishes if either $x \in B^r$ or $y \in B^r$, $q_B(t, x, y)$ is upper semicontinuous in x and y separately, q_B is jointly continuous on $(0, \infty) \times (\mathbb{R}^d \backslash \partial B) \times (\mathbb{R}^d \backslash \partial B)$, $q_B(t, x, y) > 0$ if x and y are in the same component of $(\bar{B})^c$, and $q_B(t, x, y) = 0$ if x and y are in different components of $(\bar{B})^c$.

Proof. First the semigroup property of the operators q_B^t, $t \geq 0$, will be obtained. Let f be bounded on $\mathbb{R}^d$. Then by the Markov property for s, $t \geq 0$

$$\begin{aligned} q_B^{s+t} f(x) &= E_x(f(X(s + t)); \tau_B > s + t) \\ &= E_x(f(X(t, \theta_s \omega)); \tau_B \cdot \theta_s > t, \tau_B > s) \\ &= E_x(E_{X(s)}(f(X(t)); \tau_B > t); \tau_B > s) \\ &= E_x(q_B^t f(X(s)); \tau_B > s) \\ &= (q_B^s q_B^t f)(x), \end{aligned}$$

so that $q_B^{s+t} = q_B^s q_B^t$ for s, $t \geq 0$. Consequently for s, $t > 0$

$$q_B(s + t, x, u) = \int q_B(s, x, z) q_B(t, z, u)\, dz \qquad \text{for almost all } u.$$

Thus for $0 < \varepsilon < t$

$$\int q_B(s + t - \varepsilon, x, u) p(\varepsilon, u, y)\, du = \int q_B(s, x, z) \left[\int q(t - \varepsilon, z, u) p(\varepsilon, u, y)\, du \right] dz.$$

The semigroup property (9) now follows by letting $\varepsilon \downarrow 0$ in the above equation and using Proposition 4.2 and the dominated convergence theorem.

Let $t > 0$. In order to show that $q_B(t, x, y)$ is symmetric in x and y it will first be shown that

$$(10)\quad \int_C P_x(\tau_B > t, X(t) \in A)\, dx = \int_A P_x(\tau_B > t, X(t) \in C)\, dx, \qquad A, C \subset \mathbb{R}^d.$$

By Proposition 3.7 it can be assumed that A and C are both subsets of B^c. Under this assumption (10) will be proven first when B is open, next when B is closed, and finally when $B \in \mathscr{B}$.

Suppose that B is open. Set $J_n = \{2^{-n}mt : m = 1, \ldots, 2^n\}$. Since $P_x(\tau_B = t) = 0$ and $p(s, x, y)$ is symmetric in x and y,

$$\begin{aligned}
&\int_C P_x(\tau_B > t, X(t) \in A)\, dx \\
&\quad = \int_C P_x(X(s) \notin B \text{ for } 0 < s \le t, X(t) \in A)\, dx \\
&\quad = \lim_n \int_C P_x(X(s) \notin B \text{ for } s \in J_n, X(t) \in A)\, dx \\
&\quad = \lim_n \int \cdots \int I_C(x_0) I_A(x_{2^n}) \prod_{m=1}^{2^n} [p(2^{-n}t, x_{m-1}, x_m) I_{B^c}(x_m)\, dx_0 \cdots dx_{2^n} \\
&\quad = \lim_n \int \cdots \int I_A(x_0) I_C(x_{2^n}) \prod_{m=1}^{2^n} [p(2^{-n}t, x_{m-1}, x_m) I_{B^c}(x_m)]\, dx_0 \cdots dx_{2^n} \\
&\quad = \lim_n \int_A P_x(X(s) \notin B \text{ for } s \in J_n, X(t) \in C)\, dx \\
&\quad = \int_A P_x(X(s) \notin B \text{ for } 0 < s \le t, X(t) \in C)\, dx \\
&\quad = \int_A P_x(\tau_B > t, X(t) \in C)\, dx.
\end{aligned}$$

Thus (10) holds when B is open.

Suppose next that B is closed. Set $B_n = \{x \in \mathbb{R}^d : d(x, B) < 1/n\}$. Then B_n is open, $B_n \downarrow B$ and $\bar{B}_n \downarrow B$. It follows from Proposition 3.8 that $\{\tau_{B_n} > t\} \uparrow \{\tau_B > t\}$ on $\{X(0) \in B^c\}$. Thus

$$\begin{aligned}
\int_C P_x(\tau_B > t, X(t) \in A)\, dx &= \lim_n \int_C P_x(\tau_{B_n} > t, X(t) \in A)\, dx \\
&= \lim_n \int_A P_x(\tau_{B_n} > t, X(t) \in C)\, dx \\
&= \int_A P_x(\tau_B > t, X(t) \in C)\, dx
\end{aligned}$$

and hence (10) holds when B is closed.

Suppose finally that $B \in \mathscr{B}$. Let B_n be closed sets such that $B_n \uparrow B$. Then $\{\tau_B > t\} \subset \bigcap_n \{\tau_{B_n} > t\} \subset \{\tau_B \ge t\}$ by Proposition 3.8. Since $P_x(\tau_B = t) = 0$

for all $x \in \mathbb{R}^d$, the remainder of the proof in this case is the same as that given in the previous paragraph. This completes the proof that (10) holds for all $B \in \mathscr{B}$.

Equation (10) can be written as

$$\int_A \int_C (q_B(t, x, y) - q_B(t, y, x))\, dx\, dy = 0.$$

This implies that $q_B(t, x, y) = q_B(t, y, x)$ for almost all x, y. Thus for all x, $y \in \mathbb{R}^d$ and $\varepsilon, \delta > 0$ such that $\varepsilon + \delta < t$

$$\iint p(\varepsilon, x, u) q_B(t - \varepsilon - \delta, u, v) p(\delta, v, y)\, du\, dv$$
$$= \iint p(\delta, y, v) q_B(t - \varepsilon - \delta, v, u) p(\varepsilon, u, x)\, du\, dv.$$

It follows by using Proposition 4.2 to let $\varepsilon \downarrow 0$ and then $\delta \downarrow 0$ in the above equation that $q_B(t, x, y) = q_B(t, y, x)$ for all $x, y \in \mathbb{R}^d$.

If $x \in B^r$, then $r_B(t, x, y) = p(t, x, y)$ and hence $q_B(t, x, y) = 0$. If $y \in B^r$, then $q_B(t, x, y) = q_B(t, y, x) = 0$. Since $r_B(t, x, y)$ is lower semicontinuous in y, $q_B(t, x, y)$ is upper semicontinuous in y. By symmetry it is also upper semicontinuous in x. Since $\mathring{B} \subset B^r$, $q_B(t, x, y) = 0$ if either $x \in \mathring{B}$ or $y \in \mathring{B}$. Thus q_B is jointly continuous on $(0, \infty) \times \mathring{B} \times \mathbb{R}^d$ and on $(0, \infty) \times \mathbb{R}^d \times \mathring{B}$.

It will now be shown that q_B is jointly continuous on $(0, \infty) \times (\bar{B})^c \times (\bar{B})^c$. Let $t_0 > 0$, $x_0 \in (\bar{B})^c$, and $y_0 \in (\bar{B})^c$ be given. Let C be a compact subset of $(\bar{B})^c$ containing x_0 and y_0 in its interior. Choose $\varepsilon > 0$. Since $p(t, x, y)$ is uniformly continuous on $(0, \infty) \times \bar{B} \times C$, it follows from (2) that there is a $\delta > 0$ such that

$$|r_B(t, x, y) - r_B(t_0, x, y_0)| < \varepsilon, \qquad x \in \mathbb{R}^d, \quad |t - t_0| < \delta, \quad \text{and} \quad \|y - y_0\| < \delta$$

and

$$|r_B(t_0, y_0, x) - r_B(t_0, y_0, x_0)| < \varepsilon, \qquad \|x - x_0\| < \delta.$$

Since $r_B(t, x, y)$ is symmetric in x and y, the latter inequality can be written as

$$|r_B(t_0, x, y_0) - r_B(t_0, x_0, y_0)| < \varepsilon, \qquad \|x - x_0\| < \delta.$$

Consequently

$$|r_B(t, x, y) - r_B(t_0, x_0, y_0)| < 2\varepsilon, \qquad |t - t_0| < \delta, \quad \|x - x_0\| < \delta, \quad \text{and} \quad \|y - y_0\| < \delta.$$

This shows that r_B and hence q_B are jointly continuous on $(0, \infty) \times (\bar{B})^c \times (\bar{B})^c$. Consequently q_B is jointly continuous on $(0, \infty) \times (\mathbb{R}^d \backslash \partial B) \times (\mathbb{R}^d \backslash \partial B)$.

It will now be shown that $q_B(t, x, y) = 0$ if x and y are in different components of $(\bar{B})^c$. Let D be a component of $(\bar{B})^c$ and let $x \in (\bar{B})^c \backslash D$. By continuity of paths $P_x(\tau_B > t, X(t) \in D) = 0$ and hence $q_B(t,x,y) = 0$ for almost all $y \in D$. Since $q_B(t,x,y)$ is continuous in y on D, it equals zero for all $y \in D$ as desired.

Let D be a component of $(\bar{B})^c$ and let $x \in D$. The proof of the theorem will be completed by showing that $q_B(t,x,y) > 0$ for $t > 0$ and $y \in D$. To this end set

$$U = \{y \in D : q_B(t,x,y) > 0 \text{ for all } t > 0\}.$$

By Proposition 4.1 there is a $t_0 > 0$ and an $r > 0$ such that $q_B(s,z,y) > 0$ for $0 < s \le t_0$ and $y, z \in B_r(x)$. It follows easily from the semigroup property of q_B that $q_B(t,x,y) > 0$ for $t > 0$ and $y \in B_r(x)$. Thus U is nonempty. It must be shown that $U = D$. Since D is connected it suffices to show that $\bar{U} \cap D \subset \mathring{U}$.

Choose $y_0 \in \bar{U} \cap D$. It follows from Proposition 4.1 as in the previous paragraph that there is an $r > 0$ such that $q_B(u,z,y) > 0$ for $u > 0$ and $z, y \in B_r(y_0)$. Choose $y_1 \in U \cap B_r(y_0)$ and $s > 0$. Then $q_B(s,x,y_1) > 0$, so by continuity $q_B(s,x,z) > 0$ for all z in a subset of $B_r(y_0)$ having positive measure. Thus

$$q_B(t,x,y) \ge \int_{B_r(y_0)} q_B(s,x,z) q_B(t-s,z,y)\,dz > 0, \qquad t > s \quad \text{and} \quad y \in B_r(y_0).$$

Since s can be made arbitrarily small $q_B(t,x,y) > 0$ for all $t > 0$ and $y \in B_r(y_0)$. Thus $B_r(y_0) \subset U$ and hence $y_0 \in \mathring{U}$. Consequently $\bar{U} \cap D \subset \mathring{U}$ as desired. This completes the proof of the theorem.

The dependence of $q_B(t,x,y)$ on B will now be studied.

Proposition 4.4. *Let $B, C \in \mathscr{B}$ with $B \subset C$. Then $q_B(t,x,y) \ge q_C(t,x,y)$ for $t > 0$ and $x, y \in \mathbb{R}^d$.*

Proof. Choose $t > 0$, $x \in \mathbb{R}^d$ and $A \subset \mathbb{R}^d$. Then

$$\begin{aligned}\int_A q_B(t,x,y)\,dy &= P_x(\tau_B > t, X(t) \in A) \ge P_x(\tau_C > t, X(t) \in A) \\ &= \int_A q_C(t,x,y)\,dy.\end{aligned}$$

Thus $q_B(t,x,y) \ge q_C(t,x,y)$ for almost all y. By Proposition 4.2 this inequality holds for all y.

Proposition 4.5. *If $B_n \in \mathscr{B}$ and $B_n \uparrow B$, then $q_{B_n}(t,x,y) \downarrow q_B(t,x,y)$ for $t > 0$ and $x, y \in \mathbb{R}^d$. If each B_n is closed and $B_n \downarrow B$, then $q_{B_n}(t,x,y) \uparrow q_B(t,x,y)$ for $t > 0$ and $x, y \in B^c \cup B^r$.*

Proof. Let $t > 0$. Suppose $B_n \in \mathscr{B}$ and $B_n \uparrow B$. It follows from Proposition 3.8 that $\tau_{B_n} \downarrow \tau_B$ and hence that $\{\tau_{B_n} < t\} \uparrow \{\tau_B < t\}$. Since

(11) $$q_{B_n}(t, x, y) = p(t, x, y) - E_x(p(t - \tau_{B_n}, X(\tau_{B_n}), y)); \tau_{B_n} < t),$$

Fatou's lemma implies that $\overline{\lim}_n q_{B_n}(t, x, y) \leq q_B(t, x, y)$. The first conclusion now follows immediately from Proposition 4.4.

Suppose each B_n is closed and $B_n \downarrow B$. Then B is closed. By Proposition 4.4, $q_{B_n}(t, x, y)$ is nondecreasing in n and $q_{B_n}(t, x, y) \leq q_B(t, x, y)$. If either $x \in B^r$ or $y \in B^r$, then $q_B(t, x, y) = 0$ by Theorem 4.3 and hence $q_{B_n}(t, x, y) \uparrow q_B(t, x, y)$. Suppose now that $x, y \in B^c$. Then $P_x(\tau_{B_n} \uparrow \tau_B) = 1$ by Proposition 3.8 and $P_x(\tau_B = t) = 0$. There is an $n_0 \geq 1$ and a $\delta > 0$ such that $\|z - y\| > \delta$ for $n \geq n_0$ and $z \in B_n$ and hence

$$\sup_{n \geq n_0} \sup_{s > 0, z \in B_n} p(s, z, y) < \infty.$$

It now follows from (11) and the dominated convergence theorem that $q_{B_n}(t, x, y) \uparrow q_B(t, x, y)$ as desired.

Let D be an open subset of $\mathbb{R}^d$. Then $T_D = \tau_{D^c}$ is called the *exit time* of D. If D_0 is a component of D, then $P_{.}(T_D = T_{D_0}) = 1$ on D_0. Set $Q_D^t(x, A) = q_{D^c}^t(x, A)$ for $t \geq 0$, $x \in D$ and $A \subset D$. Then Q_D^t, $t \geq 0$, are called the *transition operators for Brownian motion on* D. Note that $Q_D^t f(x) = E_x(f(X(t)); T_D > t)$ for $t \geq 0$, $x \in D$ and $f \geq 0$ on D. Set $Q_D(t, x, y) = q_{D^c}(t, x, y)$ for $t > 0$ and $x, y \in D$. Then $Q_D^t(x, A) = \int_A Q_D(t, x, y)\, dy$ for $t > 0$, $x \in D$, and $A \subset D$. The properties of q_{D^c} translate immediately into corresponding properties of Q_D. In particular Q_D is jointly continuous on $(0, \infty) \times D \times D$, $Q_D(t, x, y)$ is symmetric in x and y, and $0 \leq Q_D(t, x, y) \leq p(t, x, y)$. Also $Q_D(t, x, y)$ is positive if x and y are in the same component of D and zero if x and y are in different components of D. Moreover $Q_D(t, x, y)$ approaches zero as either x or y approaches a point in $\partial D \cap (D^c)^r$ from within D. If D_0 is a component of D, then $Q_D(t, x, y) = Q_{D_0}(t, x, y)$ for $t > 0$ and $x, y \in D_0$. The semigroup property

$$Q_D(s + t, x, y) = \int_D Q_D(s, x, z) Q_D(t, z, y)\, dz \quad \text{for} \quad s, t > 0 \quad \text{and} \quad x, y \in D$$

is valid. If each D_n is an open subset of D and $D_n \uparrow D$, then

$$Q_{D_n}(t, x, y) \uparrow Q_D(t, x, y), \quad \text{for} \quad t > 0 \quad \text{and} \quad x, y \in D.$$

The next result gives a geometric interpretation of the statement that polar sets are negligible.

Proposition 4.6. *Let D be a connected open subset of $\mathbb{R}^d$ and let B be a closed polar subset of D. Then $D \backslash B$ is connected.*

Proof. Let D_1 and D_2 be disjoint open sets such that D_1 is nonempty and $D\backslash B = D_1 \cup D_2$. Choose $x \in D_1$. Now $X(\tau_{D_2}) \in B$ on $\{X(0) = x, \tau_{D_2} < T_D\}$, so $P_x(\tau_{D_2} < T_D) \leq P_x(\tau_B < \infty) = 0$. Consequently $0 = P_x(T_D > t, X(t) \in D_2) = \int_{D_2} Q_D(t, x, y)\, dy$. Since $Q_D(t, x, \cdot) > 0$ on D it must be true that D_2 has Lebesgue measure zero. Since D_2 is open it is empty. Therefore $D\backslash B$ is connected.

5. Properties of λ-Potentials

Let $\lambda > 0$ throughout this section. The properties of λ-potentials derived here will be used in the following section to prove that $B\backslash B^r$ is polar.

Let g^λ denote the *λ-potential kernel* defined by

$$g^\lambda(x) = \int_0^\infty e^{-\lambda t} p(t, x)\, dt, \qquad x \in \mathbb{R}^d.$$

It is easily seen that $g^\lambda(x) = \infty$ if $d \geq 2$ and $x = 0$ and that $g^\lambda(x) < \infty$ otherwise. Also g^λ is positive and continuous in the extended sense. In particular g^λ is bounded away from zero on compacts. Set $g^\lambda(x, y) = g^\lambda(y - x)$ for $x, y \in \mathbb{R}^d$.

Let $B \in \mathscr{B}$. Set

$$g_B^\lambda(x, y) = \int_0^\infty e^{-\lambda t} q_B(t, x, y)\, dt, \qquad x, y \in \mathbb{R}^d.$$

Then $0 \leq g_B^\lambda(x, y) \leq g^\lambda(x, y)$, $g_B^\lambda(x, y) = 0$ if either $x \in B^r$ or $y \in B^r$ and $g_B^\lambda(x, y)$ is symmetric in x and y. These properties follow immediately from the corresponding properties of q_B.

For $x \in \mathbb{R}^d$, define $h_B^\lambda(x, \cdot)$ by

$$h_B^\lambda(x, A) = E_x(e^{-\lambda \tau_B}; \tau_B < \infty \text{ and } X(\tau_B) \in A).$$

Then $h_B^\lambda f = E_\cdot(e^{-\lambda \tau_B} f(X(\tau_B));\ \tau_B < \infty)$ for $f \geq 0$. The measure $h_B^\lambda(x, \cdot)$ is concentrated on $\bar{B}$. It follows easily from (4.2) that

$$\int_0^\infty e^{-\lambda t} r_B(t, x, y)\, dt = \int h_B^\lambda(x, dz) g^\lambda(z, y), \qquad x, y \in \mathbb{R}^d.$$

Since $p(t, x, y) = q_B(t, x, y) + r_B(t, x, y)$ it follows that

$$g^\lambda(x, y) = g_B^\lambda(x, y) + \int h_B^\lambda(x, dz) g^\lambda(z, y), \qquad x, y \in \mathbb{R}^d. \tag{1}$$

This equation is called the *fundamental identity for λ-potentials.*

Let μ be a measure on $\mathbb{R}^d$. The function $g^\lambda \mu$ on $\mathbb{R}^d$ defined by $g^\lambda \mu(x) = \int g^\lambda(x, y) \mu(dy)$ is called the *λ-potential* of μ. By Fatou's lemma, $g^\lambda \mu$ is lower semicontinuous. If $g^\lambda \mu$ is not identically infinite, then μ is a Radon measure. Thus if μ has compact support and $g^\lambda \mu$ is not identically infinite, then μ is a finite measure.

Many principles of potential theory involve extending a property of the potential of a measure from a set on which the measure is concentrated to the entire space. The following *maximum principle for λ-potentials* is one such result. It will be used to prove Proposition 6.2 below.

Theorem 5.1. *Let μ be a measure on $\mathbb{R}^d$ and let B be a set on which μ is concentrated. Then* $\sup_{x \in \mathbb{R}^d} g^\lambda \mu(x) = \sup_{x \in B} g^\lambda \mu(x)$.

Proof. It can be assumed that $M = \sup_{x \in B} g^\lambda \mu(x) < \infty$. Choose $\varepsilon > 0$ and set $A = \{x \in \mathbb{R}^d : g^\lambda \mu(x) \le M + \varepsilon\}$. Then A is closed since $g^\lambda \mu$ is lower semicontinuous. Observe that for $t > 0$

$$g^\lambda \mu \ge \int_t^\infty e^{-\lambda s} p^s \mu \, ds = e^{-\lambda t} \int_0^\infty e^{-\lambda s} p^{s+t} \mu \, ds = e^{-\lambda t} p^t g^\lambda \mu$$

and hence $g^\lambda \mu \ge (M + \varepsilon) e^{-\lambda t} P(\tau_A > t)$. Therefore $g^\lambda \mu \ge (M + \varepsilon) P_.(\tau_A > 0)$, so $g^\lambda \mu \ge M + \varepsilon$ on $(A^r)^c$ and consequently $B \subset A^r$. It now follows from the fundamental identity for λ-potentials that $g^\lambda \mu \le h_A^\lambda g^\lambda \mu \le M + \varepsilon$. Since ε can be made arbitrarily small $g^\lambda \mu \le M$, which yields the desired conclusion.

It follows from the formula $g^\lambda(x, y) = \int_0^\infty e^{-\lambda t} p(t, x, y)\, dt$ that

$$\int g^\lambda(x, y) \lambda \, dy = 1, \qquad x \in \mathbb{R}^d, \tag{2}$$

i.e., that the λ-potential of the measure μ given by $\mu(dy) = \lambda\, dy$ equals 1 on $\mathbb{R}^d$. Let $B \in \mathscr{B}$. Observe that

$$\int h_B^\lambda(x, dz) g^\lambda(z, y) = \int h_B^\lambda(y, dz) g^\lambda(z, x), \qquad x, y \in \mathbb{R}^d. \tag{3}$$

This result is trivially true if $x = y$ and follows from the fundamental identity for λ-potentials and the finiteness and symmetry of $g^\lambda(x, y)$ and $g_B^\lambda(x, y)$ in x and y if $x \ne y$. Set $\mu_B^\lambda = \int \lambda\, dy\, h_B^\lambda(y, \cdot)$. It follows easily from (2) and (3) that

$$g^\lambda \mu_B^\lambda = h_B^\lambda 1 = E_.(e^{-\lambda \tau_B}), \tag{4}$$

where $e^{-\infty} = 0$. The measure μ_B^λ, called the *λ-equilibrium measure* of B, is concentrated on $\bar{B}$ (by Theorem 6.5 below it is concentrated on $\partial B \cap B^r$) and its λ-potential $g^\lambda \mu_B^\lambda = E_.(e^{-\lambda \tau_B})$, called the *$\lambda$-equilibrium potential* of B, equals 1 exactly on the set B^r. The total measure $C^\lambda(B) = \mu_B^\lambda(\mathbb{R}^d)$ of the λ-equilibrium measure is called the *λ-capacity* of B. It follows easily from the definition of μ_B^λ that

$$C^\lambda(B) = \int E_y(e^{-\lambda \tau_B}) \lambda \, dy. \tag{5}$$

Since $g^\lambda \mu_B^\lambda < \infty$, μ_B^λ is a Radon measure. In particular if $B \in \mathscr{B}$ is bounded, μ_B^λ is a finite measure and $C^\lambda(B) < \infty$. Note that the λ-equilibrium measure,

λ-equilibrium potential, and λ-capacity of $\mathbb{R}^d$ are given, respectively, by $\mu^\lambda_{\mathbb{R}^d}(dy) = \lambda\, dy$, $g^\lambda \mu^\lambda_{\mathbb{R}^d} = 1$ and $C^\lambda(\mathbb{R}^d) = \infty$. The λ-equilibrium measure of B can be written as $\mu^\lambda_B = \int \mu^\lambda_{\mathbb{R}^d}(dy)\, h^\lambda_B(y, \cdot)$ and (5) can be written as $C^\lambda(B) = \int E_y(e^{-\lambda\tau_B})\mu^\lambda_{\mathbb{R}^d}(dy)$.

Proposition 5.2. *If $A, B \in \mathscr{B}$ and $A \subset B$, then $g^\lambda \mu^\lambda_A \le g^\lambda \mu^\lambda_B$ on $\mathbb{R}^d$ and $C^\lambda(A) \le C^\lambda(B)$. If B_n is closed for each n, $B_n \downarrow B$ and $C^\lambda(B_1) < \infty$, then $C^\lambda(B_n) \downarrow C^\lambda(B)$.*

Proof. If $A, B \in \mathscr{B}$ and $A \subset B$, then $\tau_A \ge \tau_B$, so the first conclusion follows from (4) and (5). If B_n is closed for each n and $B_n \downarrow B$, then $E_\cdot(e^{-\lambda\tau_{B_n}}) \downarrow E_\cdot(e^{-\lambda\tau_B})$ a.e. by Propositions 3.7 and 3.8. The second conclusion now follows from (5) and the dominated convergence theorem.

6. Polar Sets

Let $\lambda > 0$ throughout this section. The properties of λ-potentials will be used to show that if $B \in \mathscr{B}$, then $B \backslash B^r$ is polar. Several applications of this important result will be presented.

Proposition 6.1. *Let $B \in \mathscr{B}$. Then B is polar if and only if $C^\lambda(B) = 0$.*

Proof. It follows from (5.5) that if B is polar, then $C^\lambda(B) = 0$. Suppose conversely that $C^\lambda(B) = 0$. Then $\mu^\lambda_B = 0$, so $P_\cdot(\tau_B < \infty) = 0$ by (4) and hence B is polar.

By Proposition 2.5 if $d \ge 2$ every one-point set is polar. Here is another proof of this result using λ-potentials. Let $b \in \mathbb{R}^d$. Then $\mu^\lambda_{\{b\}}$ is concentrated at b. Since $g^\lambda(b, b) = \infty$, it follows from (5.4) that $\mu^\lambda_{\{b\}}$ is the zero measure and hence that $C^\lambda(\{b\}) = 0$. Thus $\{b\}$ is polar by Proposition 6.1.

The next result is needed to prove the theorem which follows it.

Proposition 6.2. *Suppose $B \in \mathscr{B}$ and $\sup_{x \in B} g^\lambda \mu^\lambda_B(x) < 1$. Then B is polar.*

Proof. By Propositions 2.4 and 5.2 it can be assumed that B is compact. Set $M = \sup_{x \in B} g^\lambda \mu^\lambda_B(x)$. By the maximum principle for λ-potentials $g^\lambda \mu^\lambda_B \le M$ on $\mathbb{R}^d$. Set $B_n = \{x : d(x, B) \le 1/n\}$. Then $g^\lambda \mu^\lambda_{B_n} = 1$ on B since $B \subset \mathring{B}_n \subset B^r_n$. By the symmetry of $g^\lambda(x, y)$ in x and y and Fubini's theorem

$$C^\lambda(B) = \int g^\lambda \mu^\lambda_{B_n}\, d\mu^\lambda_B = \int g^\lambda \mu^\lambda_B\, d\mu^\lambda_{B_n} \le MC^\lambda(B_n).$$

Thus $C^\lambda(B) \le MC^\lambda(B)$ by Proposition 5.2. Now $M < 1$, so $C^\lambda(B) = 0$ and hence B is polar by Proposition 6.1.

Theorem 6.3. *Let $B \in \mathscr{B}$. Then $B \backslash B^r$ is polar.*

Proof. Set $B_n = \{x \in B : g^\lambda \mu_B^\lambda(x) \le 1 - 1/n\}$. Then $B_n \in \mathscr{B}$ since $g^\lambda \mu_B^\lambda$ is lower semicontinuous. By Proposition 5.2

$$g^\lambda \mu_{B_n}^\lambda \le g^\lambda \mu_B^\lambda \le 1 - 1/n \qquad \text{on } B_n.$$

Thus B_n is polar by Proposition 6.2. Since

$$B \backslash B^r = \{x \in B : g^\lambda \mu_B^\lambda(x) < 1\} = \bigcup_n B_n,$$

it follows from Proposition 2.3 that $B \backslash B^r$ is polar.

Theorem 6.4. *Let $B \in \mathscr{B}$. Then B is polar if and only if B^r is empty.*

Proof. If B is polar then B^r is clearly empty. Conversely if B^r is empty, then $B = B \backslash B^r$ is polar by Theorem 6.3.

For a simple example of a strong Markov process in which Theorems 6.3 and 6.4 fail to hold, consider uniform motion to the right on $\mathbb{R}$. Then $\{0\}$ is nonpolar, but 0 is irregular for $\{0\}$ and hence $\{0\}^r$ is empty. The proof of Theorem 6.3 depended on the symmetry of $g_B^\lambda(x, y)$, which fails to hold in this example.

Let μ be a measure on an open subset D of $\mathbb{R}^d$. If $\mu(B) = 0$ for every polar set $B \in \mathscr{B}$ contained in D or equivalently if $\mu(B) = 0$ for every compact polar subset B of D, it is said that μ *does not charge polar sets.*

Let $B \in \mathscr{B}$. If $x \in B^r$, then $h_B(x, \cdot) = \delta_x$. In the next theorem the behavior of $h_B(x, \cdot)$ for $x \notin B^r$ is studied.

Theorem 6.5. *Let $B \in \mathscr{B}$ and $x \in (B^r)^c$. Then $h_B(x, \cdot)$ is concentrated on $\partial B \cap B^r$ and does not charge polar sets. If B is closed, then $h_{\partial B}(x, \cdot) = h_B(x, \cdot)$.*

Proof. If A is a compact polar set, then

$$h_B(x, A) = P_x(0 < \tau_B < \infty, X(\tau_B) \in A) \le P_x(\tau_A < \infty) = 0.$$

Consequently $h_B(x, \cdot)$ does not charge polar sets. Since $B \backslash B^r$ is polar by Theorem 6.3, $h_B(x, B \backslash B^r) = 0$. Observe that if $\tau_B < \infty$ and $X(\tau_B) \in B^c$, then $\tau_B \circ \theta_{\tau_B} = 0$. Thus by the strong Markov property

$$\begin{aligned} h_B(x, B^c) &= P_x(\tau_B < \infty, X(\tau_B) \in B^c, \tau_B \circ \theta_{\tau_B} = 0) \\ &= E_x(P_{X(\tau_B)}(\tau_B = 0); \tau_B < \infty, X(\tau_B) \in B^c) \\ &= P_x(\tau_B < \infty, X(\tau_B) \in B^c \cap B^r) \\ &= h_B(x, B^c \cap B^r). \end{aligned}$$

Consequently $h_B(x, B^c \backslash B^r) = 0$ and hence

$$h_B(x, (B^r)^c) = h_B(x, B \backslash B^r) + h_B(x, B^c \backslash B^r) = 0;$$

this shows that $h_B(x, \cdot)$ is concentrated on B^r. By continuity of paths $X(\tau_B) \in \partial B$ on $\{0 < \tau_B < \infty\}$, so $h_B(x, \cdot)$ is concentrated on ∂B and therefore on $\partial B \cap B^r$. Suppose B is closed. By continuity of paths $\tau_{\partial B} = \tau_B$ on $\{\tau_B > 0\}$, so $P_x(\tau_{\partial B} = \tau_B) = 1$ and hence $h_{\partial B}(x, \cdot) = h_B(x, \cdot)$. This completes the proof of the theorem.

The next result is needed for the two propositions which follow it.

Proposition 6.6. *Let $B \in \mathscr{B}$, let τ be a stopping time, and let $x \in \mathbb{R}^d$. Then $\tau_B = \tau + \tau_B \circ \theta_\tau$ a.s. (P_x) on $\{\tau \leq \tau_B, \tau < \infty\}$. In particular if $A \in \mathscr{B}$ and $A \supset B$, then $\tau_B = \tau_A + \tau_B \circ \theta_{\tau_A}$ a.s. (P_x) on $\{\tau_A < \infty\}$.*

Proof. Now $\tau_B = \tau + \tau_B \circ \theta_\tau$ on $\{\tau < \tau_B\}$. Also $X(\tau) = X(\tau_B) \in B^r$ a.s. (P_x) on $\{\tau = \tau_B < \infty\}$ by Theorem 6.5. Thus by the strong Markov property $\tau_B \circ \theta_\tau = 0$ a.s. (P_X) on $\{\tau = \tau_B < \infty\}$ and hence $\tau_B = \tau + \tau_B \circ \theta_\tau$ a.s. (P_x) on $\{\tau = \tau_B < \infty\}$. Consequently $\tau_B = \tau + \tau_B \circ \theta_\tau$ a.s. (P_x) on $\{\tau \leq \tau_B, \tau < \infty\}$ as desired.

Theorem 6.7. *Let $A, B \in \mathscr{B}$ with $A \supset B$. Then $h_B = h_A h_B = h_B h_A$.*

Proof. Let f be bounded on $\mathbb{R}^d$ and let $x \in \mathbb{R}^d$. It follows from Proposition 6.6 and the strong Markov property that

$$\begin{aligned} h_B f(x) &= E_x(f(X(\tau_B)); \tau_B < \infty) \\ &= E_x(f(X(\tau_B \circ \theta_{\tau_A}, \theta_{\tau_A}\omega)); \tau_A < \infty, \tau_B \circ \theta_{\tau_A} < \infty) \\ &= E_x(E_{X(\tau_A)}(f(X(\tau_B)); \tau_B < \infty); \tau_A < \infty) \\ &= h_A h_B f(x) \end{aligned}$$

and hence that $h_B = h_A h_B$. Now $h_B(x, \cdot)$ is concentrated on $B^r \subset A^r$ by Theorem 6.5 and $h_A f(z) = f(z)$ for $z \in A^r$. Thus

$$h_B h_A f(z) = \int h_B(x, dz) h_A f(z) = \int h_B(x, dz) f(z) = h_B f(z)$$

and hence $h_B h_A = h_B$. This completes the proof of the theorem.

Theorem 6.8. *Let $A, B \in \mathscr{B}$ with $A \supset B$. Then*

$$q_B(t, x, y) = q_A(t, x, y) + E_x(q_B(t - \tau_A, X(\tau_A), y); \tau_A < t), \qquad t > 0 \text{ and } x, y \in \mathbb{R}^d.$$

Proof. Let $t > 0$ and $x \in \mathbb{R}^d$ and let f be bounded on $\mathbb{R}^d$. By Proposition 6.6 and the strong Markov property

$$\begin{aligned} q_B^t f(x) &= E_x(f(X(t)); \tau_B > t) \\ &= E_x(f(X(t)); \tau_A > t) + E_x(f(X(t)); \tau_A < t < \tau_B) \\ &= q_A^t f(x) + E_x(f(X(t)); \tau_A < t, \tau_B \circ \theta_{\tau_A} > t - \tau_A) \\ &= q_A^t f(x) + E_x\Big(E_{X(\tau_A)}\Big(f(X(t-s)I_{\{\tau_B > t-s\}}\Big)_{s=\tau_A}; \tau_A < t\Big) \\ &= q_A^t f(x) + E_x(q_B^{t-\tau_A} f(X(\tau_A)); \tau_A < t). \end{aligned}$$

Consequently

$$q_B(t, x, y) = q_A(t, x, y) + E_x(q_B(t - \tau_A, X(\tau_A), y); \tau_A < t) \qquad \text{a.e. } y \in \mathbb{R}^d.$$

Thus for $0 < \varepsilon < t$

$$\begin{aligned} \int q_B(t - \varepsilon, x, z)p(\varepsilon, z, y)\,dz &= \int q_A(t - \varepsilon, x, z)p(\varepsilon, z, y)\,dz \\ &\quad + E_x\left(\int q_B(t - \varepsilon - \tau_A, X(\tau_A), z)p(\varepsilon, z, y)\,dz; \tau_A < t - \varepsilon\right). \end{aligned}$$

The desired conclusion now follows by letting $\varepsilon \downarrow 0$ and using Proposition 4.2 and the dominated convergence theorem.

7. Nonpolar Sets for Planar Brownian Motion

After a preliminary result for d-dimensional Brownian motion is proven, the two-dimensional case is considered.

Proposition 7.1. *Let $t > 0$, let f be a continuous $\mathbb{R}^d$-valued function on $[0, t]$, let $\varepsilon > 0$, and let $x \in \mathbb{R}^d$ satisfy $\|x - f(0)\| < \varepsilon$. Then*

$$P_x(\|X(s) - f(s)\| < \varepsilon \text{ for } 0 \le s \le t) > 0.$$

Proof. Let f be extended to a continuous function on $[0, \infty)$. It suffices to prove the stronger result that if $t > 0$, then

(1) $P_x(\|X(s) - f(s)\| < \varepsilon \text{ for } 0 \le s \le t \text{ and } \|X(t) - f(t)\| < \delta) > 0, \qquad \delta > 0.$

Suppose this result is false. Let u be the infimum of the numbers $t > 0$ which fail to satisfy (1). It follows from the positivity result in Theorem 4.3 that $u > 0$. Choose r, $0 < r < \varepsilon$, and set $D = \mathring{B}_r(f(u))$. There exist numbers t_0 and t such that $0 < t_0 < u < t$ and $f(s) \in D \subset \mathring{B}_\varepsilon(f(s))$ for $t_0 \le s \le t$. Let $\delta > 0$

and set $U = \mathring{B}_\delta(f(t)) \cap D$. Then U is an open set containing $f(t)$ and hence U has positive Lebesgue measure. Now $Q_D(t - t_0, z, y) > 0$ for $z \in D$ and $y \in U$. By hypothesis

$$P_x(\|X(s) - f(s)\| < \varepsilon \text{ for } 0 \leq s \leq t_0 \text{ and } X(t_0) \in D) > 0,$$

so by the Markov property

$$\begin{aligned} &P_x(\|X(s) - f(s)\| < \varepsilon \text{ for } 0 \leq s \leq t \text{ and } \|X(t) - f(t)\| < \delta) \\ &\quad \geq P_x(\|X(s) - f(s)\| < \varepsilon \text{ for } 0 \leq s \leq t_0,\ X(s) \in D \text{ for } t_0 \leq s \leq t \\ &\qquad \text{and } X(t) \in U) > 0. \end{aligned}$$

This violates the definition of u. Thus (1) holds for all $t > 0$, which completes the proof of the proposition.

Consider in the remainder of the section planar Brownian motion. The next result is due to Lebesgue in the context of the Dirichlet problem (see Section 4.2). The proof is taken from Itô and McKean [2]. An analytic proof will be given in Section 6.7. It follows from this result that all points on a nondegenerate continuous curve in $\mathbb{R}^2$ are regular for the curve. More generally let $B \in \mathscr{B}$ and $b \in B$ and suppose that there is a nonconstant continuous function f from $[0, 1]$ to B such that $f(0) = b$. Then b is regular for B.

Theorem 7.2. *Let $d = 2$ and let $B \in \mathscr{B}$ be a nonempty subset of $\mathbb{R}^2$ such that no component of B reduces to a single point. Then B is nonpolar and all points of B are regular for B.*

The proof of the theorem is split up into a number of lemmas. Given a point $b \in \mathbb{R}^2$ and a compact subset C of $\mathbb{R}^2$, C is said to *cut off* b if b is in the exterior of the unbounded component of $\mathbb{R}^2 \backslash C$.

Lemma 7.3. *Let C be a compact set in $\mathbb{R}^2$ which cuts off b and let B be a connected set in $\mathbb{R}^2$ such that $B \cap C = \{b\}$. Then B is disjoint from the unbounded component of $\mathbb{R}^2 \backslash C$.*

Proof. Let D denote the unbounded component of $\mathbb{R}^2 \backslash C$ and let D' denote the remainder of $\mathbb{R}^2 \backslash C$. Choose $r > 0$ such that $\mathring{B}_r(b) \subset D' \cup C$. Since $B \cap C = \{b\} \subset B \cap \mathring{B}_r(b)$,

$$\begin{aligned} B &= (B \cap D) \cup (B \cap D') \cup (B \cap C) = (B \cap D) \cup (B \cap D') \cup (B \cap \mathring{B}_r(b)) \\ &= (B \cap D) \cup [B \cap (D' \cup \mathring{B}_r(b))]. \end{aligned}$$

Now $B \cap D$ and $B \cap (D' \cup \mathring{B}_r(b))$ are disjoint relatively open subsets of the connected set B whose union is B and $b \in B \cap (D' \cup \mathring{B}_r(b))$, so that $B \cap D$ is empty as desired.

In the next lemma let $f(t)$, $0 \leq t \leq 1$, be the curve connecting the points $PQRSTQU$ shown in Figure 1 by straight lines. This curve clearly cuts off the center b of the square $QRSTQ$. It can be assumed that $P = f(0)$, $Q = f(\frac{1}{6})$, $R = f(\frac{2}{6})$, $S = f(\frac{3}{6})$, $T = f(\frac{4}{6})$, $Q = f(\frac{5}{6})$, $U = f(1)$ and that every point on the curve other than Q corresponds to a unique value of t.

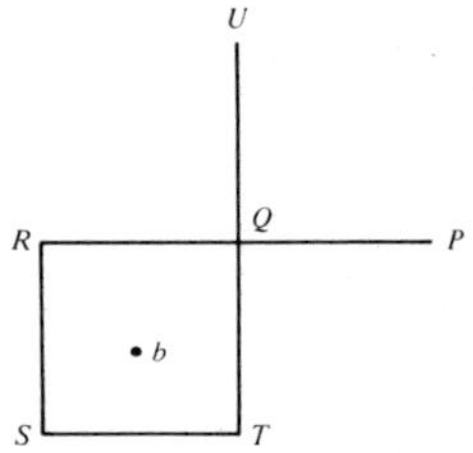

Figure 1

Lemma 7.4. *There is an* $\varepsilon > 0$ *such that if* f' *is a continuous function from* $[0,1]$ *to* $\mathbb{R}^2$ *and* $\|f' - f\| \leq \varepsilon$ *on* $[0,1]$, *then the range of* f' *cuts off* b.

Proof. Now $f(t)$, $\frac{1}{6} \leq t \leq \frac{5}{6}$, has winding number one about b (see pages 114–118 of Ahlfors [1] for the definition and elementary properties of winding numbers). Let f_n be continuous functions from $[0,1]$ to $\mathbb{R}^2$ such that $f_n \to f$ uniformly on $[0, 1]$. It suffices to show that the range of f_n cuts off b for n sufficiently large.

To this end it will first be shown that there exist numbers u_n and v_n converging to $\frac{1}{6}$ and $\frac{5}{6}$, respectively, such that $f_n(u_n) = f_n(v_n)$. Suppose this were false. Then there would be a closed curve $\gamma_1 = P'R'S'V'W'P'$ as indicated in Figure 2 that had winding number one about T' and a disjoint curve $\gamma_2 = T'U'$ from T' to a point U' in the unbounded component of $\mathbb{R}^2 \backslash \gamma_1$ (here for

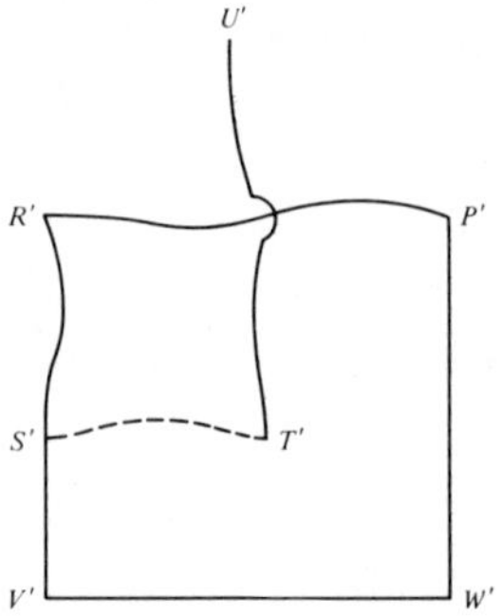

Figure 2

some sufficiently large value of n the portion of $P'R'S'$, γ_1 is the range of $f_n(t)$, $0 \le t \le \frac{3}{6}$, and γ_2 is the range of $f_n(t)$, $\frac{4}{6} \le t \le 1$). Now γ_1 has winding number zero about U'. But this is impossible since the winding number of γ_1 about points on the disjoint curve γ_2 is a continuous integer-valued function which must reduce to a constant.

Let γ_n now denote the closed curve $f_n(t)$, $u_n \le t \le v_n$. Then γ_n has winding number zero about all points in the unbounded component of $\mathbb{R}^2 \backslash \gamma_n$. But for n sufficiently large γ_n has winding number one about b and in fact about all points in $\mathring{B}_r(b)$ for some $r > 0$. Thus for n sufficiently large γ_n and hence also the range of f_n cut off b as desired.

Lemma 7.5. *$P_b(\{X(s), 0 \le s \le t\}$ cuts off b for all $t > 0) = 1$.*

Proof. Let Λ_n be the event that $\{X(s), 0 \le s \le 1/n\}$ cuts off b. By scale invariance $P_b(\Lambda_n)$ is independent of n, and Proposition 7.1 and Lemma 7.4 together show that $P_b(\Lambda_1) > 0$. Since $\Lambda_n \downarrow$,

$$P_b\left(\bigcap_n \Lambda_n\right) = \lim_n P_b(\Lambda_n) = P_b(\Lambda_1) > 0.$$

By Blumenthal's zero–one law $P_b(\bigcap_n \Lambda_n) = 1$, which is equivalent to the statement of the lemma.

With this preparation it is easy to complete the proof of the theorem. Let B be such that no component of B reduces to a single point and let $b \in B$. It must be shown that b is regular for B. Let B' be the component of B containing b and let c be a point in B' other than b. It follows from Lemma 7.5 and continuity of paths that a.s. (P_b) for $t > 0$ sufficiently small $C_t = \{X(s): 0 \le s \le t\}$ cuts off b and c lies in the unbounded component of $\mathbb{R}^2 \backslash C_t$, so that B' fails to be disjoint from the unbounded component of $\mathbb{R}^2 \backslash C_t$. Since B' is connected it follows from Lemma 7.3 that a.s. (P_b) for $t > 0$, $B' \cap C_t$ contains a point other than b and hence $\tau_B \le t$. Thus $P_b(\tau_B = 0) = 1$ and hence b is regular for B as desired.

8. Brownian Motion on an Interval

In this section linear Brownian motion is considered. The density $Q_D(t, x, y)$ is computed explicitly when D is an open interval.

Proposition 8.1. *Let D be an interval of the form $(-\infty, a)$ or (a, ∞) for some $a \in \mathbb{R}$. Then $Q_D(t, x, y) = p(t, x - a, y - a) - p(t, x - a, -(y - a))$ for $t > 0$ and $x, y \in D$.*

Proof. It suffices to consider the case $D = (0, \infty)$. Choose $t > 0$, $x \in D$, and $A \subset D$. By the strong Markov property

$$\begin{aligned} P_x(T_D \le t, X(t) \in A) &= E_x(P_0(X(t-s) \in A)\big|_{s=T_D}; T_D \le t) \\ &= E_x(P_0(X(t-s) \in -A)\big|_{s=T_D}; T_D \le t) \\ &= P_x(T_D \le t, X(t) \in -A) \\ &= P_x(X(t) = -A), \end{aligned}$$

the last equality following from continuity of paths. Consequently

$$\begin{aligned} \int_A Q_D(t,x,y)\,dy &= P_x(T_D > t, X(t) \in A) \\ &= P_x(X(t) \in A) - P_x(T_D \le t, X(t) \in A) \\ &= P_x(X(t) \in A) - P_x(X(t) \in -A) \\ &= \int_A (p(t,x,y) - p(t,x,-y))\,dy. \end{aligned}$$

Thus $Q_D(t,x,y) = p(t,x,y) - p(t,x,-y)$ for almost all $y \in D$. By continuity this equation holds for all $y \in D$ as desired.

Proposition 8.2. *Let $D = (a,b)$ be a bounded open interval. Then for $t > 0$ and x, $y \in D$*

$$\begin{aligned} &Q_D(t,x,y) \\ &\quad = \sum_{-\infty}^{\infty} [p(t, x-a, 2m(b-a) + y - a) - p(t, x-a, 2m(b-a) - y + a)] \\ &\quad = \frac{2}{b-a}\sum_{1}^{\infty} \exp\left[-\frac{n^2\pi^2 t}{2(b-a)^2}\right] \sin\left(n\pi\frac{x-a}{b-a}\right)\sin\left(n\pi\frac{y-a}{b-a}\right). \end{aligned}$$

Proof. It follows easily from the scale invariance of Brownian motion and the continuity of $Q_D(t,x,y)$ that for $t > 0$ and x, $y \in D$

$$Q_D(t,x,y) = \frac{1}{b-a} Q_{(0,1)}\left(\frac{t}{(b-a)^2}, \frac{x-a}{b-a}, \frac{y-a}{b-a}\right).$$

Thus in verifying the desired formulas it suffices to consider the special case $D = (0,1)$.

Let $t > 0$ and x, $y \in D = (0,1)$. Then $Q_D(t,x,-y) = 0$ and $Q_D(t,x,2m + y) = Q_D(t,x,2m-y) = 0$ for $m \ne 0$. Thus

$$Q_D(t,x,y) = \sum_{-\infty}^{\infty} [Q_D(t,x,2m+y) - Q_D(t,x,2m-y)]$$

and hence by (4.3)

$$(1)\qquad Q_D(t,x,y) = \sum_{-\infty}^{\infty} [p(t,x,2m+y) - p(t,x,2m-y)]$$

$$- \sum_{-\infty}^{\infty} [r_{D^c}(t,x,2m+y) - r_{D^c}(t,x,2m-y)].$$

By (4.2)

$$(2)\qquad \sum_{-\infty}^{\infty} [r_{D^c}(t,x,2m+y) - r_{D^c}(t,x,2m-y)]$$

$$= E_x\left(\sum_{-\infty}^{\infty} [p(t-T_D, X(T_D), 2m+y) - p(t-T_D, X(T_D), 2m-y)]; T_D < t \right).$$

Observe that $X(T_D) = 0$ or 1 on $\{T_D < t\}$ and

$$\sum_{-\infty}^{\infty} [p(s,z,2m+y) - p(s,z,2m-y)] = 0, \qquad s > 0 \quad \text{and} \quad z = 0 \text{ or } 1.$$

Thus the right side of (2) equals zero, so the first formula for $Q_D(t,x,y)$ follows from (1).

The second formula will be obtained from the first formula by expanding $Q_D(t,x,y)$, $0 < y < 1$, in a Fourier sine series on the interval $(0,1)$. Specifically $Q_D(t,x,y) = 2\sum_1^{\infty} a_n \sin n\pi y$ for almost all $y \in (0,1)$, where

$$\begin{aligned}
a_n &= \int_0^1 \sin(n\pi y) Q_D(t,x,y)\,dy \\
&= \int_0^1 \sin(n\pi y) \sum_{-\infty}^{\infty} [p(t,2m+y-x) - p(t,2m-y-x)]\,dy \\
&= \sum_{-\infty}^{\infty} \int_{-1}^{1} \sin(n\pi y) p(t,2m+y-x)\,dy \\
&= \int_{-\infty}^{\infty} \sin(n\pi y) p(t,y-x)\,dy \\
&= \sin(n\pi x) \int_{-\infty}^{\infty} \cos(n\pi y)(2\pi t)^{-1/2} \exp(-y^2/2t)\,dy \\
&= \sin(n\pi x) \exp(-n^2\pi^2 t/2).
\end{aligned}$$

Thus the second formula of the proposition is valid for almost all y. By continuity it is valid for all y as desired.

Proposition 8.3. *Let $D = (a, b)$ be a bounded open interval. Then for $t > 0$ and $x \in D$*

$$P_x(T_D > t, X(T_D) = b) = -\frac{2}{\pi}\sum_1^\infty \frac{\cos n\pi}{n}\exp\left[-\frac{n^2\pi^2 t}{2(b-a)^2}\right]\sin\left(n\pi\frac{x-a}{b-a}\right)$$

and

$$P_x(T_D > t, X(T_D) = a) = \frac{2}{\pi}\sum_1^\infty \frac{1}{n}\exp\left[-\frac{n^2\pi^2 t}{2(b-a)^2}\right]\sin\left(n\pi\frac{x-a}{b-a}\right).$$

Proof. By the Markov property

$$P_x(T_D > t, X(T_D) = b) = \int Q_D(t, x, y)P_y(X(T_D) = b)\,dy$$

and

$$P_x(T_D > t, X(T_D) = a) = \int Q_D(t, x, y)P_y(X(T_D) = a)\,dy.$$

The desired formulas now follow easily from Propositions 2.20 and 8.2.

Proposition 8.3 can be used to find the exit distribution of the region between two parallel hyperplanes for Brownian motion on $\mathbb{R}^d$. The method of derivation is similar to that used in obtaining Proposition 2.22 from Proposition 2.17. The answer involves Bessel functions and the details are left as an exercise for the reader.

The following consequence of Proposition 8.3 is immediate.

Proposition 8.4. *Let $D = (a, b)$ be a bounded open interval and let $x \in D$. Then*

$$\lim_{t\to\infty} \exp\left[\frac{\pi^2 t}{2(b-a)^2}\right]P_x(T_D > t, X(T_D) = b) = \frac{2}{\pi}\sin\left(\pi\frac{x-a}{b-a}\right)$$

and

$$\lim_{t\to\infty} \exp\left[\frac{\pi^2 t}{2(b-a)^2}\right]P_x(T_D > t, X(T_D) = a) = \frac{2}{\pi}\sin\left(\pi\frac{x-a}{b-a}\right).$$

Chapter 3

Potentials on the Whole Space

In the first three sections of this chapter transient Brownian motion is considered. In Section 1 some elementary properties of Newtonian potential theory related especially to the equilibrium measure and capacity of a bounded set are obtained and applied to Brownian motion. Further applications are given in Sections 2 and 3. Newtonian potential theory as an end in itself will be thoroughly developed in the more general context of Green potentials in Chapter 6. Logarithmic potentials and their application to planar Bownian motion are discussed in Section 4. Further properties of logarithmic potentials will be obtained in Section 6.7. Linear potentials and their connection with one-dimensional Brownian motion are considered in Section 5.

Set $g(x, y) = \int_0^\infty p(t, x, y)\,dt$ for $x, y \in \mathbb{R}^d$. If $d \leq 2$, g is identically infinite on $\mathbb{R}^d \times \mathbb{R}^d$ while if $d \geq 3$, $g(x, y)$ is finite or infinite according as $x \neq y$ or $x = y$.

Let $B \in \mathscr{B}$. Set $g_B(x, y) = \int_0^\infty q_B(t, x, y)\,dt$ for $x, y \in \mathbb{R}^d$. The properties of q_B obtained in Section 2.4 translate immediately into properties of g_B. In particular $0 \leq g_B \leq g$ and $g_B = g$ if B is polar. Also $g_B(x, y)$ is symmetric in x and y and equals zero if either $x \in B^r$ or $y \in B^r$. For $f \geq 0$

$$g_B f(x) = \int g_B(x, y) f(y)\,dy = \int_0^\infty q_B^t f(x)\,dt = E_x\left(\int_0^{\tau_B} f(X(t))\,dt\right).$$

Thus by Proposition 2.2.7, if $d \geq 3$ or if B is nonpolar, then $g_B(x, \cdot)$ is integrable on compacts for each $x \in \mathbb{R}^d$. If $d \geq 3$, $g_B(x, y)$ is finite when $x \neq y$. In Sections 4 and 5 of this chapter it is shown that if $d \leq 2$ and B is nonpolar, then $g_B(x, y)$ is again finite when $x \neq y$.

Let $A, B \in \mathscr{B}$ with $A \supset B$. By Theorem 2.6.8

$$g_B(x, y) = g_A(x, y) + \int h_A(x, dz) g_B(z, y), \qquad x, y \in \mathbb{R}^d \tag{1}$$

and in particular $g_B \geq g_A$.

1. Newtonian Potentials

Consider in this section d-dimensional Brownian motion with $d \geq 3$. The *Newtonian potential kernel g* on $\mathbb{R}^d$ is defined by

$$g(x) = \int_0^\infty p(t, x)\, dt = \frac{\Gamma(d/2 - 1)}{2\pi^{d/2}} \|x\|^{2-d}, \qquad x \in \mathbb{R}^d. \tag{1}$$

Note that g is integrable on compacts by Proposition 2.1.1 and that $g(x, y) = g(y - x)$ for $x, y \in \mathbb{R}^d$.

Let μ be a measure on $\mathbb{R}^d$. The *Newtonian potential* $g\mu$ of μ is defined by $g\mu(x) = \int g(x, y)\mu(dy)$ for $x \in \mathbb{R}^d$. If μ is a finite measure, then $g\mu$ is integrable on compacts and hence finite almost everywhere. For $t > 0$ let $p^t\mu$ be the function on $\mathbb{R}^d$ defined by $p^t\mu(x) = \int p(t, x, y)\mu(dy)$. Then $g\mu = \int_0^\infty p^t\mu\, dt$ and $p^t g\mu = \int_t^\infty p^s\mu\, ds \uparrow g\mu$ as $t \downarrow 0$.

Proposition 1.1. *Let μ and ν be finite measures on $\mathbb{R}^d$ such that $g\mu = g\nu$ almost everywhere. Then $\mu = \nu$.*

Proof. Choose $t > 0$. Now $g\mu - p^t g\mu = \int_0^t p^s\mu\, ds$ on the set where $g\mu < \infty$ and the same formula holds with μ replaced by ν, so $\int_0^t p^s\mu\, ds = \int_0^t p^s\nu\, ds$ almost everywhere. Let f be a nonnegative continuous function on $\mathbb{R}^d$ having compact support. By the symmetry of $p(s, x, y)$ in x and y

$$\int \left(\frac{1}{t} \int_0^t p^s f\, ds \right) d\mu = \int \left(\frac{1}{t} \int_0^t p^s f\, ds \right) d\nu.$$

Since $p^s f \to f$ as $s \to 0$ uniformly on $\mathbb{R}^d$, $\int f\, d\mu = \int f\, d\nu$. Thus $\mu = \nu$ as desired.

Recall the fundamental identity for λ-potentials

$$g^\lambda(x, y) = g_B^\lambda(x, y) + \int h_B^\lambda(x, dz) g^\lambda(z, y), \qquad x, y \in \mathbb{R}^d.$$

By the definition of h_B^λ and the monotone convergence theorem

$$\begin{aligned}\lim_{\lambda\to 0}\int h_B^\lambda(x,dz)g^\lambda(z,y) &= \lim_{\lambda\to 0} E_x(e^{-\lambda\tau_B}g^\lambda(X(\tau_B),y);\tau_B<\infty)\\ &= E_x(g(X(\tau_B),y);\tau_B<\infty)\\ &= \int h_B(x,dz)g(z,y).\end{aligned}$$

Thus

$$(2)\qquad g(x,y) = g_B(x,y) + \int h_B(x,dz)g(z,y), \qquad x,y\in\mathbb{R}^d.$$

This equation is called the *fundamental identity* for Newtonian potentials. The next result is a *uniqueness principle* for Newtonian potentials.

Proposition 1.2. *Let $B\in\mathscr{B}$ and let μ and ν be finite measures on $\mathbb{R}^d$ which are concentrated on B^r. If $g\mu = g\nu$ on B^r, then $\mu=\nu$.*

Proof. If $x\in\mathbb{R}^d$, then $g_B(x,\cdot)=0$ on B^r and $h_B(x,\cdot)$ is concentrated on B^r by Theorem 2.6.5. Thus by the fundamental identity, $g\mu = h_B g\mu = h_B g\nu = g\nu$ on $\mathbb{R}^d$ and hence $\mu=\nu$ by Proposition 1.1.

Proposition 1.3. *Let $B\in\mathscr{B}$ and $x\in\mathbb{R}^d$. Then $g_B(x,\cdot)$ is upper semicontinuous on $\mathbb{R}^d\backslash\{x\}$. In particular $\lim_{y\to b} g_B(x,y) = g_B(x,b) = 0$ for $b\in B^r\backslash\{x\}$.*

Proof. This result follows from the fundamental identity and Fatou's lemma.

Proposition 1.4. *Let $B\in\mathscr{B}$ be bounded. Then $\lim_{\|y\|\to\infty} g_B(x,y)/g(y) = P_x(\tau_B=\infty)$ uniformly for x in compacts.*

Proof. By the fundamental identity

$$\lim_{\|y\|\to\infty}\frac{g_B(x,y)}{g(y)} = 1 - P_x(\tau_B<\infty) = P_x(\tau_B=\infty)$$

uniformly on compacts.

Proposition 1.5. *Let $0<r<q$, let D denote the annulus $\{y: r<\|y\|<q\}$ and let $x\in D$. Then*

$$P_x(\|X(T_D)\|=r) = 1 - P_x(\|X(T_D)\|=q) = \frac{\|x\|^{2-d}-q^{2-d}}{r^{2-d}-q^{2-d}}.$$

Proof. Note that $P_x(T_D < \infty) = 1$ and that $g_{D^c}(x, 0) = 0$. Thus by (1), the fundamental identity and continuity of paths

$$\|x\|^{2-d} = r^{2-d} P_x(\|X(T_D)\| = r) + q^{2-d}(1 - P_x(\|X(T_D)\| = r)),$$

from which the desired result follows.

Proposition 1.6. *Let $r > 0$ and $x \in \mathbb{R}^d$. Then $P_x(\tau_{B_r} < \infty) = P_x(\tau_{S_r} < \infty) = (r/\|x\|)^{d-2} \wedge 1$.*

Proof. If $\|x\| \le r$, the result follows from Proposition 2.3.4 and continuity of paths. Suppose $\|x\| > r$. It follows from Propositions 1.5 and 2.2.11, continuity of paths and the strong Markov property that

$$P_x(\tau_{B_r} < \infty) = P_x(\tau_{S_r} < \infty) = \lim_{q \to \infty} \frac{\|x\|^{2-d} - q^{2-d}}{r^{2-d} - q^{2-d}} = \left(\frac{r}{\|x\|}\right)^{d-2}$$

as desired.

Let $r > 0$. The Newtonian potential of the uniform distribution σ_r on S_r will now be determined.

Proposition 1.7. *Let $r > 0$. Then*

$$\int g(x, y)\sigma_r(dy) = \frac{\Gamma(d/2 - 1)}{2\pi^{d/2}} (\|x\| \vee r)^{2-d}, \qquad x \in \mathbb{R}^d.$$

Also

$$\int_{B_r} g(x, y)\, dy = \frac{r^2}{d-2} - \frac{\|x\|^2}{d}, \qquad \|x\| \le r,$$

$$= \frac{2r^d \|x\|^{2-d}}{d(d-2)}, \qquad \|x\| > r.$$

Proof. It follows from the invariance of σ_r under orthogonal linear transformations that $\int g(z, y)\sigma_r(dy)$ depends on z only through $\|z\|$ and hence takes on a constant value c_r on S_r. By the fundamental identity and Proposition 1.6, for $x \in \mathbb{R}^d$

$$\int g(x, y)\sigma_r(dy) = \int h_{S_r}(x, dz) \int g(z, y)\sigma_r(dy) = c_r P_x(\tau_{S_r} < \infty)$$

$$= c_r \left(\frac{r}{\|x\|} \wedge 1\right)^{d-2}.$$

Thus by (1)

$$c_r = \int g(0, y)\sigma_r(dy) = \frac{\Gamma(d/2 - 1)}{2\pi^{d/2}} r^{2-d},$$

so the first result is valid. By Proposition 2.1.1

$$\begin{aligned}\int_{B_r} g(x, y)\,dy &= \frac{2\pi^{d/2}}{\Gamma(d/2)} \int_0^r \left(\int g(x, y)\sigma_s(dy) \right) s^{d-1}\,ds \\ &= \frac{2}{d-2} \int_0^r (\|x\| \vee s)^{2-d} s^{d-1}\,ds,\end{aligned}$$

from which the second formula follows easily.

Let $r > 0$. Then $\int_{B_r} g(x, y)\,dy$, computed above, is the expected time spent in the ball B_r by Brownian motion starting at x. For $\|x\| \le r$, $E_x(\tau_{S_r})$ is the expected time spent in the ball B_r before leaving the ball for Brownian motion starting at x. Recall that $E_0(\tau_{S_r}) = r^2/d$ by Proposition 2.2.21. This formula will now be generalized.

Proposition 1.8. *Let $r > 0$ and $\|x\| \le r$. Then $E_x(\tau_{S_r}) = (r^2 - \|x\|^2)/d$.*

Proof. Choose $c \in S_r$. By the fundamental identity

$$E_x(\tau_{S_r}) = \int_{B_r} g_{S_r}(x, y)\,dy = \int_{B_r} g(x, y) - \int_{B_r} g(c, y)\,dy.$$

The desired result now follows from Proposition 1.7.

Let μ be a measure on $\mathbb{R}^d$ such that $g\mu$ is not identically infinite. Since the Newtonian potential kernel g is bounded away from zero on compacts, μ must be a Radon measure. Thus if μ has compact support it must be a finite measure.

Let $B \in \mathscr{B}$ be bounded. By the uniqueness principle there is at most one measure on $\mathbb{R}^d$ which is concentrated on B^r and whose Newtonian potential equals one on B^r. If such a measure exists it is denoted by μ_B and called the *equilibrium measure* of B, its total measure $C(B) = \mu_B(\mathbb{R}^d)$ is called the *capacity* of B, and the Newtonian potential $g\mu_B$ is called the *equilibrium potential* of B.

Proposition 1.9. *Let $r > 0$. Then B_r has equilibrium measure*

(3) $$\mu_{B_r} = \frac{2\pi^{d/2} r^{d-2}}{\Gamma(d/2 - 1)} \sigma_r,$$

capacity $C(B_r) = 2\pi^{d/2}r^{d-2}/\Gamma(d/2 - 1)$, *and equilibrium potential given by*

$$g\mu_{B_r}(x) = P_x(\tau_{B_r} < \infty) = (r/\|x\|)^{d-2} \wedge 1, \qquad x \in \mathbb{R}^d.$$

Proof. Let μ denote the measure on the right-hand side of (3). It follows from Propositions 1.6 and 1.7 that $g\mu = P_.(\tau_{B_r} < \infty)$. Now μ is concentrated on B_r, all of whose points are regular for B_r, and $g\mu = P_.(\tau_{B_r} < \infty) = 1$ on B_r. Thus μ is the equilibrium measure of B_r. The formulas for the capacity and equilibrium potential of B_r follow immediately.

Let $B \in \mathscr{B}$. Observe that

$$\int h_B(x, dz)g(z, y) = \int h_B(y, dz)g(z, x), \qquad x, y \in \mathbb{R}^d. \tag{4}$$

For if $x = y$, (4) is trivially true and if $x \neq y$, (4) follows from the fundamental identity, the finiteness of $g(x, y)$ and the symmetry of $g(x, y)$ and $g_B(x, y)$ in x and y.

Theorem 1.10. *Let $B \in \mathscr{B}$ be bounded and let $r > 0$ be such that $B \subset B_r$. Then B has equilibrium measure $\mu_B = \int \mu_{B_r}(dy)h_B(y, \cdot)$, equilibrium potential $g\mu_B = P_.(\tau_B < \infty)$ and capacity*

$$C(B) = \int P_.(\tau_B < \infty)\, d\mu_{B_r}.$$

Moreover

$$\lim_{\|x\| \to \infty} \frac{h_B(x, A)}{g(x)} = \mu_B(A), \qquad A \subset \mathbb{R}^d.$$

Proof. Set $\mu = \int \mu_{B_r}(dy)h_B(y, \cdot)$. Then μ is concentrated on B^r by Theorem 2.6.5. By Proposition 1.9, $g\mu_{B_r} = 1$ on B_r. Thus by (4)

$$g\mu = h_B g\mu_{B_r} = h_B 1 = P_.(\tau_B < \infty),$$

which equals one on B^r. Consequently μ is the equilibrium measure of B, $P_.(\tau_B < \infty)$ is the equilibrium potential of B, and $C(B) = \int P_.(\tau_B < \infty)\, d\mu_{B_r}$ is the capacity of B.

By the fundamental identity, Proposition 1.4 and the symmetry of $g_B(x, y)$ in x and y

$$\lim_{\|x\| \to \infty} \int \frac{h_{B_r}(x, dz)}{g(x)} g(z, y) = 1 - P_y(\tau_{B_r} = \infty)$$
$$= P_y(\tau_{B_r} < \infty) = g\mu_{B_r}(y) \qquad y \in \mathbb{R}^d.$$

Thus if $x_n \in \mathbb{R}^d$, $\|x_n\| \to \infty$, and $h_{B_r}(x_n, \cdot)/g(x_n)$ converges completely to v, then $gv = g\mu_{B_r}$ on $\mathbb{R}^d \backslash S_r$ and hence $v = \mu_{B_r}$ by Proposition 1.1. Consequently $h_{B_r}(x, \cdot)/g(x)$ converges completely to μ_{B_r} as $\|x\| \to \infty$.

Choose $A \subset \mathbb{R}^d$ and let $r > 0$ be such that $\bar{B} \subset \mathring{B}_r$. Now $\lim_{t \to 0} P_{\cdot}(\tau_B \leq t) = 0$ uniformly on S_r so $\lim_{t \to 0} p^t h_B I_A = h_B I_A = h_B(\cdot, A)$ uniformly on S_r. Since $p^t h_B I_A$ is continuous on $\mathbb{R}^d$ for $t > 0$, $h_B(\cdot, A)$ is continuous on S_r. Now $h_B = h_{B_r} h_B$ by Theorem 2.6.7, so

$$\lim_{\|x\| \to \infty} \frac{h_B(x, A)}{g(x)} = \lim_{\|x\| \to \infty} \int \frac{h_{B_r}(\cdot, dz)}{g(x)} h_B(z, A) = \int \mu_{B_r}(dz) h_B(z, A)$$
$$= \mu_B(A),$$

which completes the proof of the theorem.

Newtonian capacity was first defined for an arbitrary compact set by Wiener [2].

Proposition 1.11. *Let $B \in \mathscr{B}$ be bounded. Then B is polar if and only if $C(B) = 0$. Also μ_B is supported on ∂B and does not charge polar sets. If B is compact, $\mu_{\partial B} = \mu_B$ and $C(\partial B) = C(B)$.*

Proof. Choose $r > 0$ such that $B \subset B_r$. Then $\mu_B = \int \mu_{B_r}(dy) h_B(y, \cdot)$ by Theorem 1.10. If B is polar, then $C(B) = \mu_B(\mathbb{R}^d) = \int \mu_{B_r}(dy) h_B(y, \mathbb{R}^d) = 0$. If $C(B) = 0$, then $P_{\cdot}(\tau_B < \infty) = g\mu_B = 0$ and hence B is polar. By Theorem 2.6.5 for each $y \in S_r$, $h_B(y, \cdot)$ is supported on ∂B and does not charge polar sets. Thus μ_B is supported on ∂B and does not charge polar sets. Suppose B is compact. Then $h_{\partial B}(y, \cdot) = h_B(y, \cdot)$ for $y \in S_r$ by Theorem 2.6.5. Consequently $\mu_{\partial B} = \mu_B$ and hence $C(\partial B) = C(B)$.

Let $B \in \mathscr{B}$ be bounded. By the last result of Theorem 1.10

$$\lim_{\|x\| \to \infty} \frac{P_x(\tau_B < \infty, X(\tau_B) \in A)}{g(x)} = \mu_B(A), \qquad A \subset \mathbb{R}^d. \tag{5}$$

In particular

$$\lim_{\|x\| \to \infty} \frac{P_x(\tau_B < \infty)}{g(x)} = C(B). \tag{6}$$

It follows from (5) and (6) that if B is nonpolar, then

$$\lim_{\|x\| \to \infty} P_x(X(\tau_B) \in A \,|\, \tau_B < \infty) = \frac{\mu_B(A)}{C(B)}, \qquad A \subset \mathbb{R}^d. \tag{7}$$

Thus if B is nonpolar, the normalized equilibrium measure of B can be interpreted as the conditional hitting distribution of B for Brownian motion starting at infinity.

Proposition 1.12. *The capacity $C(B)$ as B ranges over bounded sets in $\mathscr{B}$ satisfies the following properties: $C(b + B) = C(B)$ for $b \in \mathbb{R}^d$; $C(aB) = a^{d-2}C(B)$ for $a > 0$; $C(A \cup B) + C(A \cap B) \leq C(A) + C(B)$; and $C(A) \leq C(B)$ if $A \subset B$.*

Proof. Now $P_{b+\cdot}(\tau_{b+B} < \infty) = P_{\cdot}(\tau_B < \infty)$ by translation invariance, so μ_{b+B} is a translation of μ_B by Proposition 1.1 and hence $C(b + B) = \mu_{b+B}(\mathbb{R}^d) = \mu_B(\mathbb{R}^d) = C(B)$.

In proving the next result it can be assumed that $a > 0$. Observe that $g = a^{2-d}g(a^{-1}\cdot)$ and by scale invariance $P_{a^{-1}\cdot}(\tau_B < \infty) = P_{\cdot}(\tau_{aB} < \infty)$. Let ν be the measure on $\mathbb{R}^d$ defined by $\nu(aU) = a^{d-2}\mu_B(U)$ for $U \subset \mathbb{R}^d$. Then

$$g\nu = g\mu_B(a^{-1}\cdot) = P_{a^{-1}\cdot}(\tau_B < \infty) = P_{\cdot}(\tau_{aB} < \infty) = g\mu_{aB},$$

so $\mu_{aB} = \nu$ by Proposition 1.1 and hence

$$C(aB) = \mu_{aB}(\mathbb{R}^d) = \nu(\mathbb{R}^d) = a^{d-2}\mu_B(\mathbb{R}^d) = a^{d-2}C(B).$$

Let $A, B \in \mathscr{B}$ be bounded. Now $\tau_{A \cup B} = \tau_A \wedge \tau_B$ and $\tau_{A \cap B} \geq \tau_A \vee \tau_B$, so

$$P_{\cdot}(\tau_{A \cap B} < \infty) + P_{\cdot}(\tau_{A \cap B} < \infty) \leq P_{\cdot}(\tau_A < \infty) + P_{\cdot}(\tau_B < \infty).$$

Thus $C(A \cup B) + C(A \cap B) \leq C(A) + C(B)$ by the formula for capacity in Theorem 1.10. If $A \subset B$, then $P_{\cdot}(\tau_A < \infty) \leq P_{\cdot}(\tau_B < \infty)$, so $C(A) \leq C(B)$ by the same formula. This completes the proof of the proposition.

Proposition 1.13. (i) *If $B_n \in \mathscr{B}$ for $n \geq 1$ and $B_n \uparrow B$ where B is bounded, then $P_{\cdot}(\tau_{B_n} < \infty) \uparrow P_{\cdot}(\tau_B < \infty)$ on $\mathbb{R}^d$ and $C(B_n) \uparrow C(B)$.* (ii) *If B_n is compact for $n \geq 1$ and $B_n \downarrow B$, then $P_{\cdot}(\tau_{B_n} < \infty) \downarrow P_{\cdot}(\tau_B < \infty)$ on $B^r \cup B^c$ and $C(B_n) \downarrow C(B)$.*

Proof. Choose $r > 0$ such that B_n for $n \geq 1$ and B are contained in $\mathring{B}_r$. Suppose first that $B_n \in B$ for $n \geq 1$ and $B_n \uparrow B$. Then $\tau_{B_n} \downarrow \tau_B$ by Proposition 2.3.8, so $\{\tau_{B_n} < \infty\} \uparrow \{\tau_B < \infty\}$ and hence $P_{\cdot}(\tau_{B_n} < \infty) \uparrow P_{\cdot}(\tau_B < \infty)$ on $\mathbb{R}^d$. Consequently

$$C(B_n) = \int P_{\cdot}(\tau_{B_n} < \infty)\, d\mu_{B_r} \uparrow \int P_{\cdot}(\tau_B < \infty)\, d\mu_{B_r} = C(B).$$

Suppose next that B_n is compact for $n \geq 1$ and $B_n \downarrow B$. Choose $x \in B^r \cup B^c$. Then $P_x(\tau_{B_n} \uparrow \tau_B) = 1$ by Proposition 2.3.8. Since the Brownian motion process is transient, $P_x(\tau_{B_n} < \infty$ for all n and $\tau_B = \infty) = 0$ and hence $P_x(\tau_{B_n} < \infty) \downarrow P_x(\tau_B < \infty)$. Since μ_{B_r} is supported on $S_r \subset B^c$,

$$C(B_n) = \int P_{\cdot}(\tau_{B_n} < \infty)\, d\mu_{B_r} \downarrow \int P_{\cdot}(\tau_B < \infty)\, d\mu_{B_r} = C(B).$$

This completes the proof of the proposition.

For B bounded, let $\mathscr{M}(B)$ denote the collection of finite measures on $\mathbb{R}^d$ which are concentrated on B. The characterization of capacity in the next result is due to La Valleé Poussin [1].

Proposition 1.14. *Let $B \in \mathscr{B}$ be bounded. Then*

$$g\mu_B = \sup[g\mu : \mu \in \mathscr{M}(B) \text{ and } g\mu \le 1 \text{ on } \mathbb{R}^d]$$

and

$$C(B) = \sup[\mu(B) : \mu \in \mathscr{M}(B) \text{ and } g\mu \le 1 \text{ on } \mathbb{R}^d].$$

Proof. Suppose first that B is compact. Set $B_n = \{x : d(x, B) \le 1/n\}$. Then $P_.(\tau_{B_n} < \infty) \downarrow P_.(\tau_B < \infty)$ a.e. by Propositions 1.13 and 2.3.7. Choose $\mu \in \mathscr{M}(B)$ such that $g\mu \le 1$ on $\mathbb{R}^d$. Since μ is supported on B and $g_{B_n}(x, \cdot) = 0$ on $B_n^r \supset \mathring{B}_n \supset B$ for $x \in \mathbb{R}^d$, it follows from the fundamental identity that $g\mu = h_{B_n} g\mu \le P_.(\tau_{B_n} < \infty)$. Consequently $g\mu \le P_.(\tau_B < \infty)$ a.e. and therefore $p^t g\mu \le P_.(\tau_B \circ \theta_t < \infty)$ on $\mathbb{R}^d$ for $t > 0$. It follows by letting $t \downarrow 0$ that $g\mu \le P_.(\tau_B < \infty) = g\mu_B$ on $\mathbb{R}^d$. Also

$$C(B_n) \ge \int g\mu \, d\mu_{B_n} = \int g\mu_{B_n} \, d\mu = \mu(B),$$

so $\mu(B) \le C(B)$ by Proposition 1.13. Since $\mu_B \in \mathscr{M}(B)$ and $g\mu_B \le 1$ on $\mathbb{R}^d$ the two conclusions of the proposition are valid for B compact.

Consider now the general case. Let B_n be compact sets for $n \ge 1$ such that that $B_n \uparrow B$. Choose $\mu \in \mathscr{M}(B)$ such that $g\mu \le 1$ on $\mathbb{R}^d$. Let $\mu|_{B_n}$ be the measure on $\mathbb{R}^d$ defined by $\mu|_{B_n}(A) = \mu(A \cap B_n)$ for $A \subset \mathbb{R}^d$. Then $\mu|_{B_n} \in \mathscr{M}(B_n)$ and $g\mu|_{B_n} \le 1$ on $\mathbb{R}^d$, so $g\mu|_{B_n} \le P_.(\tau_{B_n} < \infty) \le P_.(\tau_B < \infty) = g\mu_B$ on $\mathbb{R}^d$ and $\mu(B_n) \le C(B_n) \le C(B)$ on $\mathbb{R}^d$. Consequently $g\mu \le g\mu_B$ on $\mathbb{R}^d$ and $\mu(B) \le C(B)$ on $\mathbb{R}^d$. Now μ_B is not necessarily concentrated on B, but $\mu_{B_n} \in \mathscr{M}(B_n) \subset \mathscr{M}(B)$ and $g\mu_{B_n} \le 1$ on $\mathbb{R}d$. Since $g\mu_{B_n} = P_.(\tau_{B_n} < \infty) \uparrow P_.(\tau_B < \infty)$ on $\mathbb{R}^d$ and $\mu_{B_n}(B) = C(B_n) \uparrow C(B)$ by Proposition 1.13, the proof is complete.

2. Asymptotic Behavior of Hitting Times

Consider in this section d-dimensional Brownian motion with $d \ge 3$. The potential theory developed in the previous section will be used to determine the asymptotic behavior of $P_x(t < \tau_B < \infty, X(\tau_B) \in A)$ as $t \to \infty$.

Let $B \in \mathscr{B}$ be bounded and hence transient. Recall from Section 2.2 the definition of the last exit time L_B from B. Now $\{L_B > t\} = \{\tau_B \circ \theta_t < \infty\}$, so by the Markov property

(1) $$P_x(L_B > t) = \int p(t, x, y) P_y(\tau_B < \infty), \qquad x \in \mathbb{R}^d \quad \text{and} \quad t > 0.$$

Since $P_.(\tau_B < \infty) = g\mu_B$, it follows from (1) that

$$P_x(L_B > t) = \int_t^\infty \left(\int p(s, x, y) \mu_B(dy) \right) ds.$$

Thus the measure $P_x(L_B \in \cdot)$ restricted to $(0, \infty)$ is absolutely continuous and has density $\int p(t, x, y)\mu_B(dy), t > 0$. Consequently $P_x(L_B = t) = 0$ for all $t > 0$ and

(2) $$E_x(e^{-\lambda L_B}; L_B > 0) = \int g^\lambda(x, y)\mu_B(dy), \qquad x \in \mathbb{R}^d \quad \text{and} \quad \lambda \geq 0.$$

The next result provides an interesting interpretation of μ_B in terms of the last time in B.

Theorem 2.1. *Let $B \in \mathscr{B}$ be bounded. Then*

$$P_x(L_B > 0, X(L_B) \in A) = \int_A g(x, y)\mu_B(dy), \qquad x \in \mathbb{R}^d \quad \textit{and} \quad A \subset \mathbb{R}^d.$$

Proof. Choose $x \in \mathbb{R}^d$. Let f be a nonnegative continuous function on $\mathbb{R}^d$ having compact support and vanishing in some neighborhood of x. Note that $L_B = t + L_B \circ \theta_t$ on $\{L_B > t\}$. Thus for $\lambda \geq 0$

$$\begin{aligned}
&\int_0^\infty e^{-\lambda t} E_x(f(X(L_B - t)); L_B > t)\, dt \\
&= E_x\left(\int_0^{L_B} e^{-\lambda t} f(X(L_B - t))\, dt\right) \\
&= E_x\left(\int_0^{L_B} e^{-\lambda(L_B - t)} f(X(t))\, dt\right) \\
&= E_x\left(\int_0^{L_B} e^{-\lambda L_B \circ \theta_t} f(X(t))\, dt\right) \\
&= \int_0^\infty E_x(e^{-\lambda L_B \circ \theta_t} f(X(t)); L_B > t)\, dt \\
&= \int_0^\infty E_x(f(X(t)) E_{X(t)}(e^{-\lambda L_B}; L_B > 0))\, dt \\
&= \int g(x, z) f(z) E_z(e^{-\lambda L_B}; L_B > 0)\, dz \\
&= \int g(x, z) f(z) \left(\int g^\lambda(z, y)\mu_B(dy)\right) dz \\
&= \int_0^\infty e^{-\lambda t}\left[\int\left(\int p(t, z, y) f(z) g(x, z)\, dz\right)\mu_B(dy)\right] dt.
\end{aligned}$$

Since the Laplace transform of a distribution on $[0, \infty)$ determines that distribution uniquely (see Theorem 1 on page 430 of Feller [1]),

(3) $$E_x(f(X(L_B - t)); L_B > t) = \int\left(\int p(t, z, y) f(z) g(x, z)\, dz\right)\mu_B(dy)$$

for almost all $t \in (0, \infty)$. It is easily seen that both sides of (3) are continuous in t, so (3) holds for all $t > 0$ (note that $f(\cdot)g(x,\cdot)$ is continuous and has compact support). It follows by letting $t \to 0$ in (3) that

$$E_x(f(X(L_B)); L_B > 0) = \int f(y)g(x,y)\mu_B(dy).$$

Thus

$$(4) \qquad P_x(L_B > 0, X(L_B) \in A) = \int_A g(x,y)\mu_B(dy)$$

holds for $A \subset \mathbb{R}^d \backslash \{x\}$. Now $P_x(L_B > 0) = P_x(\tau_B < \infty) = \int g(x,y)\mu(dy)$, so (4) holds for all $A \subset \mathbb{R}^d$ as desired. (For another proof this result, see that of Theorem 6.5.23.)

Set $r(t) = (2\pi)^{d/2}(d/2 - 1)t^{d/2-1}$ for $t > 0$. Then

$$(5) \qquad \lim_{t\to\infty} r(t)\int_t^\infty p(s,x)\,ds = 1 \text{ uniformly on compacts}$$

and

$$(6) \qquad \lim_{t\to\infty} r(t)p(t,x) = 0 \text{ uniformly on compacts.}$$

The asymptotic behavior of $P_x(L_B > t, X(L_B) \in A)$ and $P_x(t < \tau_B < \infty, X(\tau_B) \in A)$ as $t \to \infty$ will be determined in the next two results.

Proposition 2.2. *Let $B \in \mathscr{B}$ be bounded and let $A \subset \mathbb{R}^d$. Then*

$$\lim_{t\to\infty} r(t)P_x(L_B > t, X(L_B) \in A) = \mu_B(A) \textit{ uniformly on compacts.}$$

In particular

$$\lim_{t\to\infty} r(t)P_x(L_B > t) = C(B) \textit{ uniformly on compacts.}$$

Proof. For $x \in \mathbb{R}^d$ and $t > 0$

$$\begin{aligned} P_x(L_B > t, X(L_B) \in A) &= \int p(t,x,y)P_y(L_B > 0, X(L_B) \in A)\,dy \\ &= \int p(t,x,y)\left(\int_A g(y,z)\mu_B(dz)\right)dy \\ &= \int_A\left(\int_t^\infty p(s,x,z)\,ds\right)\mu_B(dz), \end{aligned}$$

so the desired result follows from (5).

The next result is due to Spitzer [2] for $d = 3$ and to Port [1] for the general case.

Theorem 2.3. *Let $B \in \mathscr{B}$ be bounded and let $x \in \mathbb{R}^d$ and $A \subset \mathbb{R}^d$. Then*

$$\lim_{t\to\infty} r(t)P_x(t < \tau_B < \infty, X(\tau_B) \in A) = \mu_B(A)P_x(\tau_B = \infty).$$

In particular

$$\lim_{t\to\infty} r(t)P_x(t < \tau_B < \infty) = C(B)P_x(\tau_B = \infty).$$

Proof. It suffices to prove the second result since the first result follows from it together with (6) and (1.7). Let $t > 0$ and $y \in \mathbb{R}^d$. Then

$$\{\tau_B \circ \theta_t < \infty\} = \{t < \tau_B < \infty\} \cup \{\tau_B \le t \text{ and } \tau_B \circ \theta_t < \infty\}$$

and hence

$$P_y(t < \tau_B < \infty) = P_y(\tau_B \circ \theta_t < \infty) - P_y(\tau_B \le t \text{ and } \tau_B \circ \theta_t < \infty).$$

Set

$$\psi(y, t) = P_y(\tau_B \circ \theta_t < \infty) = \int p(t, y, z)P_z(\tau_B < \infty)\,dz = P_y(L_B > t).$$

Then $\psi(y, t)$ is nonincreasing in t and by Proposition 2.2

(7) $$\lim_{t\to\infty} r(t)\psi(y, t) = C(B) \text{ uniformly on compacts.}$$

Observe that

(8) $$r(t)P_x(t < \tau_B < \infty) = r(t)\psi(x, t) - r(t)E_x(\psi(X(\tau_B), t - \tau_B); \tau_B \le t).$$

By (7) and (8) to prove the second result of the theorem it suffices to show that

(9) $$\lim_{t\to\infty} r(t)E_x(\psi(X(\tau_B), t - \tau_B); \tau_B \le t) = C(B)P_x(\tau_B < \infty).$$

Choose $\delta \in (0, 1)$. Now $(1 - \delta)^{d/2-1}r(t) \le r(s) \le r(t)$ for $0 \le (1 - \delta)t \le s \le t$ by the definition of $r(t)$. Thus by (7) it suffices to show that

(10) $$\lim_{t\to\infty} r(t)E_x(\psi(X(\tau_B), t - \tau_B); \delta t < \tau_B \le t) = 0.$$

Choose $\varepsilon > 0$. By (7) there is an $M > 0$ such that $\psi(y, M) \le \varepsilon$ for $y \in \bar{B}$. For $t > M$

$$\begin{aligned} &r(t)E_x(\psi(X(\tau_B), t - \tau_B); t - M < \tau_B \le t) \\ &\quad\le r(t)P_x(t - M < \tau_B \le t) \\ &\quad\le r(t)P_x(\tau_B \circ \theta_{t-M} \le M) \\ &\quad= r(t)\int p(t - M, x, y)P_y(\tau_B \le M)\,dy \\ &\quad\le r(t)p(t - M, 0)\int P_y(\tau_B \le M)\,dy. \end{aligned}$$

By (6) this yields

$$\lim_{t\to\infty} r(t)E_x(\psi(X(\tau_B), t-\tau_B); t-M<\tau_B\le t)=0 \tag{11}$$

provided that

$$\int P_y(\tau_B \le M)\,dy<\infty. \tag{12}$$

To verify (12) set $C=\{x: d(x,B)\le 1\}$. There is an $\alpha>0$ such that $P_z(X(s)\in C)\ge\alpha$ for $z\in\bar{B}$ and $0\le s\le M+1$. Now $P_y(X(M+1)\in C)\ge \alpha P_y(\tau_B\le M)$. Consequently

$$\begin{aligned}\int P_y(\tau_B\le M)\,dy &\le \alpha^{-1}\int P_y(X(M+1)\in C)\,dy\\ &=\alpha^{-1}\int dy\int p(M+1,y,z)I_C(z)\,dz=\alpha^{-1}|C|<\infty,\end{aligned}$$

where $|C|$ is the Lebesgue measure of C. This proves (12) and hence (11).
Since $P_x(\delta t<\tau_B\le t-M)\le P_x(L_B>\delta t)=\psi(x,\delta t)$, it follows from (7) that

$$\overline{\lim_{t\to\infty}}\, r(t)E_x(\psi(X(\tau_B),t-\tau_B);\delta t<\tau_B\le t-M)\le \varepsilon\delta^{1-d/2}C(B). \tag{13}$$

Since ε can be made arbitrarily small, (10) follows from (11) and (13). This completes the proof of the theorem.

3. Criteria for Regularity and Recurrence

Let $B\in\mathscr{B}$ and let $x\in\mathbb{R}^d$. In this section necessary and sufficient conditions are obtained for x to be regular for B and for B to be recurrent. These results are applied to thorns, defined below.

Let A_n, $n\ge 1$, be events. Then

$$\{A_n \text{ occurs i.o.}\}=\left\{\sum_n I_{A_n}=\infty\right\}$$

is the event that A_n occurs infinitely often. The following elementary extension of the Borel–Cantelli lemma to dependent events is required.

Proposition 3.1. *Let A_n, $n\ge 1$, be events (in an arbitrary probability space) such that for some $M>0$*

$$P(A_m\cap A_n)\le MP(A_m)P(A_n),\qquad |m-n|>1.$$

Then $P(A_n$ occurs i.o.$)>0$ if and only if $\sum_n P(A_n)=\infty$.

Proof. If $P(A_n \text{ occurs i.o.}) > 0$, then $\sum_n P(A_n) = E\sum_n I_{A_n} = \infty$. Suppose conversely that $\sum_n P(A_n) = \infty$. Then either $\sum_n P(A_{2n}) = \infty$ or $\sum_n P(A_{2n+1}) = \infty$. By this observation, in verifying that $P(A_n \text{ occurs i.o.}) > 0$ it can be assumed that

$$(1) \qquad P(A_n \cap A_m) \le MP(A_m)P(A_n), \qquad m \ne n.$$

Set $N_n = \sum_1^n I_{A_i}$ and $N = \sum_1^\infty I_{A_i}$. It must be shown that $P(N = \infty) > 0$. Now $EN_n = \sum_1^n P(A_i) \to \infty$ as $n \to \infty$, so there is a positive integer n_0 such that $(EN_n)^2 \ge EN_n > 0$ for $n \ge n_0$. Let $n \ge n_0$. Then by (1)

$$\begin{aligned} EN_n^2 &= \sum_{i=1}^n P(A_i) + 2\sum_{i=1}^{n-1}\sum_{j=i+1}^{n} P(A_i \cap A_j) \\ &\le \sum_{i=1}^n P(A_i) + M\sum_{i=1}^{n}\sum_{j=1}^{n} P(A_i)P(A_j) \\ &= EN_n + M(EN_n)^2 \\ &\le (M+1)(EN_n)^2. \end{aligned}$$

Now $E(N_n; N_n < EN_n/2) \le EN_n/2$ and hence $E(N_n; N_n \ge EN_n/2) \ge EN_n/2$. Thus by Schwarz's inequality

$$\begin{aligned} \frac{(EN_n)^2}{4} &\le \left[E\left(N_n; N_n \ge \frac{EN_n}{2}\right)\right]^2 \le EN_n^2 P\left(N_n \ge \frac{EN_n}{2}\right) \\ &\le (M+1)(EN_n)^2 P\left(N_n \ge \frac{EN_n}{2}\right). \end{aligned}$$

Therefore

$$P\left(N \ge \frac{EN_n}{2}\right) \ge P\left(N_n \ge \frac{EN_n}{2}\right) \ge \frac{1}{4(M+1)}.$$

Since $EN_n \to \infty$, $P(N = \infty) \ge \frac{1}{4}(M+1) > 0$ as desired.

Throughout the remainder of this section consider d-dimensional Brownian motion with $d \ge 3$. The necessary and sufficient condition in the next result is called *Wiener's test.*

Theorem 3.2. *Let $B \in \mathscr{B}$, $x \in \mathbb{R}^d$ and $0 < \lambda < 1$ and set $B_n = \{y \in B: \lambda^{n+1} < \|y - x\| \le \lambda^n\}$. Then $x \in B^r$ if and only if $\sum_n \lambda^{n(2-d)} C(B_n) \doteq \infty$.*

Proof. Since $\{x\}$ is polar, it follows easily from Proposition 2.2.12 and continuity of paths that $x \in B^r$ if and only if $P_x(\tau_{B_n} < \infty \text{ i.o.}) > 0$. Now $P_x(\tau_{B_n} < \infty) = \int g(y - x)\mu_{B_n}(dy)$ where μ_{B_n} is the equilibrium measure of B_n.

It follows easily from formula (1.1) for g that $\sum_n P_x(\tau_{B_n} < \infty) = \infty$ if and only if $\sum_n \lambda^{n(2-d)} C(B_n) = \infty$. Thus to complete the proof of the theorem it suffices to show that $P_x(\tau_{B_n} < \infty \text{ i.o.}) > 0$ if and only if $\sum_n P_x(\tau_{B_n} < \infty) = \infty$. By Proposition 3.1 it suffices to show that there is an $M > 0$ such that

$$(2) \quad P_x(\tau_{B_m} < \infty, \tau_{B_n} < \infty) \le M P_x(\tau_{B_m} < \infty) P_x(\tau_{B_n} < \infty), \qquad |m - n| > 1.$$

To this end, by (1.1) and the formula $P_z(\tau_{B_n} < \infty) = \int_{B_n} g(y - z)\mu_{B_n}(dy)$, there is an $M > 0$ such that $P_z(\tau_{B_n} < \infty) \le M P_x(\tau_{B_n} < \infty)/2$ for $|m - n| > 1$ and $z \in B_m$. Thus by the strong Markov property

$$P_x(\tau_{B_m} < \infty, \tau_{B_n} \circ \theta_{\tau_{B_m}} < \infty) \le \frac{M}{2} P_x(\tau_{B_m} < \infty) P_x(\tau_{B_n} < \infty), \qquad |m - n| > 1.$$

Since $\{\tau_{B_m} < \infty,\ \tau_{B_n} < \infty\} = \{\tau_{B_m} < \infty,\ \tau_{B_n} \circ \theta_{\tau_{B_m}} < \infty\} \cup \{\tau_{B_n} < \infty, \tau_{B_m} \circ \theta_{\tau_{B_n}} < \infty\}$ for $|m - n| > 1$, (2) holds as desired.

Theorem 3.3. *Let $\lambda > 1$ and $B \in \mathscr{B}$ and set $B_n = \{y \in B: \lambda^n \le \|y\| < \lambda^{n+1}\}$ for $n \ge 1$. Then B is recurrent if and only if $\sum_n \lambda^{n(2-d)} C(B_n) = \infty$.*

Proof. The proof of this result is similar to that of Theorem 3.2. The details are left to the reader.

In order to apply Theorems 3.2 and 3.3 it is necessary to bound $C(B_n)$ both above and below. The next result is useful for this purpose.

Proposition 3.4. *Let $c(L)$ denote the capacity of the cylinder $\{(x_1, \ldots, x_d): 0 \le x_1 \le L \text{ and } x_2^2 + \cdots + x_d^2 \le 1\}$ of length L and unit radius. For $L_0 > 0$ there are positive constants M and N such that if $d > 3$, then*

$$ML \le c(L) \le NL, \qquad L \ge L_0,$$

and if $d = 3$, then

$$\frac{ML}{\log L} \le c(L) \le \frac{NL}{\log L}, \qquad L \ge L_0.$$

Proof. Note that $c(L)$ is positive and nondecreasing in L for $L > 0$. Suppose first that $d > 3$. By Proposition 1.12, $c(L_1 + L_2) \le c(L_1) + c(L_2)$ for $L_1, L_2 > 0$. This yields the upper bound. In verifying the lower bound it can be assumed that L is a positive integer. For $j \ge 1$ let S_j denote the cylindrical surface

$$\{(x_1, \ldots, x_d): j - 1 \le x_1 \le j \text{ and } x_2^2 + \cdots + x_d^2 = 1\}.$$

It follows easily from the cone condition for regularity and continuity of paths that S_j has regular points and hence is nonpolar. Thus $C(S_j) > 0$. Let $\mu_j = \mu_{S_j}$ denote the equilibrium measure of S_j and set $b = (1, 0, \ldots, 0) \in \mathbb{R}^d$. Then $g\mu_{j+1}(x) = g\mu_1(x - jb)$. Set $\nu_L = \mu_1 + \cdots + \mu_L$. Then ν_L is concentrated on the cylinder of length L and unit radius, $\nu_L(\mathbb{R}^d) = C(S_1)L$ and $g\nu_L(x) = \sum_{j=0}^{L-1} g\mu_1(x - jb)$. Since $g\mu_1 \leq 1$, $\lim_{\|x\| \to \infty} \|x\|^{d-2} g\mu_1(x) < \infty$ and $\sup_x \sum_{j \geq 0} \|x - jb\|^{2-d} \wedge 1 < \infty$, there is a positive constant K such that $g\nu_L \leq K$ for all positive integers L. It now follows from Proposition 1.14 that $c(L) \geq C(S_1)L/K = ML$ where $M = C(S_1)/K$. This yields the desired lower bound.

Suppose now that $d = 3$. In verifying the lower bound it can be assumed that L is a positive integer greater than one. By an argument similar to that used for $d > 3$, there is a positive constant K independent of L such that $g\nu_L \leq K \log L$. Thus $c(L) \geq LC(S_1)/K \log L$, which yields the desired lower bound. To verify the upper bound, let S_L now denote the cylindrical surface $\{(x_1, x_2, x_3) : 0 \leq x_1 \leq L \text{ and } x_2^2 + x_3^2 = 1\}$. It suffices to show that there is a positive constant N such that $C(S_L) \leq NL/\log L$ for $L \geq 2$. Let $\mu_L = \mu_{S_L}$ denote the equilibrium measure of this surface. Then $g\mu_L \leq 1$. Set $b = (1, 0, 0) \in \mathbb{R}^3$. Then

$$L \geq \int_0^L g\mu_L(rb)\,dr = \int_0^L \left(\int_{S_L} g(y - rb)\mu_L(dy) \right) dr = \int_{S_L} \mu_L(dy) \int_0^L g(y - rb)\,dr.$$

There is a positive constant N such that $\int_0^L g(y - rb)\,dr \geq (\log L)/N$ for $L \geq 2$ and $y \in S_L$. Thus $L \geq C(S_L)(\log L)/N$ and hence $C(S_L) \leq NL/\log L$ for $L \geq 2$ as desired. This completes the proof of the proposition.

Let $h(r)$, $r \geq 0$, be a continuous function such that $h(r) > h(0) = 0$ for $r > 0$, $h(r)/r$ is nondecreasing in r for r sufficiently small and $h(r)/r$ is nonincreasing in r for r sufficiently large. Let T_h denote the *thorn*

$$\{(x_1, \ldots, x_d) : x_1 \geq 0 \text{ and } x_2^2 + \cdots + x_d^2 \leq h^2(x_1)\}.$$

Proposition 3.5. *For $d > 3$, 0 is regular for the thorn T_h if and only if*

$$\int_0^1 \left(\frac{h(r)}{r} \right)^{d-3} \frac{dr}{r} = \infty. \tag{3}$$

For $d = 3$, 0 is regular for the thorn if and only if

$$\int_0^1 \left| \log \frac{h(r)}{r} \right|^{-1} \frac{dr}{r} = \infty.$$

Proof. It will be shown that the condition for $d > 3$ is sufficient, the proofs of the remaining parts of the proposition being similar and left to the reader. If $\lim_{r \to 0} h(r)/r > 0$ the cone condition for regularity is satisfied, so 0 is regular for the thorn. Suppose $\lim_{r \to 0} h(r)/r = 0$. Set $B_n = \{y \in T_h: 2^{-n-1} < \|y\| \le 2^{-n}\}$. By Theorem 3.2 it must be shown that $\sum_n 2^{n(d-2)} C(B_n) = \infty$. Let S_n denote the cylinder

$$\{(x_1, \ldots, x_d): \tfrac{4}{3} 2^{-n-1} \le x_1 \le \tfrac{3}{4} 2^{-n} \text{ and } x_2^2 + \cdots + x_d^2 \le h^2(2^{-n-1})\}.$$

Then $S_n \subset B_n$ for n sufficiently large, so it sufficies to show that

$$\sum_n 2^{n(d-2)} C(S_n) = \infty.$$

By Propositions 3.4 and 1.12 there is a positive constant K such that $C(S_n) \ge K 2^{-n} (h(2^{-n}))^{d-3}$ for all n. Thus it suffices to show that $\sum_n h(2^{-n})/2^{-n})^{d-3} = \infty$. Since this follows from (3) and the monotonicity of $h(r)/r$ near $r = 0$, the proof is complete.

Some applications of Proposition 3.5 will now be described. Suppose $d = 3$. If $h(r) = r^\alpha$ for r sufficiently small, where $\alpha \ge 1$, then 0 is regular for the thorn T_h. If $h(r) = r^{|\log r|}$ for r sufficiently small, 0 is irregular for the thorn. Suppose $d > 3$ and $h(r) = r|\log r|^{-\alpha}$ for r sufficiently small, where $\alpha \ge 0$. Then 0 is regular for the thorn if $\alpha \le 1/(d-3)$ and irregular if $\alpha > 1/(d-3)$.

Proposition 3.6. *For $d > 3$ the thorn T_h is recurrent if and only if*

$$\int_1^\infty \left(\frac{h(r)}{r}\right)^{d-3} \frac{dr}{r} = \infty.$$

For $d = 3$ the thorn is recurrent if and only if

$$\int_1^\infty \left|\log \frac{h(r)}{r}\right|^{-1} \frac{dr}{r} = \infty.$$

Proof. The proof of this result uses Theorem 3.3 instead of Theorem 3.2 and is otherwise similar to that of Proposition 3.5. The details are left to the reader.

Suppose $d = 3$. If $h(r) = r^{-\alpha}$ for r sufficiently large, where $\alpha > 0$, the thorn T_h is recurrent. If $h(r) = r^{-\log r}$ for r sufficiently large the thorn is transient. Suppose $d > 3$ and $h(r) = r(\log r)^{-\alpha}$ for r sufficiently large, where $\alpha \ge 0$. The thorn is recurrent if $\alpha \le 1/(d-3)$ and transient if $\alpha > 1/(d-3)$. These results follow from Proposition 3.6.

Suppose $d = 3$. By Propositions 3.4 and 1.12 there is an $N > 0$ such that if $L > 0$ and $0 < \varepsilon \leq \frac{1}{2}$ the cylinder with radius εL and length L has capacity at most $NL/|\log \varepsilon|$. By Proposition 1.9 for $r > 0$ the closed ball B_r of radius r has capacity $2\pi r$. These observations can be used to show that a curve in $\mathbb{R}^3$ of bounded variation has zero capacity and hence is polar for three-dimensional Brownian motion. The details are left to the reader.

Lebesgue used thorns to give an example of a simply connected bounded open set in $\mathbb{R}^3$ such that the Dirichlet problem on D is not solvable for some continuous function on the boundary of D (see Section 4.2). Wiener [3] developed the test in Theorem 3.2 as a necessary and sufficient condition for $x \in \partial D$ to be regular for the Dirichlet problem on D; by Theorem 4.2.2 below, this is true if and only if $x \in (D^c)^r$. The planar version of Wiener's test is contained in Theorem 6.7.35. Wiener's test also yields a necessary and sufficient condition for B not to be "thin" at x (see Section 5.2). The necessary and sufficient condition for recurrence in Theorem 3.5 is due to Itô and McKean [1], [2]. The method of using a Borel–Cantelli lemma for dependent events to prove Theorems 3.2 and 3.3 is due to Lamperti [1]. The connection between these two theorems will be made more explicit in Section 6.5.

4. Logarithmic Potentials

Consider in this section planar Brownian motion. Analogs of the results in Sections 1 and 2 are obtained.

Observe that $\lim_{\lambda \to 0} g^\lambda(x) = \infty$ for $x \in \mathbb{R}^2$. To obtain a finite limit, let u denote a fixed point in $\mathbb{R}^2$ with $\|u\| = 1$. Set $k^\lambda(\cdot) = g^\lambda(\cdot) - g^\lambda(u)$ and $k^\lambda(x, y) = k^\lambda(y - x)$ for $x, y \in \mathbb{R}^2$. Then $k^\lambda(x, y)$ is symmetric in x and y. Note that

$$k^\lambda(x) = \int_0^\infty e^{-\lambda t}(p(t, x) - p(t, u))\, dt = \frac{1}{2\pi}\int_0^\infty e^{-\lambda t}(e^{-\|x\|^2/2t} - e^{-1/2t})\frac{dt}{t}.$$

Set

$$k(x) = \int_0^\infty (p(t, x) - p(t, u))\, dt = \frac{1}{2\pi}\int_0^\infty (e^{-\|x\|^2/2t} - e^{-1/2t})\frac{dt}{t}.$$

If $\|x\| = 1, k^\lambda(x) = k(x) = 0$ for all $\lambda > 0$. If $\|x\| < 1, k^\lambda(x) > 0$ and $k^\lambda(x) \uparrow k(x)$ as $\lambda \downarrow 0$. If $\|x\| > 1, k^\lambda(x) < 0$ and $k^\lambda(x) \downarrow k(x)$ as $\lambda \downarrow 0$.

Observe that

$$k(x) = \frac{1}{2\pi}\int_0^\infty (e^{-\|x\|^2 t} - e^{-t})\frac{dt}{t} = \frac{1}{2\pi}\int_0^\infty \left(\int_{\|x\|^2 t}^t e^{-s}\, ds\right)\frac{dt}{t}.$$

It follows by first changing the order of integration that

(1) $$k(x) = \frac{1}{\pi}\log\left(\frac{1}{\|x\|}\right), \qquad x \in \mathbb{R}^2.$$

The function $k(\cdot)$ is called the *logarithmic potential kernel*. By Proposition 2.1.1 it is integrable on compacts. Set $k(x, y) = k(y - x)$ for $x, y \in \mathbb{R}^2$. Then $k(x, y)$ is symmetric in x and y. Note that

$$(2) \qquad \lim_{\|y\| \to \infty} (k(x, y) - k(0, y)) = 0 \text{ uniformly on compacts.}$$

Let μ be a finite measure on $\mathbb{R}^2$ having compact support. Its *logarithmic potential* $k\mu$ is defined by $k\mu(\cdot) = \int k(\cdot, y)\mu(dy)$. This function is lower semicontinuous and integrable on compacts. It follows easily from the formula $k(x) = \int_0^\infty (p(s, x) - p(s, u))\, ds$ that $p^t k\mu$ is finite-valued and $k\mu = p^t k\mu + \int_0^t p^s \mu\, ds$ for $t > 0$.

Theorem 4.1. *Let μ, ν be finite measures on $\mathbb{R}^2$ having compact support and let $\alpha \in \mathbb{R}$.* If $k\mu = k\nu + \alpha$ a.e., then $\mu = \nu$.

Proof. Note that for $t > 0$

$$\int_0^t p^s \mu\, ds = k\mu - p^t k\mu = k\nu - p^t k\nu = \int_0^t p^s \nu\, ds \text{ a.e.}$$

The remainder of the proof is the same as that of Proposition 1.1.

Let $B \in \mathscr{B}$. Recall the fundamental identity for λ-potentials

$$(3) \qquad g^\lambda(x, y) = g_B^\lambda(x, y) + \int h_B^\lambda(x, dz) g^\lambda(z, y), \qquad x, y \in \mathbb{R}^2.$$

Set $W_B^\lambda(\cdot) = g^\lambda(u)(1 - E_\cdot(e^{-\lambda \tau_B}))$. Then $W_B^\lambda \geq 0$ and $W_B^\lambda = 0$ on B^r. By (3)

$$(4) \qquad k^\lambda(x, y) = g_B^\lambda(x, y) + \int h_B^\lambda(x, dz) k^\lambda(z, y) - W_B^\lambda(x), \qquad x, y \in \mathbb{R}^2.$$

According to the next theorem, if B is nonpolar one can let $\lambda \downarrow 0$ in (4) and obtain (5) below, called the *fundamental identity for logarithmic potentials.*

Theorem 4.2. *Let $B \in \mathscr{B}$ be nonpolar. Then $W_B(x) = \lim_{\lambda \to 0} W_B^\lambda(x)$ exists, is nonnegative, equals zero on B^r, and is bounded on compacts; if B is bounded the convergence is uniform on compacts. For $x \neq y$, $g_B(x, y) < \infty$ and $\int h_B(x, dz)|k(z, y)|\, dy < \infty$. Also*

$$(5) \qquad k(x, y) = g_B(x, y) + \int h_B(x, dz) k(z, y) - W_B(x), \qquad x, y \in \mathbb{R}^2.$$

Proof. Let C be a compact subset of $\mathbb{R}^2$ having positive Lebesgue measure $|C|$. By (4)

$$(6) \qquad \int_C k^\lambda(x, y)\, dy = g_B^\lambda(x, C) + \int h_B^\lambda(x, dz) \int_C k^\lambda(z, y)\, dy - |C|\, W_B^\lambda(x).$$

Now $\sup_y g_B(y, C) < \infty$ by Proposition 2.2.7 and for $s > 0$

$$0 \le g_B(x, C) - g_B^\lambda(x, C) = E_x \int_0^{\tau_B} (1 - e^{-\lambda t}) I_C(X(t))\, dt$$
$$\le \sup_y G_B(y, C)(1 - e^{-\lambda s} + P_x(\tau_B \ge s)).$$

Thus by Proposition 2.2.2, $\lim_{\lambda \to 0} g_B^\lambda(x, C) = g_B(x, C)$ uniformly on compacts. It follows easily from the formula

$$\int_C k^\lambda(x, y)\, dy = \frac{1}{2\pi} \int_0^\infty e^{-\lambda t} \left(\int_C (e^{-\|y - x\|^2/2t} - e^{-1/2t})\, dy \right) \frac{dt}{t}$$

and a similar formula for $\int_C k(x, y)\, dy$ that $\lim_{\lambda \to 0} \int_C k^\lambda(x, y)\, dy = \int_C k(x, y)\, dy$ uniformly on compacts.

Suppose B is bounded. By Proposition 2.2.2 and the definition of h_B^λ

$$\lim_{\lambda \to 0} \int h_B^\lambda(x, dz) \int_C k^\lambda(z, y)\, dy = \int h_B(x, dz) \int_C k(z, y)\, dy$$

uniformly on compacts. These observations together show that $W_B(x) = \lim_{\lambda \to 0} W_B^\lambda(x)$ exists and is finite and that the convergence is uniform on compacts. Since $W_B^\lambda \ge 0$ and $W_B^\lambda = 0$ on B^r, W_B satisfies these two properties. By the definition of h_B^λ and the monotone convergence theorem

$$\lim_{\lambda \to 0} \int h_B^\lambda(x, dz) k^\lambda(z, y) = \int h_B(x, dz) k(z, y), \qquad x, y \in \mathbb{R}^2. \tag{7}$$

(Consider separately the integrals over $\|z - y\| \le 1$ and $\|z - y\| > 1$.) Equation (5) now follows from (4). It follows from (5) that if $x \ne y$ then $g_B(x, y) < \infty$ and $\int h_B(x, dz) |k(z, y)| < \infty$ (note that k is bounded below on compacts). This completes the proof of the theorem when B is bounded.

Suppose now that B is unbounded. Set $A = \{z : d(z, C) \le 1\}$. Since A is compact

$$\lim_{\lambda \to 0} \int_A h_B^\lambda(x, dz) \int_C k^\lambda(z, y)\, dy = \int_A h_B(x, dz) \int_C k(z, y)\, dy,$$

the right-hand side of this equation being bounded in x. Now

$$0 \ge \int_C k^\lambda(z, y)\, dy \downarrow \int_C k(z, y)\, dy \qquad \text{for } z \in A^c.$$

It follows easily from the definition of h_B^λ and the monotone convergence theorem that

$$\int_{A^c} h_B^\lambda(x, dz) \int_C k^\lambda(z, y)\, dy \downarrow \int_{A^c} h_B(x, dz) \int_C k(z, y)\, dy \qquad \text{as } \lambda \downarrow 0.$$

Consequently

$$\lim_{\lambda \to 0} \int h_B^\lambda(x, dz) \int_C k^\lambda(z, y)\, dy = \int h_B(x, dz) \int_C k(z, y)\, dy,$$

the right-hand side of this equation being bounded above. Since $W_B^\lambda \geq 0$ it now follows from (6) that $W_B(x) = \lim_{\lambda \to 0} W_B^\lambda(x)$ exists and is bounded on compacts. As in the bounded case, $W_B \geq 0$ and $W_B = 0$ on B^r.

To see that $g_B(x, y) < \infty$ for $x \neq y$, observe that by Proposition 2.2.4 there is a compact nonpolar subset A of B. Now $g_A(x, y) < \infty$ for $x \neq y$ by what has already been shown, so $g_B(x, y) \leq g_A(x, y) < \infty$ for $x \neq y$ by (1) in the introduction to this chapter.

It will now be shown that

$$\text{(8)} \qquad \int_{B_1(y)} h_B(x, dz) k(z, y)\, dy < \infty, \qquad x \neq y.$$

Set $A = B \cap B_2(y)$. If A is polar, (8) clearly holds. Suppose A is nonpolar. Then

$$\text{(9)} \qquad \int_{B_1(y)} h_A(x, dz) k(z, y) < \infty, \qquad x \neq y,$$

since the theorem holds when applied to the bounded set A. Observe that if $X(\tau_B) \in B_1(y)$, then $\tau_A = \tau_B$ and hence $X(\tau_A) = X(\tau_B)$. Consequently the restriction of the measure $h_B(x, \cdot)$ to $B_1(y)$ is dominated by $h_A(x, \cdot)$, so (8) follows from (9).

If $x \neq y$, then $g_B(x, y) < \infty$ as we have already seen, so (5) follows from (4) and (8). In particular $\int h_B(x, dz)|k(z, y)| < \infty$ for $x \neq y$. Thus by (1), $\int h_B(x, dz) k^-(z, y) < \infty$ for all $x, y \in \mathbb{R}^2$. Therefore (5) also follows from (4) if $x = y$. This completes the proof of the theorem.

The next result is a *uniqueness principle for logarithmic potentials.*

Theorem 4.3. *Let $B \in \mathscr{B}$ be bounded and nonpolar and let μ and ν be finite measures on $\mathbb{R}^2$ which are concentrated on B^r and such that $\mu(\mathbb{R}^2) = \nu(\mathbb{R}^2)$ and $k\mu = k\nu + \alpha$ on B^r for some $\alpha \in \mathbb{R}$. Then $\mu = \nu$.*

Proof. Now $h_B(x, \cdot)$ is concentrated on B^r for $x \in \mathbb{R}^2$ by Theorem 2.6.5. Thus by the fundamental identity for logarithmic potentials

$$k\mu = h_B k\mu - W_B \mu(\mathbb{R}^2) = h_B k\nu + \alpha - W_B \nu(\mathbb{R}^2) = k\nu + \alpha \qquad \text{on } \mathbb{R}^2.$$

Consequently $\mu = \nu$ by Theorem 4.1.

Proposition 4.4. *Let $B \in \mathscr{B}$ be bounded and nonpolar. Then $\lim_{\|y\| \to \infty} g_B(x, y) = W_B(x)$ uniformly on compacts.*

Proof. By the fundamental identity for logarithmic potentials

$$k(x, y) - k(0, y) = g_B(x, y) + \int h_B(x, dz)(k(z, y) - k(0, y)) - W_B(x), \qquad x \neq y,$$

so the desired result follows from (2).

Proposition 4.5. *Let $B \in \mathscr{B}$ be nonpolar, let C be compact and let $\delta > 0$. Then* $\sup[g_B(x, y): x \in C,\ y \in \mathbb{R}^2$ *and* $\|y - x\| \geq \delta] < \infty$.

Proof. Let A be a compact nonpolar subset of B. Since $g_B \leq g_A$ it suffices to prove the result for A. In other words, without loss of generality it can be assumed that B is compact. Then W_B is bounded on compacts by Theorem 4.2. Thus by Proposition 4.4 it suffices to show that $\sup[g_B(x, y): x, y \in C$ and $\|y - x\| \geq \delta] < \infty$. But this follows easily from the inequality

$$g_B(x, y) \leq k(x, y) + W_B(x) - \int_{\|z-y\| \geq 1} h_B(x, dz) k(z, y),$$

which is a consequence of the fundamental identity for logarithmic potentials.

Proposition 4.6. *Let $B \in \mathscr{B}$ be nonpolar and let $x \in \mathbb{R}^2$. Then $g_B(x, \cdot)$ is upper semicontinuous on $\mathbb{R}^2 \setminus \{x\}$. In particular $\lim_{y \to b} g_B(x, y) = g_B(x, b) = 0$ for $b \in B^r \setminus \{x\}$.*

Proof. By the fundamental identity for logarithmic potentials it suffices to show that $\int h_B(x, dz) k(z, \cdot)$ is lower semicontinuous on $\mathbb{R}^2$. Choose $y_0 \in \mathbb{R}^2$ and let $f_1(y)$, $f_2(y)$, and $f_3(y)$ denote, respectively, the integrals of $h_B(x, dz)k(z, y)$ over $\|z - y_0\| \leq \frac{1}{2}$, $\frac{1}{2} < \|z - y_0\| \leq 2$, and $\|z - y_0\| > 2$. By Fatou's lemma, f_1 is lower semicontinuous at y_0. By the dominated convergence theorem, f_2 is continuous at y_0. Now $\int_{\|z-y_0\| \geq 2} h_B(x, dz) k(z, y_0) < \infty$ by Theorem 4.2 and, for y sufficiently close to y_0, $0 \geq k(z, y) \geq 2k(z, y_0)$ for $\|z - y_0\| \geq 2$. Thus f_3 is continuous at y_0 by the dominated convergence theorem. This completes the proof of the proposition.

Proposition 4.7. *Let $B \in \mathscr{B}$ be nonpolar. If B^c is bounded, then $W_B = 0$ on $\mathbb{R}^2$. If B is closed, then $W_B = 0$ on each bounded component of B^c.*

Proof. Suppose B^c is bounded. Choose $x \in \mathbb{R}^2$. Then $E_x \tau_B < \infty$ by Proposition 2.2.8. Thus

$$\lim_{\lambda \to 0} \frac{1 - E_x(e^{-\lambda \tau_B})}{\lambda} = \lim_{\lambda \to 0} \int_0^\infty e^{-\lambda t} P_x(\tau_B \geq t)\, dt = \int_0^\infty P_x(\tau_B \geq t)\, dt$$
$$= E_x \tau_B < \infty.$$

Now $p(t, u)$ is bounded in t and approaches zero as $t \to \infty$, so

$$\lim_{\lambda \to 0} \lambda g^\lambda(u) = \lim_{\lambda \to 0} \int_0^\infty \lambda e^{-\lambda t} p(t, u)\, dt = 0.$$

Consequently

$$W_B(x) = \lim_{\lambda \to 0} W_B^\lambda(x) = \lim_{\lambda \to 0} \lambda g^\lambda(u)\left(\frac{1 - E_x(e^{-\lambda \tau_B})}{\lambda}\right) = 0.$$

Thus $W_B = 0$ on $\mathbb{R}^2$. Suppose B is closed. Choose x in a bounded component D of B^c. Then $P_x(\tau_B = \tau_{D^c}) = 1$ and hence $W_B^\lambda(x) = W_{D^c}^\lambda(x)$. Consequently $W_B(x) = W_{D^c}(x) = 0$, which completes the proof of the proposition.

Proposition 4.8. *Let $0 < r < q$, let D denote the annulus $\{y : r < \|y\| < q\}$ and let $x \in D$. Then*

$$P_x(\|X(T_D)\| = r) = 1 - P_x(\|X(T_D)\| = q) = \frac{\log q - \log\|x\|}{\log q - \log r}.$$

Proof. Set $B = D^c$. Then $W_B(x) = 0$ by Proposition 4.7 and $g_B(x, 0) = 0$. Thus by (1) and the fundamental identity for logarithmic potentials

$$\log\|x\| = P_x(\|X(T_D)\| = r)\log r + (1 - P_x(\|X(T_D)\| = r))\log q,$$

from which the desired result follows.

Proposition 4.9. *Let $r > 0$ and $x \in \mathbb{R}^2$. Then $W_{B_r}(x) = \pi^{-1}\log^+(\|x\|/r)$ and $\int k(x, y)\sigma_r(dy) = \pi^{-1}\log(1/(\|x\| \vee r))$. Also*

$$\int_{B_r} k(x, y)\,dy = \frac{(r^2 - \|x\|^2)^+}{2} + r^2 \log\left(\frac{1}{\|x\| \vee r}\right).$$

Proof. The indicated formula for W_{B_r} will first be verified. If $\|x\| \le r$, then x is regular for B_r by Proposition 2.3.3 so $W_{B_r}(x) = 0$ by Theorem 4.2. Suppose $\|x\| > r$. It follows by setting $y = 0$ in the fundamental identity for logarithmic potentials that

$$W_{B_r}(x) = \pi^{-1}\log\|x\| - \pi^{-1}\log r = \pi^{-1}\log(\|x\|/r).$$

Thus in general $W_{B_r}(x) = \pi^{-1}\log^+(\|x\|/r)$ as desired.

To prove the second result choose $c \in S_r$. Suppose first that $\|x\| \le r$. By the fundamental identity for logarithmic potentials

$$\int k(x, y)\sigma_r(dy) = \int h_{S_r}(x, dz)\int k(z, y)\sigma_r(dy) = \int k(c, y)\sigma_r(dy)$$

independently of x. Thus

$$\int k(x, y)\sigma_r(dy) = \int k(0, y)\sigma_r(dy) = k(c) = \pi^{-1}\log(1/r).$$

Suppose now that $\|x\| > r$. Then by the fundamental identity for logarithmic potentials

$$\int k(x, y)\sigma_r(dy) = \int h_{B_r}(x, dz) \int k(z, y)\sigma_r(dy) - W_{B_r}(x)$$
$$= \pi^{-1}\log(1/r) - \pi^{-1}\log(\|x\|/r) = \pi^{-1}\log(1/\|x\|),$$

so the second result is valid. The last result follows easily from the second result and Proposition 2.1.1.

Let $r > 0$ and $\|x\| \leq r$. The formula $E_x(\tau_{S_r}) = (r^2 - \|x\|^2)/d$ has been verified for $d \geq 3$ in Proposition 1.8 and for $d = 1$ in Proposition 2.2.20. The next result shows that this formula is valid in the remaining case $d = 2$.

Proposition 4.10. *Let $r > 0$ and $\|x\| \leq r$. Then $E_x(\tau_{S_r}) = (r^2 - \|x\|^2)/2$.*

Proof. Set $B = \{y : \|y\| \geq r\}$ and choose $c \in S_r$. Now $W_B = 0$ by Proposition 4.7, so by the fundamental identity for logarithmic potentials and continuity of paths

$$E_x(\tau_{S_r}) = E_x(\tau_B) = \int_{B_r} g_B(x, y)\, dy = \int_{B_r} k(x, y)\, dy - \int_{B_r} k(c, y)\, dy.$$

The desired result now follows easily from Proposition 4.9.

Let $B \in \mathscr{B}$ be bounded and nonpolar. By the uniqueness principle for logarithmic potentials there is at most one probability measure μ_B on $\mathbb{R}^2$ which is concentrated on B^r and whose logarithmic potential is constant on B^r (the constant is necessarily finite by the fundamental identity and the local integrability of k). If such a measure μ_B exists it is called the *equilibrium measure* of B, the constant value $R(B)$ of $k\mu_B$ on B^r is called the *Robin constant* of B and $k\mu_B$ is called the *equilibrium potential* of B.

Proposition 4.11. *Let $r > 0$. Then $\mu_{B_r} = \sigma_r$ and $R(B_r) = \pi^{-1}\log(1/r)$.*

Proof. This result follows from Proposition 4.9.

Let $B \in \mathscr{B}$ be nonpolar. Then

$$\text{(10)} \quad \int h_B(x, dz)k(z, y) - W_B(x) = \int h_B(y, dz)k(z, x) - W_B(y), \qquad x, y \in \mathbb{R}^2.$$

For if $x = y$, then (10) is trivially true and if $x \neq y$, then (10) follows from the fundamental identity for logarithmic potentials, the finiteness of $k(x, y)$, and the symmetry of $k(x, y)$ and $g_B(x, y)$ in x and y.

According to the next result, every bounded nonpolar set $B \in \mathscr{B}$ has an equilibrium measure μ_B, which can be interpreted as the "hitting distribution of B for Brownian motion starting at infinity." The analytic version of the next result is due to La Vallée Poussin [1].

Theorem 4.12. *Let $B \in \mathscr{B}$ be bounded and nonpolar and let $r > 0$ be such that $B \subset B_r$. Then B has equilibrium measure $\mu_B = \int \sigma_r(dy) h_B(y, \cdot)$, Robin constant*

$$R(B) = \pi^{-1} \log(1/r) + \int W_B \, d\sigma_r = \lim_{\|x\| \to \infty} (k(x) + W_B(x)) \tag{11}$$

and equilibrium potential $k\mu_B = R(B) - W_B$. The function W_B is upper semicontinuous on $\mathbb{R}^2$. Finally

$$\lim_{\|x\| \to \infty} h_B(x, A) = \mu_B(A), \qquad A \subset \mathbb{R}^2. \tag{12}$$

Proof. Set $\mu = \int \sigma_r(dy) h_B(y, \cdot)$. Then μ is concentrated on B^r by Theorem 2.6.5 and μ is a probability measure by Proposition 2.2.10. It follows from (10) and Proposition 4.11 that

$$\begin{aligned} k\mu(x) &= \int \sigma_r(dy) \int h_B(y, dz) k(z, x) \\ &= \int \sigma_r(dy) \left[\int h_B(x, dz) k(z, y) + W_B(y) - W_B(x) \right] \\ &= \int h_B(x, dz) k\sigma_r(z) + \int W_B \, d\sigma_r - W_B(x) \\ &= \pi^{-1} \log(1/r) + \int W_B \, d\sigma_r - W_B(x). \end{aligned}$$

Now $W_B = 0$ on B^r, so μ is the equilibrium measure of B, $R(B) = \pi^{-1} \log(1/r) + \int W_B \, d\sigma_r$ is the Robin constant of B and $R(B) - W_B$ is the equilibrium potential of B. Since $k\mu_B$ is lower semicontinuous on $\mathbb{R}^2$, $W_B = R(B) - k\mu_B$ is upper semicontinuous on $\mathbb{R}^2$. By (2)

$$R(B) = \lim_{\|x\| \to \infty} (k\mu_B(x) + W_B(x)) = \lim_{\|x\| \to \infty} (k(x) + W_B(x)).$$

Observe that

$$k(x, y) = g_{B_r}(x, y) + \int h_{B_r}(x, dz) k(z, y) - W_{B_r}(x), \qquad x, y \in \mathbb{R}^2,$$

$$\lim_{\|x\| \to \infty} g_{B_r}(x, y) = W_{B_r}(y), \qquad y \in \mathbb{R}^2,$$

and

$$\lim_{\|x\|\to\infty}(k(x,y)+W_{B_r}(x)) = \lim_{\|x\|\to\infty}(k(x)+W_{B_r}(x)) = R(B_r).$$

Consequently

$$\lim_{\|x\|\to\infty}\int h_{B_r}(x,dz)k(z,\cdot) = R(B_r) - W_{B_r} = k\mu_{B_r} = k\sigma_r \qquad \text{on } \mathbb{R}^2.$$

Thus if $x_n \in \mathbb{R}^2$, $\|x_n\| \to \infty$, and $h_{B_r}(x_n,\cdot)$ converges completely to v, then $kv = k\sigma_r$ on $\mathbb{R}^2\backslash S_r$ and hence $v = \sigma_r$ by Theorem 4.1. Therefore $h_{B_r}(x,\cdot)$ converges completely to σ_r as $\|x\| \to \infty$.

Choose $A \subset \mathbb{R}^2$. Let r be such that $\bar{B} \subset \mathring{B}_r$. Now $h_B(\cdot, A)$ is continuous on S_r and $h_B(\cdot, A) = \int h_{B_r}(\cdot, dz)h_B(z, A)$, as was seen in the proof of Theorem 1.10. Thus

$$\lim_{\|x\|\to\infty} h_B(x,A) = \lim_{\|x\|\to\infty}\int h_{B_r}(x,dz)h_B(z,A) = \int \sigma_r(dz)h_B(z,A) = \mu_B(A),$$

which completes the proof of the theorem.

Theorem 4.13. *Let $B \in \mathscr{B}$ be bounded and nonpolar. Then μ_B is supported supported on ∂B and does not charge polar sets. If B is compact, then $\mu_{\partial B} = \mu_B$ and $R(\partial B) = R(B)$.*

Proof. Choose $r > 0$ such that $B \subset B_r$. Then $\mu_B = \int \sigma_r(dz)h_B(z,\cdot)$ by Theorem 4.12 so it follows from Theorem 2.6.5 that μ_B is supported on ∂B and does not charge polar sets and that if B is compact, then $\mu_{\partial B} = \mu_B$. Suppose now that B is compact and $x \notin B$. Then $W_{\partial B}^{\lambda}(x) = W_B^{\lambda}(x)$ for $\lambda > 0$ and hence $W_{\partial B}(x) = W_B(x)$. Consequently by Theorem 4.12

$$R(\partial B) = k\mu_{\partial B}(x) - W_{\partial B}(x) = k\mu_B(x) - W_B(x) = R(B),$$

which completes the proof of the theorem.

If $B \in \mathscr{B}$ is bounded and polar, set $R(B) = \infty$ and set $W_B = \infty$ on $\mathbb{R}^2$.

Theorem 4.14. *Let $A, B \in \mathscr{B}$ be bounded. Then $R(b + B) = R(B)$ for $b \in \mathbb{R}^2$; $R(aB) = \pi^{-1}\log(1/a) + R(B)$ for $a \geq 0$; and*

$$R(A \cup B) + R(A \cap B) \geq R(A) + R(B).$$

If $A \subset B$, then $W_A \geq W_B$ on $\mathbb{R}^2$ and $R(A) \geq R(B)$.

Proof. The first result is an obvious consequence of translation invariance. The second result is obvious if $a = 0$ or B is polar. Suppose $a > 0$ and B is nonpolar. Note that $k = \pi^{-1}\log(1/a) + k(a^{-1}\cdot)$ on $\mathbb{R}^2$. Note also that $P_{a\cdot}(\tau_{aB} = 0) = P_{\cdot}(\tau_B = 0)$ by scale invariance (Proposition 2.2.14), so $(aB)^r = aB^r$ and hence $W_B(a^{-1}\cdot) = 0$ on $(aB)^r$ by Theorem 4.2. Let ν be the probability measure on $\mathbb{R}^2$ defined by $\nu(aU) = \mu_B(U)$ for $U \subset \mathbb{R}^2$. Then ν is concentrated on $aB)^r = (aB)^r$. Also

$$k\nu = \pi^{-1}\log(1/a) + k\mu_B(a^{-1}\cdot) = \pi^{-1}\log(1/a) + R(B) + W_B(a^{-1}\cdot).$$

Thus $k\nu = \pi^{-1}\log(1/a) + R(B)$ on $(aB)^r$, so $\nu = \mu_{aB}$ and $R(aB) = \pi^{-1}\log(1/a) + R(B)$.

In proving the third result it can be assumed that $A \cap B$ is nonpolar. Note that $\tau_{A\cup B} = \tau_A \wedge \tau_B$ and $\tau_{A\cap B} \geq \tau_A \vee \tau_B$. Thus for $\lambda > 0$

$$e^{-\lambda\tau_{A\cup B}} + e^{-\lambda\tau_{A\cap B}} \leq e^{-\lambda\tau_A} + e^{-\lambda\tau_B},$$

so

$$E_{\cdot}(e^{-\lambda\tau_{A\cup B}}) + E_{\cdot}(e^{-\lambda\tau_{A\cap B}}) \leq E_{\cdot}(e^{-\lambda\tau_A}) + E_{\cdot}(e^{-\lambda\tau_B})$$

and hence

$$W^\lambda_{A\cup B} + W^\lambda_{A\cap B} \geq W^\lambda_A + W^\lambda_B.$$

Consequently

$$W_{A\cup B} + W_{A\cap B} \geq W_A + W_B.$$

It now follows easily from (11) that the third result is valid.

Suppose $A \subset B$. If A is polar the last result is trivially true, so suppose A is nonpolar. Then B is also nonpolar. Now $W^\lambda_A \geq W^\lambda_B$ for $\lambda > 0$, so $W_A \geq W_B$. Choose $r > 0$ such that $B \subset B_r$. Then by (11)

$$R(A) = \pi^{-1}\log(1/r) + \int W_A\,d\sigma_r \geq \pi^{-1}\log(1/r) + \int W_B\,d\sigma_r = R(B),$$

which completes the proof of the theorem.

Theorem 4.15. (i) *If $B_n \in B$ for $n \geq 1$ and $B_n \uparrow B$ where B is bounded, then $W_{B_n} \downarrow W_B$ on $\mathbb{R}^2$ and $R(B_n) \downarrow R(B)$.* (ii) *If B_n is compact for $n \geq 1$ and $B_n \downarrow B$, then $W_{B_n} \uparrow W_B$ on $B^r \cup B^c$ and $R(B_n) \uparrow R(B)$.*

Proof. Let f be a continuous nonnegative function having compact support and Lebesgue integral equal to one. Define kf on $\mathbb{R}^2$ by $kf(x) = \int k(x, y)f(y)\,dy = \int k(y)f(x + y)\,dy$ and observe that kf is continuous on $\mathbb{R}^2$. Choose $r > 0$ such that $\mathring{B}$ contains $\bar{B}$ and $\bar{B}_n$ for $n \geq 1$.

In proving the theorem it can be assumed without loss of generality that each set B_n is nonpolar. Choose $x \in \mathbb{R}^2$, with $x \in B^r \cup B^c$ in proving (ii). By the fundamental identity for logarithmic potentials

(13) $$kf(x) = g_{B_n}f(x) + h_{B_n}kf(x) - W_{B_n}(x).$$

It follows from Proposition 2.3.8 that

$$P_x\left(\lim_n \tau_{B_n} = \tau_B\right) = 1. \tag{14}$$

Since $g_{B_n}f(x) = E_x \int_0^{\tau_{B_n}} f(X(t))\,dt$ and the same formula holds with B_n replaced by B, it follows from Proposition 2.2.7, the dominated convergence theorem, and the monotone convergence theorem that

$$\lim_n g_{B_n}f(x) = g_B f(x). \tag{15}$$

Suppose now that B is nonpolar. Then $P_x(\tau_B < \infty) = 1$ by Proposition 2.2.10 so it follows from (14) and the continuity of kf that

$$\lim_n h_{B_n}kf(x) = h_B kf(x). \tag{16}$$

By (13) (applied to both B_n and B), (15) and (16), $\lim_n W_{B_n}(x) = W_B(x)$. By (11) and the dominated convergence theorem

$$\begin{aligned} \lim_n R(B_n) &= \pi^{-1}\log(1/r) + \lim_n \int W_{B_n}\,d\sigma_r \\ &= \pi^{-1}\log(1/r) + \int W_B\,d\sigma_r = R(B). \end{aligned} \tag{17}$$

Suppose instead that B_n is a compact nonpolar set for $n \geq 1$ and that $B_n \downarrow B$ where B is polar. Then by (15)

$$\lim_n g_{B_n}f(x) = \infty. \tag{18}$$

Now $h_{B_n}kf(x)$ is bounded below in n so by (13) and (18), $\lim_n W_{B_n}(x) = \infty = W_B(x)$. By (11) and the monotone convergence theorem, (17) holds in this case also. This completes the proof of the theorem.

The *logarithmic capacity* $C(B)$ of a bounded set $B \in \mathscr{B}$ is defined by $C(B) = e^{-R(B)}$. By Theorems 4.13, 4.14, and 4.15 logarithmic capacity satisfies the following properties: (i) if B is compact, $C(\partial B) = C(B)$; (ii) if $A, B \in \mathscr{B}$ are bounded and $A \subset B$, then $C(A) \leq C(B)$; (iii) if $B_n \in \mathscr{B}$ for each n and $B_n \uparrow B$ where B is bounded, then $C(B_n) \uparrow C(B)$; (iv) if B_n is compact for each n and $B_n \downarrow B$, then $C(B_n) \downarrow C(B)$. A bounded set $B \in \mathscr{B}$ is polar if and only if its logarithmic capacity is zero (this is true by definition, but it is the only definition which is consistent with the definition of $C(B)$ for B compact and nonpolar and properties (ii)–(iv)). In particular if B is compact, then $P_\cdot(\tau_B < \infty)$ equals zero or one according as B has zero or positive logarithmic capacity; this result is due to Kakutani [1].

The capacity of an arbitrary compact subset B of $\mathbb{R}^2$ was first defined by Wiener [2], [3] as $R(B)^{-1}$ when $R(B) > 0$, which is true if B lies inside an open disk $\mathring{B}_1(a)$ (the definition in Wiener [2] ignored the requirement $R(B) > 0$; this was corrected in Wiener [3]).

In the next result the asymptotic behavior of $P_x(\tau_B > t, X(\tau_B) \in A)$ as $t \to \infty$ is determined. The result is due to Hunt [1]; for an interesting application see Spitzer [2].

Theorem 4.16. *Let $B \in \mathscr{B}$ be bounded and nonpolar and let $x \in \mathbb{R}^2$. Then*

$$\lim_{t\to\infty} \frac{1}{2\pi} \log t P_x(\tau_B > t, X(\tau_B) \in A) = W_B(x)\mu_B(A), \qquad A \subset \mathbb{R}^2. \tag{19}$$

In particular

$$\lim_{t\to\infty} \frac{1}{2\pi} \log t P_x(\tau_B > t) = W_B(x). \tag{20}$$

Proof. Observe first that

$$W_B^\lambda(x) = g^\lambda(u)(1 - E_x(e^{-\lambda\tau_B})) = \lambda g^\lambda(u) \int_0^\infty e^{-\lambda\tau} P_x(\tau_B > t)\, dt.$$

Now $p(t, u)$ is bounded in t and $\lim_{t\to\infty} tp(t, u) = 1/2\pi$ so

$$\lim_{\lambda\to 0} |\log \lambda|^{-1} g^\lambda(u) = \lim_{\lambda\to 0} |\log \lambda|^{-1} \int_1^\infty \frac{e^{-\lambda t}}{2\pi t}\, dt = \frac{1}{2\pi}.$$

Thus by Theorem 4.2

$$\lim_{\lambda\to 0} \frac{\lambda|\log \lambda|}{2\pi} \int_0^\infty e^{-\lambda t} P_x(\tau_B > t)\, dt = W_B(x).$$

Consequently by a standard Tauberian theorem (Example (c) on page 447 of Feller [1]), (20) holds. Since $\lim_{t\to\infty} \log t P_x(\|X(t)\| \le r) = 0$ for all $r > 0$, (19) follows from (12) and (20).

5. Linear Potentials

Consider in this section linear Brownian motion. Analogs of the results in Section 4 are obtained.

Set $k^\lambda(\cdot) = g^\lambda(\cdot) - g^\lambda(0)$ and $k^\lambda(x, y) = k^\lambda(y - x)$ for $x, y \in \mathbb{R}$. Then $k^\lambda(x, y)$ is symmetric in x and y. Observe that

$$k^\lambda(x) = \int_0^\infty e^{-\lambda t}(p(t, x) - p(t, 0))\, dt = (1/\sqrt{2\pi}) \int_0^\infty e^{-\lambda t}(e^{-x^2/2t} - 1)t^{-1/2}\, dt.$$

Set

$$k(x) = \int_0^\infty (p(t,x) - p(t,0))\,dt = (1/\sqrt{2\pi}) \int_0^\infty (e^{-x^2/2t} - 1)t^{-1/2}\,dt.$$

Then $0 \geq k^\lambda(x) \downarrow k(x)$ for $x \in \mathbb{R}$ as $\lambda \downarrow 0$.

It follows by first making the substitution $u = x^2/2t$ and then changing the order of integration that

$$k(x) = (|x|/2\sqrt{\pi}) \int_0^\infty (e^{-u} - 1)u^{-3/2}\,du = (-|x|/2\sqrt{\pi}) \int_0^\infty \left(\int_0^u e^{-s}\,ds\right)u^{-3/2}\,du$$
$$= (-|x|/\sqrt{\pi}) \int_0^\infty s^{-1/2}e^{-s}\,ds.$$

Consequently

(1) $$k(x) = -|x|, \qquad x \in \mathbb{R}.$$

The function $k(\cdot)$ is called the *linear potential kernel*. Set $k(x, y) = k(y - x)$ for $x, y \in \mathbb{R}^2$. Then $k(x, y)$ is symmetric in x and y. Note that

(2) $$\lim_{y \to \pm\infty} (k(x, y) - k(0, y)) = \pm x \text{ uniformly on compacts.}$$

Let μ be a finite measure on $\mathbb{R}$ having compact support. Its *linear potential* $k\mu$ is defined by $k\mu(\cdot) = \int k(\cdot, y)\mu(dy)$. Note that $k\mu$ is a continuous function. It follows by the proof of Theorem 4.1 that if μ and ν are finite measures on $\mathbb{R}$ having compact support, $\alpha \in \mathbb{R}$ and $k\mu = k\nu + \alpha$ on $\mathbb{R}$, then $\mu = \nu$.

Let $B \in \mathscr{B}$. Recall from Proposition 2.2.19 that $P.(\tau_B = \tau_{\bar{B}}) = 1$, from Proposition 2.3.2 that $B^r = \bar{B}$, and from Proposition 2.2.18 that B is nonpolar if and only if it is nonempty. Recall also the fundamental identity for λ-potentials

(3) $$g^\lambda(x, y) = g_B^\lambda(x, y) + \int h_B^\lambda(x, dz)g^\lambda(z, y), \qquad x, y \in \mathbb{R}.$$

Set $W_B^\lambda(\cdot) = g^\lambda(0)(1 - E.(e^{-\lambda\tau_B}))$. Then $W_B^\lambda \geq 0$ and $W_B^\lambda = 0$ on $\bar{B}$. By (3)

(4) $$k^\lambda(x, y) = g_B^\lambda(x, y) + \int h_B^\lambda(x, dz)k^\lambda(z, y) - W_B^\lambda(x), \qquad x, y \in \mathbb{R}.$$

Choose $b \in \mathbb{R}$. It follows by applying (4) with $B = \{b\}$ and $y = b$ that

$$W_{\{b\}}(x) = \lim_{\lambda \to 0} W_{\{b\}}^\lambda(x) = -k(x, b) = |x - b|, \qquad x \in \mathbb{R}.$$

Thus by another application of (4) together with the definition of h_B^λ and the monotone convergence theorem

(5) $$g_{\{b\}}(x, y) = |x - b| + |y - b| - |y - x|, \qquad x, y \in \mathbb{R}.$$

Let $B \in \mathscr{B}$ be nonempty and choose $b \in B$. Then $g_B(x, y) \le g_{\{b\}}(x, y) < \infty$ for $x, y \in \mathbb{R}$. Consequently for every bounded set A

$$\sup[g_B(x, y) : x \in A \text{ and } y \in \mathbb{R}] < \infty. \tag{6}$$

Also $g_B(x, y) = 0$ for $x, y \in \bar{B}$. By (4), $W_B(x) = \lim_{\lambda \to 0} W_B^\lambda(x)$ exists and is finite and

$$k(x, y) = g_B(x, y) + \int h_B(x, dz) k(z, y) - W_B(x), \qquad x, y \in \mathbb{R}. \tag{7}$$

This equation is called the *fundamental identity for linear potentials*. The uniqueness principle (Theorem 4.3) holds with $\mathbb{R}^2$ replaced by $\mathbb{R}$.

Explicit formulas for g_B and W_B will now be obtained when $B \in \mathscr{B}$ is nonempty. Suppose first that x is contained in the closure of a bounded component (a, b) of $(\bar{B})^c$. By Propositions 2.2.20 and 2.3.1 $h_B(x, \{a\}) = (b - x)/(b - a)$ and $h_B(x, \{b\}) = (x - a)/(b - a)$. Thus by (7) with $y = b$

$$-(b - x) = -\frac{b - x}{b - a}(b - a) - W_B(x)$$

and hence $W_B(x) = 0$. It follows by another application of (7) that $g_B(x, y) = 0$ for $y \notin (a, b)$ and

$$g_B(x, y) = \frac{(b - x)(y - a) + (x - a)(b - y)}{b - a} - |y - x|, \qquad a < y < b. \tag{8}$$

Suppose next that $\infty > x \ge b_0 = \sup[x : x \in B]$. Then $h_B(x, \{b_0\}) = 1$. By (7), $W_B(x) = x - b_0$. Thus by another application of (7), $g_B(x, y) = 0$ for $y \le b_0$ and

$$g_B(x, y) = x + y - 2b_0 - |y - x|, \qquad y > b_0. \tag{9}$$

In particular $g_B(x, y) = 2W_B(x)$ for $y \ge x$ and hence $\lim_{y \to \infty} g_B(x, y) = 2W_B(x)$. Suppose finally that $-\infty < x \le a_0 = \inf[x : x \in B]$. Then $h_B(x, \{a_0\}) = 1$ and $W_B(x) = a_0 - x$. Also $g_B(x, y) = 0$ for $y \ge a_0$ and

$$g_B(x, y) = 2a_0 - x - y - |y - x|, \qquad y < a_0. \tag{10}$$

In particular $g_B(x, y) = 2W_B(x)$ for $y \le x$ and hence $\lim_{y \to -\infty} g_B(x, y) = 2W_B(x)$. Note that $g_B(x, y)$ is jointly continuous in x and y on $\mathbb{R} \times \mathbb{R}$. It follows from the formulas for $W_B(x)$ and from Propositions 2.2.17 and 2.8.4 that

$$\lim_{t \to \infty} (\pi t/2)^{1/2} P_x(\tau_B > t) = W_B(x), \qquad x \in \mathbb{R}. \tag{11}$$

Let $B \in \mathscr{B}$ be bounded and nonempty. Set $a_0 = \inf[x : x \in B]$, $b_0 = \sup[x : x \in B]$ and $\mu_B = (\delta_{a_0} + \delta_{b_0})/2$. Then μ_B is the unique probability measure on $\mathbb{R}$ which is supported on $\bar{B}$ and whose linear potential is constant on $\bar{B}$. The measure μ_B is called the *equilibrium measure* of B, the constant value $R(B) = -(b_0 - a_0)/2$ of $k\mu_B$ on $\bar{B}$ is called the *Robin constant* of B, and $k\mu_B$ is called the *equilibrium potential* of B. It follows from the fundamental identity for linear potentials that $k\mu_B = R(B) - W_B$ on $\mathbb{R}$. Thus by an easy computation

$$\tfrac{1}{2}\left[\lim_{x\to-\infty}(k(x) + W_B(x)) + \lim_{x\to\infty}(k(x) + W_B(x))\right] = R(B).$$

See Chacon [1] for some interesting applications of linear potentials.

Chapter 4

Harmonic Functions

Throughout this chapter D is a nonempty open subset of $\mathbb{R}^d$. Harmonic functions on D are studied and in particular the Dirichlet problem is solved.

1. Basic Properties of Harmonic Functions

A function on D is said to be *locally integrable* on D if it is integrable on every compact subset of D. A function f on D is said to be *harmonic* on D if it is locally integrable on D and satisfies the following averaging property: if $x \in D$ and $r > 0$ are such that $B_r(x) \subset D$, then $f(x) = \int f(y)\sigma_r(x, dy)$. Clearly constant functions are harmonic and a finite linear combination of harmonic functions is harmonic.

Proposition 1.1. *Let f be harmonic on D. Then f is infinitely differentiable (and in particular continuous) on D.*

Proof. Let $x_0 \in D$. Choose $\delta > 0$ such that $B_{2\delta}(x_0) \subset D$. Let ψ denote a nonnegative infinitely differentiable function on $[0, \infty)$ such that $\psi = 0$ on $[\delta^2, \infty)$ and ψ is not identically zero on $[0, \delta^2]$. Then $\int_D \psi(\|y - x\|^2)f(y)\,dy$

defines an infinitely differentiable function on $\mathring{B}_\delta(x_0)$. By Proposition 2.1.1

$$\begin{aligned}\int_D \psi(\|y - x\|^2) f(y)\, dy &= \int_{B_\delta} \psi(\|y\|^2) f(x + y)\, dy \\ &= \frac{2\pi^{d/2}}{\Gamma(d/2)} \int_0^\delta \psi(r^2)\left(\int f(y)\sigma_r(x, dy)\right) r^{d-1}\, dr \\ &= \frac{2\pi^{d/2}}{\Gamma(d/2)} f(x) \int_0^\delta \psi(r^2) r^{d-1}\, dr,\end{aligned}$$

so that f is infinitely differentiable on $\mathring{B}_\delta(x_0)$. Since x_0 is arbitrary, f is infinitely differentiable on D as desired.

If f is twice differentiable on D, its *Laplacian* Δf is the function on D defined by

$$\Delta f(x) = \sum_{i=1}^{d} \partial^2 f/\partial x_i^2, \qquad x = (x_1, \ldots, x_d) \in D.$$

A function f is commonly defined to be harmonic on D if it is twice continuously differentiable on D and satisfies Laplace's equation $\Delta f = 0$ on D. The next result shows that this definition of harmonicity is equivalent to the previous definition.

Proposition 1.2. *A function f is harmonic on D if and only if f is twice continuously differentiable on D and $\Delta f = 0$ on D.*

Proof. By Proposition 1.1 it can be assumed that f is twice continuously differentiable on D. Choose $x \in D$ and let $\delta > 0$ be such that $B_\delta(x) \subset D$. Observe that $\int f(y)\sigma_r(x, dy) = \int f(x + ry)\sigma_1(dy)$. By the divergence theorem of advanced calculus

$$\frac{d}{dr}\int f(x + ry)\sigma_1(dy) = \int [f'(x + ry) \cdot y]\sigma_1(dy) = \frac{r}{s_d}\int_{B_1} \Delta f(x + ry)\, dy$$

for $0 < r \le \delta$. Here s_d denotes the surface area of the unit sphere $S_1 = \partial B_1$ and f' denotes the first derivative (gradient) of f. Suppose $\Delta f = 0$. Then $\int f(x + ry)\sigma_1(dy)$ is constant in r for $0 < r \le \delta$ and the constant value (obtained by letting $r \to 0$) is $f(x)$. Thus f is harmonic on D. Suppose conversely that f is harmonic on D. Then $\int_{B_1} \Delta f(x + ry)\, dy = 0$ for $0 < r \le \delta$ and it follows by letting $r \to 0$ that $\Delta f(x) = 0$. Thus $\Delta f = 0$. This completes the proof of the proposition.

Suppose $d = 1$. Then $\Delta f = f''$. Now D is a finite or countable union of disjoint open intervals, so a function on D is harmonic if and only if it is linear on each such interval. In particular the linear potential kernel k given by $k(x) = -|x|$ is harmonic on $\mathbb{R}\backslash\{0\}$.

Suppose $d \geq 2$ and $f(x) = \psi(\|x\|)$, $x \in \mathbb{R}^d\backslash\{0\}$, where ψ is twice continuously differentiable on $(0, \infty)$. It is easily seen that

$$\Delta f(x) = (d-1)\|x\|^{-1}\psi'(\|x\|) + \psi''(\|x\|) \qquad \text{for} \quad x \in \mathbb{R}^d\backslash\{0\}.$$

It follows directly from this formula that the logarithmic potential kernel k is harmonic on $\mathbb{R}^2\backslash\{0\}$ and that for $d \geq 3$ the Newtonian potential kernel g is harmonic on $\mathbb{R}^d\backslash\{0\}$.

According to the next result, a harmonic function on a connected open set having an interior maximum or minimum is necessarily constant.

Proposition 1.3. *Suppose D is connected and let f be harmonic on D. If either $f \leq f(x_0)$ on D or $f \geq f(x_0)$ on D for some $x_0 \in D$, then f is constant on D.*

Proof. Suppose $x_0 \in D$ and $f \leq f(x_0) = M$ on D. Set $A = \{x \in D: f(x) = M\}$. Choose $x \in A$ and let $\delta > 0$ be such that $B_\delta(x) \subset D$. Then $M = f(x) = \int f(y)\sigma_r(x, dy)$ for $0 < r \leq \delta$ and hence $f = M$ on $B_\delta(x)$. Thus A is open. Since A is nonempty and $D\backslash A = \{x \in D: f(x) < M\}$ is open, it follows by connectivity that $A = D$ and hence that $f = M$ on D. A similar argument shows that if $f \geq f(x_0) = M$ on D, then $f = M$ on D. Thus in either case f is constant on D.

Finally it will be shown that a function which is harmonic on a bounded open set D and continuous on $\bar{D}$ assumes its maximum and minimum values on ∂D.

Proposition 1.4. *Suppose D is bounded and let f be harmonic on D and continuous on $\bar{D}$. Then $\min_{y \in \partial D} f(y) \leq f \leq \max_{y \in \partial D} f(y)$ on D.*

Proof. Without loss of generality it can be assumed that D is connected (otherwise look separately at each component of D). Since f is continuous on the compact set $\bar{D}$ there is an $x_0 \in \bar{D}$ such that $f \leq f(x_0)$ on $\bar{D}$. If $x_0 \in \partial D$, then $f \leq \max_{y \in \partial D} f(y)$ on D. If $x_0 \in D$, then f is constant on D by Proposition 1.3, so the last inequality again holds. Similarly $f \geq \min_{y \in \partial D} f(y)$ on D, so the desired conclusion holds.

2. Dirichlet Problem

Let f be defined on D. Then f is said to have *boundary value* v at $b \in \partial D$ if $f(x) \to v$ as $x \to b$ within D. Given a subset A of ∂D and a function φ on A, f is said to have *boundary function* φ on A if f has boundary value $\varphi(b)$ at b for each $b \in A$. For this to be true it is necessary that φ be continuous on A.

Let φ be a continuous function on ∂D. A solution to the *Dirichlet problem* for φ, posed by Gauss in 1840, is a harmonic function on D having boundary function φ on ∂D. Gauss thought he had solved the problem by using the "Dirichlet principle," but his reasoning was invalid (see Monna [1]). Indeed Zaremba in 1909 and Lebesgue in 1913 gave examples which will be discussed below, in which D is bounded and there is a continuous function φ on ∂D such that the Dirichlet problem for φ has no solution. Wiener [2] formulated a generalized Dirichlet problem which always has a solution and determined when the solution to the generalized problem also solves the original problem. He did not point out any connection with Brownian motion (the subject of Wiener [1] the year before). This connection was later observed by Kakutani [1] and Doob [1]. In the intervening period Courant, Friedrichs, and Lewy [1] noted an indirect connection between the two subjects. Namely, they noted that the solution to the discrete version of the Dirichlet problem (which converges to the solution to the original problem under appropriate conditions as the grid size tends to zero) has an interpretation in terms of simple random walk. In a footnote to this material in Section 3 of their paper they state that "The present treatment is essentially different from the familiar treatments which can be carried through, say, for example, in the case of Brownian motion for molecules. The difference lies precisely in the way in which the boundary of the region enters." Kellogg [1] discussed the historical development of the analytical solution to the Dirichlet and generalized Dirichlet problems.

In order to derive Wiener's result probabilistically, further definitions are required. Given $x \in D$, the *exit distribution* $H_D(x, \cdot)$ of D for Brownian motion starting at x is defined by

$$H_D(x, A) = h_{D^c}(x, A) = P_x(T_D < \infty, X(T_D) \in A),$$

where $T_D = \tau_{D^c}$ is the exit time of D. By Theorem 2.6.5, $H_D(x, \cdot)$ is concentrated on $\partial D \cap (D^c)^r$ and does not charge polar sets and $H_D(x, \cdot) = h_{\partial D}(x, \cdot)$. By Theorem 2.6.7 if D_0 is an open subset of D, then

$$H_D(x, A) = \int H_{D_0}(x, dy) H_D(y, A) \qquad \text{for} \quad x \in D_0.$$

If D_0 is a component of D, then $H_D(x, \cdot) = H_{D_0}(x, \cdot)$ for $x \in D_0$. A subset N of ∂D is said to be H_D-*null* if $H_D(\cdot, N) = 0$ on D. A polar subset of D is clearly H_D-null, but an H_D-null set need not be polar. For let $d = 2$ and let D be the open subset of $\mathbb{R}^2$ which remains after the process of removing line segments from a rectangle as illustrated in Figure 1 is continued indefinitely. Then the lower half of the left edge of the rectangle is a subset of ∂D which is H_D-null but not polar.

Figure 1

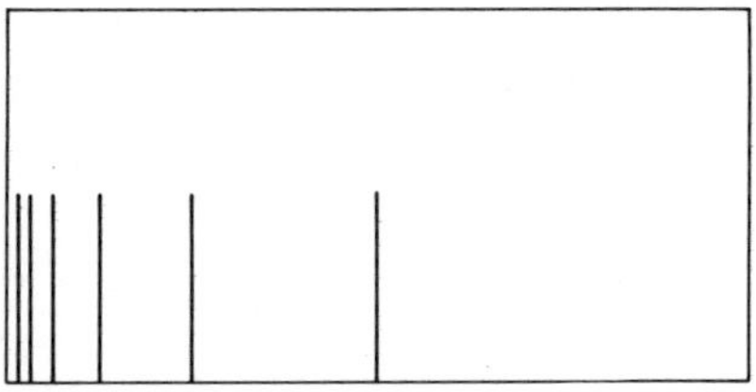

A function φ on ∂D is said to be essentially continuous on ∂D if there is an H_D-null subset N of ∂D such that φ is continuous on $\partial D \backslash N$. According to the next result, if φ is bounded and essentially continuous on ∂D, then $H_D\varphi$ solves the generalized Dirichlet problem for φ, as formulated below.

Proposition 2.1. *Let φ be essentially bounded on ∂D. Then $H_D\varphi$ is bounded and harmonic on D and has boundary value $\varphi(b)$ at b for every $b \in \partial D \cap (D^c)^r$ at which φ is defined and continuous.*

Proof. Clearly $H_D\varphi$ is bounded on D. Choose $x \in D$ and let $r > 0$ be such that $B_r(x) \subset D$. Then

$$H_D\varphi(x) = \int h_{S_r(x)}(x, dy) H_D\varphi(y) = \int \sigma_r(x, dy) H_D\varphi(y)$$

by Theorem 2.6.7. Thus $H_D\varphi$ is harmonic on D. The remainder of the conclusion follows from Proposition 2.3.6.

A point $b \in \partial D$ is said to be *regular for the Dirichlet problem* (on D) if $H_D\varphi$ has boundary value $\varphi(b)$ at b for every bounded continuous function φ on ∂D. The set D itself is said to be *regular* if every point on its boundary is regular for the Dirichlet problem. The next result is due to Doob [1].

Theorem 2.2. *A point $b \in \partial D$ is regular for the Dirichlet problem if and only if $b \in (D^c)^r$. In particular D is regular if and only if $\partial D \subset (D^c)^r$.*

Proof. The first result will be proven, from which the second result follows immediately. Let $b \in \partial D$. If $b \in (D^c)^r$ then b is regular for the Dirichlet problem by Proposition 2.1. Suppose conversely that b is regular for the Dirichlet problem. Define φ on ∂D by $\varphi(x) = (1 - \|x - b\|)^+$. Then φ is a continuous function on ∂D having compact support, $\varphi(b) = 1$ and $\varphi < 1$ on $\partial D \backslash \{b\}$. By hypothesis $H_D\varphi(x) \to \varphi(b) = 1$ as $x \to b$ within D. Consequently $H_D(x, \cdot)$ converges completely to δ_b as $x \to b$ within D. Thus $b \in (D^c)^r$ by Proposition 2.3.6.

Poincare showed that D is regular if every point in ∂D lies on a sphere whose interior is disjoint from D. Zaremba showed more generally that D is regular if the cone condition in Proposition 2.3.3 is satisfied for $B = D^c$ and each $x \in \partial D$.

Theorem 2.3. *The set of points on the boundary of D which are irregular for the Dirichlet problem is polar.*

Proof. By Theorem 2.2 the indicated set is $\partial D \backslash (D^c)^r$ which is contained in $D^c \backslash (D^c)^r$. By Theorem 2.6.3 the latter set is polar.

Theorem 2.4. *Suppose D is regular and bounded and φ is continuous on ∂D. Then $H_D\varphi$ is the unique solution to the Dirichlet problem for φ.*

Proof. By Proposition 2.1, $H_D\varphi$ is a solution to the Dirichlet problem for φ. Let f be any solution to the Dirichlet problem for φ. Then $f - H_D\varphi$ extends to a function which is harmonic on D, continuous on $\bar{D}$ and zero on ∂D. By Proposition 1.4 $f - H_D\varphi = 0$ on D. Thus $H_D\varphi$ is the unique solution to the Dirichlet problem for φ.

Let φ be essentially continuous on ∂D. A solution to the *generalized Dirichlet problem* for φ is a harmonic function on D having boundary value $\varphi(b)$ for every $b \in \partial D$ which is both regular for the Dirichlet problem and a point at which φ is defined and continuous. If D is regular and φ is continuous on ∂D, the generalized Dirichlet problem for φ is equivalent to the original Dirichlet problem for this boundary function.

Let φ be essentially bounded and essentially continuous on ∂D. By Proposition 2.1, $H_D\varphi$ is a solution to the generalized Dirichlet problem for φ. In particular $P_.(T_D < \infty) = H_D 1$ is a solution to the generalized Dirichlet problem for the boundary function 1 on ∂D and hence $P_.(T_D = \infty) = 1 - P_.(T_D < \infty)$ is a solution to the generalized Dirichlet problem for the boundary function zero on ∂D. Consequently, for every $\alpha \in \mathbb{R}$, $H_D\varphi + \alpha P_.(T_D = \infty)$ is a solution to the generalized Dirichlet problem for φ. By Proposition 2.2.9, $P_.(T_D = \infty) = 0$ on D if and only if D^c is recurrent. Thus by Proposition 2.2.11 if $d \geq 3$ and D^c is bounded (which is true for "exterior" Dirichlet problems) the generalized Dirichlet problem for φ fails to have a unique solution. It will be shown below that every bounded solution to the generalized Dirichlet problem for φ is of the form $H_D\varphi + \alpha P_.(T_D = \infty)$ for some $\alpha \in \mathbb{R}$. First some preliminary results are required.

A subset B of D is said to be a *relatively compact* subset of D if $\bar{B}$ is a compact subset of D or equivalently if B is bounded and $\bar{B} \subset D$.

Proposition 2.5. *$D_n = \{x \in D : \|x\| < n$ and $d(x, D^c) > 1/n\}$ is a relatively compact regular open subset of D and $D_n \uparrow D$.*

Proof. It is clear that D_n is a relatively compact open subset of D and that $D_n \uparrow D$. It remains to show that D_n is regular. Choose $b \in \partial D_n$. If $\|b\| = n$, then b clearly satisfies the cone condition for being regular for D_n^c. Suppose $d(b, D^c) = 1/n$. It is left to the reader to show that b again satisfies the cone condition for being regular for D_n^c. (Note that there is an $a \in D^c$ such that $\|b - a\| = 1/n$.)

Recall that if D is bounded or if $d \leq 2$ and D^c is nonpolar, then D^c is recurrent.

Proposition 2.6. *Suppose D^c is recurrent and that f is bounded and harmonic on D and continuous on $\bar{D} \backslash N$ for some H_D-null set N. Then $f = H_D f$ on D.*

Proof. If D is bounded and regular and f is harmonic on D and continuous on $\bar{D}$, then $f = H_D f$ on D by Theorem 2.4. Consider now the general case. By Proposition 2.5 there exist relatively compact regular open subsets D_n of D such that $D_n \uparrow D$. Choose $x \in D$. For n sufficiently large $x \in D_n$ and hence $f(x) = H_{D_n} f(x) = E_x(f(X(T_{D_n})))$ by the first paragraph of this proof. Now $P_.(T_{D_n} \uparrow T_D < \infty) = 1$ on D by Proposition 2.3.8 and the recurrence of D^c. Thus by continuity of paths and the continuity of f

$$f = \lim_n E_. f(X(T_{D_n})) = E_. f(X(T_D)) = H_D f \qquad \text{on } D$$

as desired.

It follows easily from the Markov property that if $f = \alpha P_.(T_D = \infty)$ on D for some $\alpha \in \mathbb{R}$, then $Q_D^t f = f$ on D for all $t \geq 0$. A converse to this result, which extends Proposition 2.1.4, will now be obtained. In this particular result the assumption that D is open is irrelevant.

Proposition 2.7. *Let $t > 0$ and suppose that f is bounded on D and $Q_D^t f = f$ on D. Then $f = \alpha P_.(T_D = \infty)$ on D for some $\alpha \in \mathbb{R}$.*

Proof. It follows from the semigroup property of the operators $Q_D^t, t > 0$, that $Q_D^{nt} f = f$ on D for every positive integer n. Let M be such that $|f| \leq M$ on D. Then for $x \in D$

$$|f(x)| = |Q_D^{nt} f(x)| = |E_x(f(X(nt)); T_D > nt)| \leq M P_x(T_D > nt)$$

and hence $|f| \leq M P_.(T_D = \infty)$ on D. Thus $f + M P_.(T_D = \infty)$ is nonnegative on D. Consequently in showing that f is of the desired form it can be assumed to be nonnegative on D. This being the case, extend f to all of $\mathbb{R}^d$ by setting $f = 0$ on D^c. Let m be a positive integer. Then $p^t f \geq Q_D^t f = f$ on D and $p^t f \geq 0 = f$ on D^c, so $p^t f \geq f$ on $\mathbb{R}^d$. Consequently $p^{(n+1)t} f = p^{nt} p^t f \geq p^{nt} f$ for $n \geq 1$. Therefore $p^{nt} f$ increases to some function on $\mathbb{R}^d$ as $n \to \infty$. By Proposition 2.1.3 this function must be a nonnegative constant α. Thus $\lim_n p^{nt} f = \alpha$. Clearly $\alpha \geq f$ since $p^{nt} f \geq f$ for each $n \geq 1$; so $f = Q_D^{nt} f \leq \alpha P_{.}(T_D > nt)$ on D and hence $f \leq \alpha P_{.}(T_D = \infty)$ on D. To obtain the opposite inequality observe that

$$p^{nt} f = Q_D^{nt} + E_{.}(f(X(nt); T_D \leq nt) \leq f + \alpha P_{.}(T_D \leq nt) \qquad \text{on } D$$

and hence that $\alpha \leq f + \alpha P_{.}(T_D < \infty)$ on D. This shows that $f \geq \alpha P_{.}(T_D = \infty)$ on D and therefore that $f = \alpha P_{.}(T_D = \infty)$ on D as desired.

Proposition 2.8. *Let f be a bounded harmonic function on D and let D_n be relatively compact open subsets of D such that $D_n \uparrow D$. Then there is an $\alpha \in \mathbb{R}$ such that*

$$f = \alpha P_{.}(T_D = \infty) + \lim_n E_{.}(f(X(T_{D_n})); T_D < \infty) \qquad \textit{on } D.$$

Proof. Let $t > 0$. By Proposition 2.6

$$\begin{aligned} Q_{D_n}^t f &= Q_{D_n}^t H_{D_n} f \\ &= E_{.}(f(X(T_{D_n})); T_{D_n} > t) \\ &= H_{D_n} f - E_{.}(f(X(T_{D_n})); T_{D_n} \leq t) \\ &= f - E_{.}(f(X(T_{D_n})); T_{D_n} \leq t) \end{aligned}$$

on D_n and hence

$$f = E_{.}(f(X(t)); T_{D_n} > t) + E_{.}(f(X(T_{D_n})); T_{D_n} \leq t) \qquad \text{on } D_n.$$

Now $P_{.}(T_{D_n} \uparrow T_D) = 1$ on D by Proposition 2.3.8, so

$$f = E_{.}(f(X(t)); T_D > t) + \lim_n E_{.}(f(X(T_{D_n})); T_D \leq t) \qquad \text{on } D.$$

Consequently $\lim_n E_{.}(f(X(T_{D_n})); T_D < \infty)$ exists on D and

(1) $$\lim_{t\to\infty} Q_D^t f = f - \lim_n E_{.}(f(X(T_{D_n})); T_D < \infty) \qquad \text{on } D.$$

Set $h = \lim_{t\to\infty} Q_D^t f$ on D. It follows from the dominated convergence theorem that $h = \lim_{s\to\infty} Q_D^t Q_D^s f = Q_D^t h$ on D for $t > 0$. Thus by Proposition 2.7 there is an $\alpha \in \mathbb{R}$ such that $h = \alpha P_{.}(T_D = \infty)$ on D. In other words

(2) $$\lim_{t\to\infty} Q_D^t f = \alpha P_{.}(T_D = \infty) \qquad \text{on } D.$$

The conclusion of the proposition follows from (1) and (2).

The behavior of $P_{\cdot}(T_D = \infty)$ will now be determined when D^c is transient. ($P_{\cdot}(T_D = \infty) = 0$ on D if D^c is recurrent.)

Proposition 2.9. *Suppose D^c is transient. Then $P_{\cdot}(T_D = \infty)$ is harmonic on D and has boundary value zero on $\partial D \cap (D^c)^r$, and $\{x \in D: P_x(T_D = \infty) > 0\}$ is an unbounded component of D.*

Proof. By Proposition 2.1, $P_{\cdot}(T_D = \infty) = 1 - H_D 1$ is harmonic on D and has boundary value zero on $\partial D \cap (D^c)^r$. By Proposition 1.3 on any component of D, $P_{\cdot}(T_D = \infty)$ is either strictly positive or identically zero. By Proposition 2.2.9 there is a necessarily unbounded component D_0 of D on which $P_{\cdot}(T_D = \infty) = P_{\cdot}(T_{D_0} = \infty)$ is strictly positive. Now D_0 is recurrent by Proposition 2.2.9, so $P_{\cdot}(\tau_{D_0} < \infty) = 1$ on D. Choose $x \in D \backslash D_0$. Then $T_D = \tau_{\partial D} \leq \tau_{D_0}$ on $\{X(0) = x\}$, so $P_x(T_D < \infty) = 1$. Thus $\{x \in D: P_x(T_D = \infty) > 0\} = D_0$, which is an unbounded component of D.

Theorem 2.10. *Let φ be a bounded and essentially continuous function on ∂D. Then f is a bounded solution to the generalized Dirichlet problem for φ if and only if there is an $\alpha \in \mathbb{R}$ such that $f = H_D\varphi + \alpha P_{\cdot}(T_D = \infty)$ on D. The generalized Dirichlet problem for φ has a unique bounded solution if and only if D^c is recurrent.*

Proof. By Proposition 2.1 the indicated functions are solutions. To see that they are the only solutions let D_n be relatively compact open subsets of D such that $D_n \uparrow D$. Then $P_{\cdot}(T_{D_n} \uparrow T_D) = 1$ on D by Proposition 2.3.8. The first conclusion now follows from Proposition 2.8. The second conclusion follows from the first conclusion and Proposition 2.2.9.

Let f be defined on an unbounded subset A of $\mathbb{R}^d$. Then f is said to have *limit α at infinity* if for every $\varepsilon > 0$ there is an $M > 0$ such that $|f(x) - \alpha| < \varepsilon$ for $x \in A$ and $\|x\| \geq M$.

Theorem 2.11. *Suppose D^c is transient. Let φ be a bounded and essentially continuous function on ∂D and, if ∂D is unbounded, let φ have limit α at infinity. Then $H_D\varphi + \alpha P_{\cdot}(T_D = \infty)$ is the unique bounded solution to the generalized Dirichlet problem for φ having limit α at infinity.*

Proof. Suppose first that $d \leq 2$. Then D^c is polar by Proposition 2.2.10, so the desired result follows immediately from Theorem 2.10. Suppose next that $d \geq 3$. Let f be a bounded solution to the generalized Dirichlet problem for φ. Then $f = H_D\varphi + \beta P_{\cdot}(T_D = \infty)$ on D for some $\beta \in \mathbb{R}$ by Theorem 2.10.

By Proposition 2.2.11, f has limit α at infinity if and only if the function $\alpha P_{\cdot}(T_D < \infty) + \beta P_{\cdot}(T_D = \infty)$ on D has limit α at infinity. By Proposition 2.2.9 this is true if and only if $\beta = \alpha$.

Consider now the original Dirichlet problem. The next result which extends Theorem 2.4 to unbounded sets, follows from Theorem 2.10.

Theorem 2.12. *Suppose D is regular. Let φ be bounded and continuous on ∂D. Then f is a bounded solution to the Dirichlet problem for φ if and only if there is an $\alpha \in \mathbb{R}$ such that $f = H_D\varphi + \alpha P_{\cdot}(T_D = \infty)$ on D. The Dirichlet problem for φ has a unique bounded solution if and only if D^c is recurrent.*

The following result was obtained by Lebesgue in 1907.

Theorem 2.13. *Suppose $d = 2$ and that D^c is nonempty and no component of D^c reduces to a single point. Let φ be bounded and continuous on ∂D. Then $H_D\varphi$ is the unique bounded solution to the Dirichlet problem for φ.*

Proof. By Theorem 2.7.2, D is regular and D^c is nonpolar. By Proposition 2.2.10, D^c is recurrent. The desired result now follows from Theorem 2.12.

The next result will be used in this section and again in Chapter 5, to obtain analytic criteria for regularity.

Proposition 2.14. *Let $b \in \partial D \backslash (D^c)^r$. Then there is a compact subset B of D such that*

$$\overline{\lim_{\substack{x \to b \\ x \in D}}}\, P_x(\tau_B < T_D) > 0.$$

Proof. There is a $t > 0$ such that $P_b(T_D > 2t) > 0$. For $0 < s < t$

$$P_b(T_D > 2t) \le \int_D Q_D(s, b, x) P_x(T_D > t)\, dx.$$

It follows by letting $s \to 0$ that

$$\delta = \overline{\lim_{\substack{x \to b \\ x \in D}}}\, P_x(T_D > t) > 0.$$

Let B_n be compact subsets of D such that $B_n \uparrow D$. Then the Lebesgue measure of $C \cap (D \backslash B_n)$ decreases to zero for every compact subset C of $\mathbb{R}^d$. Therefore $\lim_n P_{\cdot}(X(t) \in D \backslash B_n) = 0$ uniformly on compacts. Consequently there is an $\varepsilon > 0$ and a positive integer n_0 such that $P_x(X(t) \in D \backslash B_{n_0}) \le \delta/2$ for $x \in D$ and

$\|x - b\| < \varepsilon$. Set $B = B_{n_0}$. Since

$$\{\tau_B < T_D\} \supset \{T_D > t\} \backslash \{X(t) \in D \backslash B\},$$

$$\overline{\lim_{\substack{x \to b \\ x \in D}}} P_x(\tau_B < T_D) \geq \delta/2 > 0$$

as desired.

Here is an analytic criterion for regularity in terms of harmonic functions.

Theorem 2.15. *Let $b \in \partial D$. Then $b \in (D^c)^r$ if and only if there is a positive harmonic function on D having boundary value zero at b.*

Proof. Suppose $b \in (D^c)^r$. If $d = 1$, $f(x) = |x - b|$ defines a positive harmonic function on D having zero boundary value at b. Consider now $d \geq 2$. Let φ be a bounded continuous function on ∂D such that $\varphi(b) = 0$ and $\varphi > 0$ on $\partial D \backslash \{b\}$. Then $H_D \varphi$ is a nonnegative harmonic function on D which by Proposition 2.1 has boundary value zero at b. Since $b \in (D^c)^r$, D^c is nonpolar and hence $H_D(\cdot, \partial D) = P_\cdot(\tau_{D^c} < \infty) > 0$ on D by Proposition 2.2.6. Now $\{b\}$ is polar by Proposition 2.2.5, so $H_D(\cdot, \partial D \backslash \{b\}) > 0$ on D and hence $H_D \varphi > 0$ on D.

Suppose next that $b \in \partial D \backslash (D^c)^r$. By Proposition 2.14 there is a compact subset B of D such that

(3) $$\overline{\lim_{\substack{x \to b \\ x \in D}}} P_x(\tau_B < T_D) > 0.$$

Let f be a positive harmonic function on D. There is a $\delta > 0$ such that $f \geq \delta$ on B. Let D_n be relatively compact open subsets of D such that $D_n \uparrow D$. It follows from Proposition 2.6 that $f = H_{D_n \backslash B} f$ on $D_n \backslash B$ and hence that $f \geq \delta P_\cdot(\tau_B < T_{D_n})$ on $D_n \backslash B$. Consequently $f \geq \delta P_\cdot(\tau_B < T_D)$ on $D \backslash B$ by Proposition 2.3.8. Therefore by (3)

$$\overline{\lim_{\substack{x \to b \\ x \in D}}} f(x) > 0$$

and hence f fails to have boundary value zero at b. This completes the proof of the theorem.

Zaremba gave the first example of a Dirichlet problem which was not solvable. In his example $d \geq 2$ and D is the punctured open ball $\mathring{B}_1 \backslash \{0\}$. The function φ on ∂D is given by $\varphi = 0$ on S_1 and $\varphi(0) = 1$. Then φ is bounded and continuous on ∂D. Since D is bounded and $\{0\}$ is polar, the unique

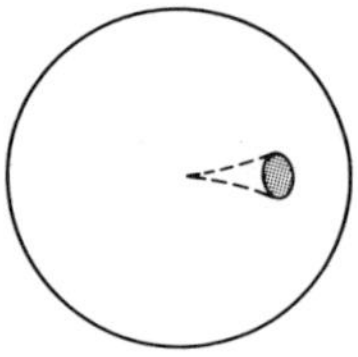

Figure 2

bounded solution to the generalized Dirichlet problem is given by $f = H_D\varphi = 0$ on D. Thus f fails to have boundary value $\varphi(0) = 1$ at 0. Consequently the Dirichlet problem for φ has no solution.

In Zaremba's example $\{0\}$ is an isolated point of ∂D. Suppose D is bounded and ∂D is connected. If $d = 2$ Dirichlet problems on D are solvable by Theorem 2.13; but if $d \geq 3$ this is not true in general, as Lebesgue showed by means of the following example.

Suppose $d \geq 3$ and that D is the open ball $\mathring{B}_1(0)$ punctured by a thorn having vertex at the origin as shown in Figure 2. If the thorn is sufficiently sharp, 0 is irregular for D^c by Proposition 3.3.5. Let this be the case and let φ be a nonnegative continuous function on ∂D which satisfies $\varphi(0) = 0$ but which is not identically zero on ∂D. Then the unique bounded solution $f = H_D\varphi$ to the generalized Dirichlet problem for φ is positive on D by Proposition 1.3 and hence by Theorem 2.15 cannot have boundary value $\varphi(0) = 0$ at 0 (this is true even if φ vanishes on the entire thorn). Consequently the Dirichlet problem for φ has no solution.

Wiener [3] gave a necessary and sufficient condition, called Wiener's test, for a point $b \in \partial D$ to be regular for the Dirichlet problem. For $d \geq 3$ this is the test described in Theorem 3.3.2; see Theorem 6.7.35 for the corresponding test when $d = 2$.

The next result yields an analytic criterion for transience and hence also for recurrence.

Theorem 2.16. *The set D^c is transient if and only if there is a bounded harmonic function on D which has boundary value zero on $\partial D \cap (D^c)^r$ but which is not identically zero on D.*

Proof. By Theorem 2.10, f is a bounded harmonic function on D having boundary value zero on $\partial D \cap (D^c)^r$ if and only if $f = \alpha P_.(T_D = \infty)$ on D for some $\alpha \in \mathbb{R}$. By Proposition 2.2.9, D^c is transient if and only if $P_.(T_D = \infty)$ is not identically zero on D. The desired conclusion follows from these two observations.

In order to obtain an analytic characterization of the exit distribution, let $\mathscr{C}_c(\partial D)$ denote the collection of continuous functions on ∂D having compact support. Note that for each $x \in D$, $H_D(x, \cdot)$ is uniquely determined by values of $H_D\varphi(x)$, $\varphi \in \mathscr{C}_c(\partial D)$. Suppose D^c is recurrent and let $x \in D$. Then $H_D(x, \cdot)$ is the unique measure μ_x on ∂D satisfying the following property: if $\varphi \in \mathscr{C}_c(\partial D)$ and f is the bounded solution to the generalized Dirichlet problem for φ, then $f(x) = \int \varphi(y)\mu_x(dy)$. Suppose instead that D^c is transient and let $x \in D$. Then $H_D(x, \cdot)$ is the unique measure μ_x on ∂D satisfying the following property: if $\varphi \in \mathscr{C}_c(\partial D)$ and f is the bounded solution to generalized Dirichlet problem for φ having limit zero at infinity, then $f(x) = \int \varphi(y)\mu_x(dy)$. These results also yield an analytic characterization of $P_{\cdot}(T_D = \infty) = 1 - H_D 1$.

The problem of determining when there are nonconstant bounded harmonic functions on D will now be considered. Suppose first that $d = 1$. Then the components of D are open intervals and a function on D is harmonic if and only if it is harmonic on each such interval. If there is more than one such interval or if D is a bounded open interval, there clearly are nonconstant bounded harmonic functions on D. But if D is an unbounded interval there are no nonconstant bounded harmonic functions on D. For $d \geq 2$ the solution to the problem yields an analytic criterion for D^c to be nonpolar.

Theorem 2.17. *Let $d \geq 2$. Then D^c is nonpolar if and only if there is a nonconstant bounded harmonic function on D.*

Proof. Suppose first that D^c is polar. It follows easily from Theorem 2.10 that if f is a bounded harmonic function on D, then $f = \alpha P_{\cdot}(T_D = \infty) = \alpha$ on D for some $\alpha \in \mathbb{R}$. Thus there are no nonconstant bounded harmonic functions on D.

Suppose next that D^c is nonpolar. Then $P_{\cdot}(T_D < \infty) > 0$ on D by Proposition 2.2.6. Thus $\partial D \cap (D^c)^r$ is nonpolar by Theorem 2.6.5. This set must contain two distinct points a and b since one-point sets are polar by Proposition 2.2.5. Let φ be a bounded continuous function on ∂D such that $\varphi(a) \neq \varphi(b)$. Then $H_D\varphi$ is a bounded harmonic function on D having boundary values $\varphi(a)$ at a and $\varphi(b)$ at b. Thus $H_D\varphi$ is a nonconstant bounded harmonic function on D. This completes the proof of the theorem.

It is easily seen that if $d \geq 2$, every bounded harmonic function on the punctured ball $\mathring{B}_1 \backslash \{0\}$ extends to a harmonic function on $\mathring{B}_1$. The next proposition extends this result of Bôcher [1]. The necessity of the condition that $D \backslash D_0$ be polar in this proposition will be obtained in Theorem 5.4.11.

Theorem 2.18. *Let D_0 be an open subset of D such that $D \backslash D_0$ is polar. Suppose f is harmonic on D_0 and bounded on $C \cap D_0$ for every compact subset C of D. Then f extends uniquely to a harmonic function on D.*

Proof. Since $D \backslash D_0$ is polar it has Lebesgue measure zero. Thus D_0 is dense in D. Consequently f has at most one continuous extension to D and hence at most one harmonic extension to D. In proving that f does have a harmonic extension to D it suffices to prove this result locally; that is, it suffices to prove that if D_1 is a relatively compact open subset of D, then f extends to a harmonic function on D_1.

To this end set $D_2 = D_1 \cap D_0$ and let N denote the polar subset $\partial D_2 \backslash D_0$ of ∂D_2. Now f is bounded and harmonic on D_2 and continuous on $\bar{D}_2 \backslash N$. Thus $f = H_{D_2} f$ on D_2 by Proposition 2.6. But $H_{D_2} f = H_{D_1} f$ on D_2 since $D_1 \backslash D_2$ is polar. Thus $f = H_{D_1} f$ on D_2 and hence f extends to the harmonic function $H_{D_1} f$ on D_1. This completes the proof of the theorem.

Let φ be bounded on ∂D and let $\alpha \in \mathbb{R}$. Then $f = H_D \varphi + \alpha P_{\cdot}(T_D = \infty)$ is bounded and harmonic on D by Proposition 2.1. It is natural to conjecture that under some appropriate restriction on D, every bounded harmonic function on D is of this form.

To see that some restriction is required let $d = 2$ and let D denote the open ball $\mathring{B}_1(0)$ punctured by a horizontal line segment L from the origin to a point on the boundary of the ball. Let f be defined on D as follows: for $x \in D$ let $f(x)$ be the probability that Brownian motion starting at x first hits ∂D on L and does so by approaching L from above. It is left to the reader to show that f cannot be written in the indicated form (reflect about the horizontal axis).

The set D is said to be *strongly starshaped about the origin* if $\overline{rD} \subset D$ for $0 < r < 1$ and *strongly starshaped* if $D - b$ is strongly starshaped about the origin for some $b \in \mathbb{R}^d$. The set D described in the previous paragraph is starshaped about the origin (i.e., $rD \subset D$), but not strongly starshaped.

Let D be strongly starshaped. Suppose D^c is polar. Let f be a bounded harmonic function on D. Then $f = \alpha$ for some $\alpha \in \mathbb{R}$ by Theorem 2.17 and $P_{\cdot}(T_D = \infty) = 1$ on D, so $f = \alpha P_{\cdot}(T_D = \infty)$ on D. The alternative that D^c is nonpolar is considered next.

Theorem 2.19. *Suppose D is strongly starshaped and D^c is nonpolar. Let f be a bounded harmonic function on D. Then there is a bounded function φ on D such that $f = H_D \varphi$ on D.*

Proof. Since D is starshaped and D^c is nonpolar, D^c is recurrent by Proposition 2.2.16. By translation invariance it can be assumed that D is strongly starshaped about the origin, which is then necessarily in D. Now for $x \in D$ the line segment from the origin to x lies entirely within D. Consequently D is connected. By Proposition 1.3 for $A \subset \partial D$, $H_D(\cdot, A)$ is either positive on D

or identically zero on D. Thus the measures $H_D(x,\cdot)$, $x \in D$, are all absolutely continuous with respect to each other. Let μ denote the measure $H_D(0,\cdot)$, which is a probability measure supported on ∂D. For $x \in D$ let ψ_x be a nonnegative function on ∂D such that $H_D(x,dy) = \psi_x(y)\mu(dy)$. Then $\int\psi_x(y)\mu(dy) = P_x(T_D < \infty) = 1$.

Let f be a bounded harmonic function on D. By adding a constant to f if necessary it can be assumed that f is nonnegative on D. Let M be an upper bound to f on D and let $0 < r < 1$. Since $(rD)^c$ is recurrent and f is harmonic on rD and continuous on $\overline{rD}$, it follows from Theorem 2.10 that

$$f(rx) = \int H_{rD}(rx, dz)f(z), \qquad x \in D.$$

By scale invariance (Proposition 2.2.14) the right-hand side of this equation equals $\int H_D(x, dz)f(rz)$. Consequently

$$f(rx) = \int \psi_x(z)f(rz)\mu(dz), \qquad x \in D. \tag{4}$$

Let μ_r denote the measure on $\mathbb{R}^d$ given by $\mu_r(dz) = f(rz)\mu(dz)$. Then there is a sequence $\{r_n\}$ of numbers in $(0,1)$ such that $r_n \uparrow 1$ and a finite measure ν on ∂D such that μ_{r_n} converges completely to ν as $n \to \infty$. Now $\nu(U) \leq \underline{\lim}_{r\to 1}\mu_r(U) \leq M\mu(U)$ for all relatively open subsets U of ∂D, so $\nu(A) \leq M\mu(A)$ for all $A \subset \partial D$. Thus ν is absolutely continuous with respect to μ and has a density $\varphi = d\nu/d\mu$ which is bounded above by M.

Choose $x \in D$ and $\varepsilon > 0$. There is a bounded continuous function ρ on ∂D such that $\int|\psi_x(z) - \rho(z)|\mu(dz) \leq \varepsilon$. Note that

$$\lim_n \int \rho(z)f(r_n z)\mu(dz) = \int \rho(z)\varphi(z)\mu(dz).$$

Since ε can be made arbitrarily small

$$\lim_n \int \psi_x(z)f(r_n z)\mu(dz) = \int \psi_x(z)\varphi(z)\mu(dz) = H_D\varphi(x), \qquad x \in D. \tag{5}$$

Clearly $\lim_n f(r_n x) = f(x)$, so it follows from (4) and (5) that $f = H_D\varphi$ on D. This completes the proof of the theorem.

3. Poisson's Formula

Throughout this section it is assumed that $d \geq 2$. Inversions with respect to spheres are defined and some of their basic properties are obtained. Kelvin transformations of harmonic functions are defined in terms of such inversions and used to derive Poisson's formula for the hitting distribution of a sphere. This formula is in turn used to derive Harnack's inequality and other properties of harmonic functions.

Let $b \in \mathbb{R}^d$ and $\rho > 0$. The homeomorphism of $\mathbb{R}^d \backslash \{b\}$ onto itself which takes x into

$$x^* = b + \frac{\rho^2}{\|x - b\|^2}(x - b)$$

is called *inversion* relative to the sphere $S_\rho(b)$. Given any subset A of $\mathbb{R}^d \backslash \{b\}$, let A^* denote its image under this inversion.

Observe that

$$\|x^* - b\|\,\|x - b\| = \rho^2 \tag{1}$$

and

$$x = b + \frac{\rho^2}{\|x^* - b\|^2}(x^* - b), \tag{2}$$

so that $x^{**} = x$. Also

$$\begin{aligned}
&(y^* - x^*) \cdot (y^* - x^*) \\
&= \rho^4\left(\frac{y - b}{\|y - b\|^2} - \frac{x - b}{\|x - b\|^2}\right) \cdot \left(\frac{y - b}{\|y - b\|^2} - \frac{x - b}{\|x - b\|^2}\right) \\
&= \frac{\rho^4}{\|x - b\|^2\|y - b\|^2}(\|x - b\|^2 - 2(x - b) \cdot (y - b) + \|y - b\|^2) \\
&= \frac{\rho^4\|y - x\|^2}{\|x - b\|^2\|y - b\|^2}
\end{aligned}$$

and hence

$$\|y^* - x^*\| = \frac{\rho^2\|y - x\|}{\|x - b\|\,\|y - b\|}. \tag{3}$$

(For a geometrical proof of (3) observe that $\|y^* - x^*\|/\|y - x\| = \|x^* - b\|/\|y - b\|$ by the law of proportions for similar triangles; (3) follows from this identity and (1).)

Consider the image $S_r^*(c)$ of a sphere $S_r(c)$, where $\|c - b\| > r$ (so that $B_r(c) \subset \mathbb{R}^d \backslash \{b\}$). The equation $\|x - c\|^2 = r^2$ of the sphere can be written as

$$\|x - b\|^2 - 2(x - b) \cdot (c - b) + \|c - b\|^2 - r^2 = 0. \tag{4}$$

If (1) and (2) are solved for $\|x - b\|$ and $(x - b)$ in terms of $\|x^* - b\|$ and $(x^* - b)$, respectively, and the results are substituted into (4) the equation

$$\|x^* - b\|^2 - \frac{2\rho^2}{\|c - b\|^2 - r^2}(x^* - b) \cdot (c - b) + \frac{\rho^4}{\|c - b\|^2 - r^2} = 0 \tag{5}$$

for $S_r^*(c)$ is obtained. Thus $S_r^*(c)$ is also a sphere. The image of the inside of $S_r(c)$ is just the inside of $S_r^*(c)$. It follows from (4) and (5) that $S_r^*(c) = S_r(c)$ if and only if $r^2 + \rho^2 = \|c - b\|^2$ (which is true if and only if $S_r(c)$ and $S_\rho(b)$ are orthogonal to each other).

Choose x such that $0 < \|x - c\| < r$. The sphere $S_\rho(b)$ will now be found such that the corresponding inversion preserves $S_r(c)$ and takes x to c. By the preceding paragraph it is necessary that

$$r^2 + \rho^2 = \|c - b\|^2. \tag{6}$$

By (1) it is necessary that

$$\|c - b\| \, \|x - b\| = \rho^2. \tag{7}$$

By (6), c, x, and b must lie in order on a straight line, so

$$\|c - b\| = \|c - x\| + \|x - b\|. \tag{8}$$

It follows by solving (7) for $\|x - b\|$, substituting the result into (8), and using (6) that

$$\|c - b\| \, \|x - c\| = r^2 \tag{9}$$

and hence that b must be the image of x under inversion relative to $S_r(c)$. Suppose conversely that b is the image of x under inversion relative to $S_r(c)$ and that (6) holds. Then inversion relative to $S_\rho(b)$ preserves $S_r(c)$. Also (9) holds and (8) continues to hold. It follows from (6), (8), and (9) that (7) holds and hence that inversion relative to $S_\rho(b)$ maps x to c.

If $\|c - b\| < r$ the image of $S_r(c)$ under inversion relative to $S_\rho(b)$ is again a sphere but now the outside of $S_r(c)$ is mapped onto the inside of the image sphere. In this case $S_r^*(c) = S_r(c)$ if and only if $S_\rho(b) = S_r(c)$. If $\|c - b\| = r$, the image of $S_r(c)\backslash\{b\}$ is a hyperplane. Let P be a hyperplane. If $b \notin P$, P^* is a sphere. If $b \in P$, the image of $P\backslash\{b\}$ is itself.

Consider inversion relative to $S_\rho(b)$. Let D be an open set in $\mathbb{R}^d\backslash\{b\}$ and let f be a harmonic function on D. Define f^* on D^* by

$$f^*(x^*) = \frac{\rho^{d-2}}{\|x^* - b\|^{d-2}} f(x), \qquad x \in D,$$

where x is defined in terms of x^* according to (2). It follows by a straightforward but tedious computation (see page 232 of Kellogg [1] or Section 13 of Wermer [1]) that the Laplacian of f^* vanishes on D^* and hence that f is harmonic on D^*. The transformation that sends the harmonic function f on D to the harmonic function f^* on D^* is called the *Kelvin transformation.* It is easily seen that $f^{**} = f$ on D.

Consider now the hitting distribution $h_{S_r(c)}(x,\cdot)$ of a sphere $S_r(c)$ for Brownian motion starting at x. If $x \in S_r(c)$, $h_{S_r(c)}(x,\cdot) = \delta_x$ since points of the sphere are regular for the sphere. If $x = c$, $h_{S_r(c)}(x,\cdot) = \sigma_r(x)$ by Proposition 2.2.21 and translation invariance. Kelvin transformations will now be used to reduce the general case $x \notin S_r(c)$ to the special case $x = c$.

Theorem 3.1. *Let $c \in \mathbb{R}^d$, $r > 0$ and $x \notin S_r(c)$. Then $h_{S_r(c)}(x,\cdot)$ is given by*

$$h_{S_r(c)}(x, dy) = \frac{r^{d-2}|r^2 - \|x - c\|^2|}{\|y - x\|^d}\,\sigma_r(c, dy).$$

Proof. This result is true for $x = c$. Suppose now that $0 < \|x - c\| < r$. Let $S_\rho(b)$ be the sphere such that inversion relative to $S_\rho(b)$ preserves $S_r(c)$ and sends x to c. Let * denote inversion relative to $S_\rho(b)$.

Let φ be a continuous function on $S_r(c)$ and let f be the solution to the Dirichlet problem on $\mathring{B}_r(c)$ with boundary function φ. Then the Kelvin transformation f^* of f is the solution to the Dirichlet problem on $\mathring{B}_r(c)$ with boundary function φ^* given by

$$\varphi^*(y^*) = \frac{\rho^{d-2}}{\|y^* - b\|^{d-2}}\,\varphi(y), \qquad y \in S_r(c).$$

Thus $f^*(c) = \int \varphi^*(y^*)\sigma_r(c, dy^*)$ by Theorem 2.4. It is easily seen that

$$dy^* = \frac{\rho^2}{\|y - b\|^2}\,Q(y)dy,$$

where $Q(y)$ is an orthogonal linear transformation for each $y \neq b$. (Specifically Q is given by

$$Qv = v - \frac{2v\cdot(y - b)}{\|y - b\|^2}(y - b), \qquad v \in \mathbb{R}^d.)$$

Consequently

$$\sigma_r(c, dy^*) = \left(\frac{\rho^2}{\|y - b\|^2}\right)^{d-1}\sigma_r(c, dy).$$

Therefore

$$f(x) = \frac{\rho^{d-2}}{\|x - b\|^{d-2}}\,f^*(c) = \frac{\rho^{4d-6}}{\|x - b\|^{d-2}}\int \frac{\varphi(y)}{\|y - b\|^{2d-2}\|y^* - b\|^{d-2}}\,\sigma_r(c, dy).$$

Now $\|y - b\|\,\|y^* - b\| = \rho^2$ by (1), so

$$f(x) = \frac{\rho^{2d-2}}{\|x - b\|^{d-2}}\int \frac{\varphi(y)}{\|y - b\|^d}\,\sigma_r(c, dy).$$

Inversion with respect to $S_r(c)$ sends x to b and leaves y fixed for $y \in S_r(c)$. It follows by applying (3) to this inversion that

$$\|y - b\| = \frac{r\|y - x\|}{\|x - c\|}, \qquad y \in S_r(c). \tag{10}$$

Also $\|b - c\| = \|x - c\| + \|x - b\|$ and $\|x - c\|\,\|b - c\| = r^2$, so that

$$\|x - b\| = \frac{r^2 - \|x - c\|}{\|x - c\|}. \tag{11}$$

By (6), $\rho^2 = \|b - c\|^2 - r^2 = (r^4/\|x - c\|^2) - r^2$ and hence

$$\rho^2 = \frac{r^2(r^2 - \|x - c\|^2)}{\|x - c\|^2}. \tag{12}$$

It follows from the last formula for $f(x)$ together with (10)–(12) that

$$f(x) = \int \varphi(y) \frac{r^{d-2}(r^2 - \|x - c\|^2)}{\|y - x\|^d} \sigma_r(c, dy).$$

The conclusion of the theorem for $0 < \|x - c\| < r$ now follows by the analytical characterization of the exit distribution of $\mathring{B}_r(c)$.

Suppose now that $\|x - c\| > r$. Set $D = \{y : \|y - c\| > r\}$ and let $*$ correspond to inversion relative to $S_r(c)$. Then $D^* = \mathring{B}_r(c)\backslash\{c\}$. Let φ be a continuous function on $S_r(c)$ and let M be an upper bound to $|\varphi|$. Set $f = H_D\varphi = h_{S_r(c)}\varphi$ on D and let f^* denote the Kelvin transformation of f. Then f^* is harmonic on $\mathring{B}_r(c)\backslash\{c\}$ and

$$f^*(x^*) = \frac{r^{d-2}}{\|x^* - c\|^{d-2}} f(x) = \frac{r^{d-2}}{\|x^* - c\|^{d-2}} f\left(c + \frac{r^2}{\|x^* - c\|^2}(x^* - c)\right).$$

By translation invariance $|f(c + z)| \le MP_z(\tau_{S_r} < \infty)$ for $c + z \in D$. It is now clear if $d = 2$ and it follows from Proposition 3.1.6 if $d \ge 3$ that $|f^*| \le M$ on $\mathring{B}_r(c)\backslash\{c\}$. By Theorem 2.18, f^* can be extended to a harmonic function on $\mathring{B}_r(c)$ having boundary function φ. Thus by what has already been shown

$$f^*(x^*) = \int \varphi(y) \frac{r^{d-2}(r^2 - \|x^* - c\|^2)}{\|y - x^*\|^d} \sigma_r(c, dy).$$

It follows from (1) and (3) applied to inversion relative to $S_r(c)$ that $\|x^* - c\| = r^2/\|x - c\|$ and $\|y - x^*\| = r\|y - x\|/\|x - c\|$ for $y \in S_r(c)$. Therefore

$$f^*(x^*) = \int \varphi(y) \frac{\|x - c\|^{d-2}(\|x - c\|^2 - r^2)}{\|y - x\|^d} \sigma_r(c, dy).$$

Now $f(x) = r^{d-2}\|x - c\|^{2-d} f^*(x^*)$ and hence

$$f(x) = \int \varphi(y) \frac{r^{d-2}(\|x - c\|^2 - r^2)}{\|y - x\|^d} \sigma_r(c, dy).$$

Since $f(x) = \int \varphi(y) h_{S_r(c)}(x, dy)$ and φ is an arbitrary continuous function on $S_r(c)$, the conclusion of the theorem holds if $\|x - c\| > r$. This completes the proof of the theorem.

The following immediate consequence of Theorems 2.4 and 3.1 is called *Poisson's integral formula.*

Theorem 3.2. *Let $c \in \mathbb{R}^d$ and $r > 0$. If f is continuous on $B_r(c)$ and harmonic on $\mathring{B}_r(c)$, then*

$$f(x) = \int f(y) \frac{r^{d-2}(r^2 - \|x - c\|^2)}{\|y - x\|^d} \sigma_r(c, dy), \qquad x \in \mathring{B}_r(c).$$

The next result is called *Harnack's inequality.*

Theorem 3.3. *Let $c \in \mathbb{R}^d$ and $r > 0$, let f be a positive harmonic function on $\mathring{B}_r(c)$, and let $0 < \rho < r$. Then*

$$\frac{f(y)}{f(x)} \le \frac{r^2}{r^2 - \rho^2} \left(\frac{r + \rho}{r - \rho} \right)^d, \qquad x, y \in \mathring{B}_\rho(c).$$

Proof. Choose $x, y \in \mathring{B}_\rho(c)$ and $s \in (\rho, r)$. By Poisson's integral formula

$$f(x) = \int f(z) \frac{s^{d-2}(s^2 - \|x - c\|^2)}{\|z - x\|^d} \sigma_s(c, dz)$$

and the same formula holds with x replaced by y. Now

$$\frac{s^2 - \|y - c\|^2}{s^2 - \|x - c\|^2} \le \frac{s^2}{s^2 - \rho^2}$$

and

$$\left(\frac{\|z - x\|}{\|z - y\|} \right)^d \le \left(\frac{s + \rho}{s - \rho} \right)^d, \qquad z \in S_r(c).$$

Consequently

$$\frac{f(y)}{f(x)} \le \frac{s^2}{s^2 - \rho^2} \left(\frac{s + \rho}{s - \rho} \right)^d.$$

The desired now follows by letting $s \uparrow r$.

It is obvious for $d = 1$ and follows from Theorem 2.17 for $d \geq 2$ that every bounded harmonic function on $\mathbb{R}^d$ is constant. This result will now be strengthened.

Theorem 3.4. *A harmonic function on $\mathbb{R}^d$ which is bounded below or above is constant.*

Proof. It suffices to prove that a positive harmonic function on $\mathbb{R}^d$ is constant. But this result follows from Harnack's inequality by letting $r \uparrow \infty$.

The next consequence of Theorem 3.3 is also called *Harnack's inequality.*

Theorem 3.5. *Suppose D is connected and let C be a compact subset of D. There is a constant $c > 0$ such that if f is a positive harmonic function on D, then $f(y)/f(x) \leq c$ for $x, y \in C$.*

Proof. Choose $x \in D$. Let D_x consist of all points $y \in D$ such that

$$\sup\left[\frac{f(y)}{f(x)} : f \text{ is a positive harmonic function on } D\right] = \infty.$$

It follows easily from Theorem 3.3 that D_x and $D \backslash D_x$ are open. Since D is connected and $x \in D \backslash D_x$, D_x is empty.

Suppose the conclusion of the theorem were false. Then there would be sequences $\{x_j\}$ and $\{y_j\}$ of points in C and a sequence $\{f_j\}$ of positive harmonic functions on D such that $f_j(y_j)/f_j(x_j) \geq j$ for $j \geq 1$. Let x and y be limit points of the sequences $\{x_j\}$ and $\{y_j\}$, respectively. It follows from Theorem 3.3 that $y \in D_x$. But this is impossible since D_x is empty. Thus the conclusion of the theorem is valid.

The next result is called *Harnack's theorem.*

Theorem 3.6. *Suppose D is connected. Let f_n be harmonic functions on D such that $f_n \uparrow f$ on D. Then either f is harmonic on D or $f = \infty$ on D.*

Proof. Suppose f is not identically infinite on D. By looking at $\{f_n - f_1 + 1\}$ and $f - f_1 + 1$ if necessary it can be assumed that $f_n > 0$ and D for $n \geq 1$. It now follows from Theorem 3.5 that f is locally bounded and hence locally integrable on D. Choose $x \in D$ and let $r > 0$ be such that $B_r(x) \subset D$. Then

$$\int f(y)\sigma_r(x, dy) = \lim_n \int f_n(y)\sigma_r(x, dy) = \lim_n f_n(x) = f(x).$$

Consequently f is harmonic on D.

Theorem 3.7. *Suppose D is connected. Let φ be a nonnegative function on ∂D. Then either $H_D\varphi$ is harmonic on D or $H_D\varphi = \infty$ on D.*

Proof. Set $\varphi_j = \varphi \wedge j$. Then $H_D\varphi_j$ is harmonic on D and $H_D\varphi_j \uparrow H_D\varphi$. The desired result now follows immediately from Harnack's theorem.

Proposition 3.8. *Let $c \in \mathbb{R}^d$ and $r > 0$ and let f be continuous on $B_r(c)$ and harmonic on $\mathring{B}_r(c)$. Then*

$$|f'(c)| \leq \frac{d}{r} \max_{y \in S_r(c)} |f(y)|.$$

Proof. It follows easily from Poisson's integral formula that

$$f'(c) = \frac{d}{r^2} \int (y - c) f(y) \sigma_r(c, dy),$$

which in turn yields the desired conclusion.

The next result is also due to Harnack.

Theorem 3.9. *Let $\{f_n\}$ be a sequence of harmonic functions on D which are uniformly bounded on compact subsets of D. Then there is a subsequence $\{f_{n_j}\}$ and a function f on D such that $f_{n_j} \to f$ on D. The function f is necessarily harmonic on D and the convergence is necessarily uniform on compacts.*

Proof. Let C be a compact subset of D. There is a compact subset C_1 of D containing C in its interior. Let M be an upper bound to $|f_n|$, $n \geq 1$, on C_1 and let $r > 0$ be such that $B_r(x) \subset C_1$ for $x \in C$. Then $|f'_n| \leq Md/r$ on C by Proposition 3.8. Thus $\{f_n\}$ is equicontinuous on C. By the Arzelá–Ascoli theorem there is a subsequence $\{f_{n_j}\}$ and a function f such that $f_{n_j} \to f$ uniformly on C. Since D is the limit of an increasing sequence of compact sets, it follows by the usual diagonal argument that there is a subsequence $\{f_{n_j}\}$ and a function f such that $f_{n_j} \to f$ on D. The convergence is necessarily uniform on compacts since the original sequence is equicontinuous. It follows from the definition of harmonicity that f is harmonic on D.

The next result is preliminary to the theorem which follows it.

Proposition 3.10. *Let $c \in \mathbb{R}^d$ and $0 < \rho < r$ and suppose f is continuous on $B_r(c)$ and harmonic on $\mathring{B}_r(c)$ and vanishes on $\mathring{B}_\rho(c)$. Then f vanishes on $B_{r/3}(c)$.*

Proof. It can be assumed that $c = 0$. By Poisson's integral formula

$$f(x) = \int \frac{f(y)r^{d-2}(r^2 - x \cdot x)}{(r^2 - 2x \cdot y + x \cdot x)^{d/2}} \sigma_r(dy), \qquad \|x\| < r.$$

Suppose $\|x\| \leq r/3$. Then $|2x \cdot y| + x \cdot x < r^2$ for $y \in S_r$ and hence $f(x)$ can be written as an absolutely convergent infinite series of the form

$$f(x) = \sum_{n=1}^{\infty} h_n(x),$$

where $h_n(bx) = b^n h_n(x)$ for $0 \leq b \leq 1$ (the constant term vanishes since $f(0) = 0$). Thus

$$f(bx) = \sum_{n=1}^{\infty} b^n h_n(x) \qquad \text{for} \quad 0 \leq b \leq 1.$$

Since $f(bx)$ vanishes for b sufficiently small, $h_n(x) = 0$ for all $n \geq 1$ and hence $f(x) = 0$ as desired.

Theorem 3.11. *Suppose D is connected. Let f be a harmonic function on D which vanishes on a nonempty open subset of D. Then f vanishes on D.*

Proof. Let D_0 denote the set of points $x \in D$ such that f vanishes in some neighborhood of x. Then D_0 is a nonempty open subset of D. It follows easily from Proposition 3.10 that D_0 is relatively closed in D. Since D is connected, $D_0 = D$ as desired.

It follows from Theorem 3.11 that if D is connected and f is a harmonic function on D which is constant on some nonempty open subset of D, then f is constant on D. The next result, called the *principle of harmonic continuation* is also an immediate consequence of Theorem 3.11.

Theorem 3.12. *Suppose $D = D_1 \cup D_2$ where D_1 and D_2 are open sets whose intersection is a nonempty open connected set. Let f_1 and f_2 be harmonic functions on D_1 and D_2, respectively, such that $f_1 = f_2$ on a nonempty open subset of $D_1 \cap D_2$. Then $f_1 = f_2$ on $D_1 \cap D_2$ and the function f on D defined by $f = f_1$ on D_1 and $f = f_2$ on D_2 is harmonic on D.*

4. Nonnegative Harmonic Functions on an Open Ball

Assume $d \geq 2$ in this section. Let $c \in \mathbb{R}^d$ and $r > 0$. A Poisson integral representation is obtained for all nonnegative harmonic functions on $\mathring{B}_r(c)$. First a preliminary result is required.

Proposition 4.1. *Let $y \in S_r(c)$. Then*

$$\frac{r^{d-2}|r^2 - \|x - c\|^2|}{\|y - x\|^d}$$

is harmonic in x on $\mathbb{R}^d \backslash S_r(c)$. Also

$$\int \frac{r^{d-2}|r^2 - \|x - c\|^2|}{\|y - x\|^d} \sigma_\rho(c, dx) = \begin{cases} 1 & \text{if } 0 < \rho < r \\ (r/\rho)^{d-2} & \text{if } \rho > r. \end{cases} \tag{1}$$

Proof. For $\delta > 0$ define φ_δ on $S_r(c)$ by $\varphi_\delta(z) = I_{B_\delta(y)}(z)/\sigma_r(c, B_\delta(y))$ and set $f_\delta = h_{S_r(c)}\varphi_\delta$. Then f_δ is harmonic on $\mathbb{R}^d \backslash S_r(c)$ and

$$f_\delta(x) = \int_{B_\delta(y)} \frac{r^{d-2}|r^2 - \|x - c\|^2|}{\|z - x\|^d} \frac{\sigma_r(c, dz)}{\sigma_r(c, B_\delta(y))}$$

for $x \in \mathbb{R}^d \backslash S_r(c)$ by Theorem 3.1. Now

$$\lim_{\delta \to 0} f_\delta(x) = \frac{r^{d-2}|r^2 - \|x - c\|^2|}{\|y - x\|^d} \tag{2}$$

uniformly for x in compact subsets of $\mathbb{R}^d \backslash S_r(c)$. It now follows from the definition of harmonicity that the right-hand side of (2) is harmonic in x on $\mathbb{R}^d \backslash S_r(c)$.

Let $\rho > 0$ and $\rho \neq r$. By Theorem 3.1

$$\begin{aligned} P_y(\tau_{S_\rho(c)} < \infty) &= \int \frac{\rho^{d-2}|\rho^2 - \|y - c\|^2|}{\|y - x\|^d} \sigma_\rho(c, dx) \\ &= \left(\frac{\rho}{r}\right)^{d-2} \int \frac{r^{d-2}|r^2 - \|x - c\|^2|}{\|y - x\|^d} \sigma_\rho(c, dx). \end{aligned}$$

If $d = 2$, then $P_y(\tau_{S_\rho(c)} < \infty)$, $(\rho/r)^{d-2}$ and the right-hand side of (1) all equal one so (1) holds. If $d \geq 3$ and $0 < \rho < r$, then $P_y(\tau_{S_\rho(c)} < \infty) = (\rho/r)^{d-2}$ by Proposition 3.1.6, so (1) again holds. Finally, if $d \geq 3$ and $\rho > r$, then $P_y(\tau_{S_\rho(c)} < \infty) = 1$ so the left side of (1) equals $(r/\rho)^{d-2}$ as desired. This completes the proof of the proposition.

The desired Poisson integral representation for the nonnegative harmonic functions on a ball will now be obtained.

Theorem 4.2. *Let f be defined on $\mathring{B}_r(c)$. Then f is a nonnegative harmonic function on $\mathring{B}_r(c)$ if and only if there is a measure μ on $S_r(c)$ such that*

$$f(x) = \int_{S_r(c)} \frac{r^{d-2}(r^2 - \|x - c\|^2)}{\|y - x\|^d} \mu(dy), \qquad x \in \mathring{B}_r(c), \tag{3}$$

in which case the measure μ on $S_r(c)$ is finite and uniquely determined by f.

Proof. Suppose first that f is nonnegative and harmonic on $\mathring{B}_r(c)$. By the Poisson formula, for $0 < \rho < r$

$$(4) \qquad f(x) = \int_{S_\rho(c)} \frac{\rho^{d-2}(\rho^2 - \|x-c\|^2)}{\|y-x\|^d} f(y)\sigma_\rho(c, dy), \qquad x \in \mathring{B}_\rho(c).$$

Let μ_ρ denote the measure on $\mathbb{R}^d$ defined by $\mu_\rho(dy) = f(y)\sigma_\rho(c, dy)$. Then μ_ρ is supported on $S_\rho(c)$ and has total measure $f(c)$. Thus there is a sequence $\rho_n \uparrow r$ and a measure μ on $\mathbb{R}^d$ supported on $S_r(c)$ and having total measure $f(c)$ such that μ_{ρ_n} converges completely to μ as $n \to \infty$. It follows from (4) that (3) holds for this choice of μ.

Suppose now that μ is a finite measure on $S_r(c)$ and let f be defined on $\mathring{B}_r(c)$ by (3). Then f is locally bounded and hence locally integrable on $\mathring{B}_r(c)$. It follows from Proposition 4.1, Fubini's theorem and the definition of harmonicity that f is a nonnegative harmonic function on $\mathring{B}_r(c)$.

Finally it will be shown that μ is uniquely determined by f. For $0 < \rho < r$ let μ_ρ be defined as above by $\mu_\rho(dx) = f(x)\sigma_\rho(c, dx)$. Let A be a closed subset of $\mathbb{R}^d$. By (3)

$$(5) \qquad \mu_\rho(A) = \int_{S_r(c)} \mu(dy) \int_A \frac{r^{d-2}(r^2 - \|x-c\|^2)}{\|y-x\|^d} \sigma_\rho(c, dx).$$

Now μ_ρ and μ have total measure $f(c)$. By (1)

$$(6) \qquad \overline{\lim_{\rho \uparrow r}} \int_A \frac{r^{d-2}(r^2 - \|x-c\|^2)}{\|y-x\|^d} \sigma_\rho(c, dx) \le I_A(y), \qquad y \in S_r(c).$$

It follows from (5) and (6) that $\limsup_{\rho \uparrow r} \mu_\rho(A) \le \mu(A)$. Consequently (see Chung [1], p. 91), μ_ρ converges completely to μ as $\rho \uparrow r$. Thus μ is uniquely determined by f.

Given a nonnegative harmonic function f on $\mathring{B}_r(c)$, let μ_f denote the measure on $S_r(c)$ corresponding to f as in (3). Given a finite measure μ on $S_r(c)$, let f_μ be the harmonic function corresponding to μ according to (3).

Proposition 4.3. *Let μ be a signed measure of finite total variation on $S_r(c)$. If*

$$\int \frac{r^{d-2}(r^2 - \|x-c\|^2}{\|y-x\|^d} \mu(dy) = 0, \qquad x \in B_r(c),$$

then $\mu = 0$.

Proof. Let $\mu = \mu^+ - \mu^-$ be the decomposition of μ into its positive and negative parts. Then $f_{\mu^+} = f_{\mu^-}$ by hypothesis. Consequently $\mu^+ = \mu^-$ by Theorem 4.2 and hence $\mu = 0$ as desired.

Proposition 4.4. *Let f_1 and f_2 be nonnegative harmonic functions on $\mathring{B}_r(c)$. Then $f_1 \geq f_2$ on $\mathring{B}_r(c)$ if and only if $\mu_{f_1} \geq \mu_{f_2}$.*

Proof. This result follows easily from Theorem 4.2 and Proposition 4.3. The details are left to the reader.

Proposition 4.5. *Let f be a nonnegative harmonic function on $\mathring{B}_r(c)$. Then there is a nonnegative function φ on $S_r(c)$ such that $f = H_{\mathring{B}_r(c)}\varphi$ on $\mathring{B}_r(c)$ if and only if μ_f is absolutely continuous with respect to $\sigma_r(c, \cdot)$, in which case $d\mu_f/d\sigma_r(c, \cdot) = \varphi$ a.e. $(\sigma_r(c, \cdot))$.*

Proof. This result follows immediately from Theorems 3.1 and 4.2 and Proposition 4.3.

Proposition 4.6. *Let f be a nonnegative harmonic function on $\mathring{B}_r(c)$ and let A be a relatively open subset of $S_r(c)$. Then $\mu_f(A) = 0$ if and only if f has boundary value zero on A.*

Proof. It follows from Theorem 4.2 that if $\mu_f(A) = 0$, then f has boundary value zero on A. Suppose conversely that f has boundary value zero on A. In order to show that $\mu_f(A) = 0$ it suffices to show that if C is a compact subset of A and v is the measure μ_f restricted to C, then $v = 0$ or equivalently $f_v = 0$. Now f_v is harmonic on $\mathring{B}_r(c)$ and $f_v \leq f$ by Proposition 4.4. Thus f_v has boundary value zero on C. By the first part of this proof f_v has boundary value zero on $S_r(c)\backslash C$. Consequently $f_v = 0$ by Proposition 1.4.

Proposition 4.7. *Let f be a nonnegative harmonic function on $\mathring{B}_r(c)$. Set $\varphi(b) = \overline{\lim}_{x \to b} f(x)$ for $b \in S_r(c)$. Then μ_f is absolutely continuous with respect to $\sigma_r(c, \cdot)$ on $A = \{b : \varphi(b) < \infty\}$ and $d\mu_f/d\sigma_r(c, \cdot) \leq \varphi$ a.e. $(\sigma_r(c, \cdot))$ on A. In particular, if μ_f is singular with respect to $\sigma_r(c, \cdot)$, then $\overline{\lim}_{x \to b} f(x) = \infty$ a.e. (μ_f).*

The proof of this result is left to the reader.

Martin [1] obtained a representation for the nonnegative harmonic functions on an arbitrary open set which generalizes Theorem 4.2 (see Brelot [8], Helms [1], Kemeny, Snell, and Knapp [1] and Meyer [1]).

5. Green Function

Set $G_D(x, y) = g_{D^c}(x, y)$ for $x, y \in D$. If G_D is not identically infinite on $D \times D$, D is said to be *Greenian* and G_D is called the *Green function* for D. In this section the Green function is characterized analytically and computed explicitly when D is a half space, ball, or exterior of a ball.

Proposition 5.1. *D is Greenian if and only if $d \geq 3$ or D^c is nonpolar.*

Proof. If $d \geq 3$ or D^c is nonpolar, then D is Greenian by Proposition 2.2.7. If $d \leq 2$ and D^c is polar, then $g_{D^c}(x, y) = g(x, y) = \infty$ for all $x, y \in \mathbb{R}^2$ so D is not Greenian.

Suppose D is Greenian. Then

$$G_D(x, y) = \int_0^\infty Q_D(t, x, y)\, dt, \qquad x, y \in D.$$

The properties of q_{D^c} obtained in Sections 2.4 and 2.6 and the properties of g_{D^c} obtained in Sections 3.1, 3.4, and 3.5 and in the introduction to Chapter 3 easily yield the following basic properties of the Green function. First, $0 \leq G_D(x, y) \leq g(x, y)$ and $G_D(x, y) = G_D(y, x)$ for $x, y \in D$ and $G_D(x, y) < \infty$ for $x \neq y$; also $G_D(x, y) > 0$ if x and y are in the same component of D and $G_D(x, y) = 0$ otherwise. For each $x \in D$, $G_D(x, \cdot)$ is locally integrable and has boundary value zero on $\partial D \cap (D^c)^r$. If A is a bounded subset of D and $\delta > 0$, then

$$\text{(1)} \qquad \sup[G_D(x, y) : x \in A, \|y - x\| > \delta] < \infty.$$

If D_0 is an open subset of D, then D_0 is Greenian and

$$\text{(2)} \qquad G_D(x, y) = G_{D_0}(x, y) + \int_D H_{D_0}(x, dz) G_D(z, y), \qquad x, y \in D_0.$$

(This follows from (1) in the introduction to Chapter 3 since by Theorems 2.6.3 and 2.6.5, $H_{D_0}(x, \cdot)$ assigns measure zero to $\{z \in D^c : g_{D^c}(z, y) > 0\}$.) If D_0 is a component of D, then $G_D(x, y) = G_{D_0}(x, y)$ for $x, y \in D_0$. If D_n are open subsets of D such that $D_n \uparrow D$, then $G_{D_n}(x, y) \uparrow G_D(x, y)$ for $x, y \in D$ (this follows from Proposition 2.4.5).

Proposition 5.2. *Suppose D is Greenian and let $y \in D$. If $d \geq 3$, the function $\int H_D(\cdot, dz) g(z, y)$ is harmonic on D. If $d \leq 2$, $\int H_D(\cdot, dz) k(z, y)$ and W_{D^c} are harmonic on D.*

Proof. Suppose $d \geq 3$. Then $g(\cdot, y)$ is bounded on ∂D, so it follows from Proposition 2.1 that $\int H_D(\cdot, dz)g(z, y)$ is harmonic on D. Suppose $d = 1$. It follows from Proposition 2.2.20 that $\int H_D(\cdot, dz)k(z, y)$ is linear on each component interval of D and hence harmonic on D and it follows similarly from the discussion in Section 3.5 that W_{D^c} is harmonic on D. Suppose finally that $d = 2$. It follows from Theorem 3.7 and Theorem 3.4.2 that $\int H_D(\cdot, dz)k(z, y)$ is harmonic on D. Choose $x_0 \in D\backslash\{y\}$ and $r > 0$ such that $B_r(x_0) \subset D\backslash\{y\}$. Then

$$G_D(x, y) = \int h_{S_r(x_0)}(x, dz)G_D(z, y), \qquad x \in \mathring{B}_r(x_0).$$

Thus $G_D(\cdot, y)$ is harmonic on $\mathring{B}_r(x_0)$ and hence it is harmonic on $D\backslash\{y\}$. Since $k(\cdot, y)$ is harmonic on $D\backslash\{y\}$ it now follows from the fundamental identity for logarithmic potentials that W_{D^c} is harmonic on $D\backslash\{y\}$. Since this is true for each $y \in D$, W_{D^c} is harmonic on all of D. This completes the proof of the proposition.

Proposition 5.3. *Suppose D is Greenian. Then G_D is continuous in the extended sense on $D \times D$.*

Proof. If $d = 1$, G_D is continuous in the usual sense (i.e., finite and continuous) as was shown in Section 3.5. Suppose $d = 2$ (a similar proof works for $d \geq 3$). Then

$$k(x, y) = G_D(x, y) + \int H_D(x, dz)k(z, y), \qquad x, y \in D,$$

by the fundamental identity for logarithmic potentials and Proposition 3.4.7; so $\int H_D(x, dz)k(z, y)$ is symmetric in x and y on $D \times D$. To prove the desired result it suffices to show that $\int H_D(x, dz)k(z, y)$ is continuous on $D \times D$. Now this function is finite on $D \times D$ by Proposition 5.2. Choose $x_0, y_0 \in D$ and $\varepsilon > 0$. There is a $\delta > 0$ such that

$$|k(z, y) - k(z, y_0)| \leq \varepsilon/2 \qquad \text{and} \qquad |k(z, x) - k(z, x_0)| \leq \varepsilon/2$$

for $\|x - x_0\| < \delta$, $\|y - y_0\| < \delta$, and $z \in \partial D$. Choose $x, y \in D$ such that $\|x - x_0\| < \delta$ and $\|y - y_0\| < \delta$. Then

$$\left|\int H_D(x, dz)k(z, y) - \int H_D(x_0, dz)k(z, y_0)\right|$$

$$\leq \int H_D(x, dz)|k(z, y) - k(z, y_0)| + \int H_D(y_0, dz)|k(z, x) - k(z, x_0)| \leq \varepsilon.$$

Therefore $\int H_D(x, dz)k(z, y)$ is continuous on $D \times D$ as desired.

Let D be Greenian and let $d \geq 3$. The next result asserts that $G_D(x,\cdot) - g(x,\cdot)$ is harmonic on $D\backslash\{x\}$ and can be extended to a harmonic function on D. This is summarized by saying that $G_D(x,\cdot) - g(x,\cdot)$ is harmonic on D. The same convention is used in other similar situations.

Proposition 5.4. *Suppose D is Greenian and let $x \in D$. If $d \geq 3$, then $G_D(x,\cdot) - g(x,\cdot)$ is harmonic on D and if $d \leq 2$, then $G_D(x,\cdot) - k(x,\cdot)$ is harmonic on D.*

Proof. This result follows from Proposition 5.2, symmetry and the fundamental identities for Newtonian, logarithmic and linear potentials.

It follows from Proposition 5.4 that if $d \geq 2$, then $G_D(x,x) = \infty$ for $x \in D$. If $d = 1$ and $D \neq \mathbb{R}$, then $G_D(x,x) < \infty$ for $x \in D$.

Let $\mathscr{G}_D$ denote the collection of nonnegative functions G on $D \times D$ such that if $d \geq 3$, then $G(x,\cdot) - g(x,\cdot)$ is harmonic on D for each $x \in D$ and if $d \leq 2$, then $G(x,\cdot) - k(x,\cdot)$ is harmonic on D for each $x \in D$. Suppose D is Greenian. Then $G_D \in \mathscr{G}_D$ by Proposition 5.4, so in particular $\mathscr{G}_D$ is nonempty. If D is bounded and regular, G_D is the unique $G \in \mathscr{G}_D$ such that $G(x,\cdot)$ has boundary value zero on ∂D for each $x \in D$. An analytic characterization of G_D which covers the general case will now be given.

Theorem 5.5. *D is Greenian if and only if $\mathscr{G}_D$ is nonempty, in which case G_D is the smallest function in $\mathscr{G}_D$.*

Proof. It suffices to show that $G \geq G_D$ for $G \in \mathscr{G}_D$. Let D_n be relatively compact regular open subsets of D such that $D_n \uparrow D$. Then $G_{D_n} \uparrow G_D$ on $D \times D$. Choose $G \in \mathscr{G}_D$ and let $x \in D_n$. Then $G(x,\cdot) - G_{D_n}(x,\cdot)$ is harmonic on D_n and has nonnegative boundary values on ∂D_n. Thus $G(x,\cdot) \geq G_{D_n}(x,\cdot)$ on D_n by Proposition 1.4. Consequently $G \geq G_D$ on $D \times D$ as desired.

It is easy to derive explicit formulas for the Green function of a half-space, ball, or the exterior of a ball. Consider first a hyperplane P and let D be one of the two half spaces (components of $\mathbb{R}^d\backslash P$) determined by P. Let x' denote the reflection of x about P. If $d \leq 2$, then $G_D(x,y) = k(y-x) - k(y-x')$ for $x, y \in D$. To see this set $G(x,y) = k(y-x) - k(y-x')$. Choose $x \in D$. Then $G_D(x,\cdot) - G(x,\cdot)$ is a bounded harmonic function on D (recall (1)) having boundary value zero on $\partial D = P$. Since D^c is recurrent, this function equals zero by Theorem 2.12. A similar proof shows that if $d \geq 3$, then $G_D(x,y) = g(y-x) - g(y-x')$ for $x, y \in D$.

Consider next an open ball $D = \mathring{B}_r(c)$. Let * denote inversion relative to $S_r(c)$. By (3.3)

$$\frac{\|x - c\| \, \|y - x^*\|}{r} = \|y - x\|, \qquad x \in \mathbb{R}^d \backslash \{c\} \quad \text{and} \quad y \in S_r(c). \tag{3}$$

If $d \geq 3$, then

$$G_D(x, y) = g(y - x) - g\left(\frac{\|x - c\|(y - x^*)}{r}\right) \tag{4}$$

for $x \in D \backslash \{c\}$ and $y \in D$. To see this let $G(x, y)$ denote the right-hand side of (4). Choose $x \in D \backslash \{c\}$. By (3), $G_D(x, \cdot) - G(x, \cdot)$ is a bounded harmonic function on D having boundary value zero on ∂D. Thus this function equals zero on D and hence (4) holds. It follows by a similar argument that if $d = 2$, then

$$G_D(x, y) = k(y - x) - k\left(\frac{\|x - c\|(y - x^*)}{r}\right) = \frac{1}{\pi} \log\left(\frac{\|x - c\| \, \|y - x^*\|}{r\|y - x\|}\right) \tag{5}$$

for $x, y \in D$ with $x \neq c$. It is left to the reader to show that (4) and (5) are also valid if $D = \mathbb{R}^d \backslash B_r(c)$.

6. Poisson's Equation

Let f be defined on D. For $x \in D$ set

$$\tilde{\Delta} f(x) = \lim_{r \to 0} \frac{2d}{r^2} \left[\int f(y) \sigma_r(x, dy) - f(x) \right]$$

provided that the indicated limit exists. The operator $\tilde{\Delta}$ is called the *generalized Laplacian.*

Let ρ be a function on D. In this section the *Poisson equation* $\Delta f = -2\rho$ on D and the *generalized Poisson equation* $\tilde{\Delta} f = -2\rho$ on D are considered. By the next result a twice continuously differentiable solution to the generalized Poisson equation also satisfies the Poisson equation.

Theorem 6.1. *Let f be twice continuously differentiable on D. Then $\tilde{\Delta} f = \Delta f$ on D.*

Proof. Choose $x \in D$. Define q on D by $q(y) = \|y - x\|^2/2d$. Then $\Delta q = 1$. Choose s and t such that $s < \Delta f(x) < t$. Choose $\delta > 0$ such that $B_\delta(x) \subset D$ and $s < \Delta f < t$ on $B_\delta(x)$. Let $0 < r \leq \delta$ and note that $\Delta(f - tq) < 0$ on $B_r(x)$. Let h be the harmonic function on $\mathring{B}_r(x)$ having boundary function $f - tq$ on $S_r(x)$. Then $\Delta(f - tq - h) < 0$ on $\mathring{B}_r(x)$ and hence f cannot have a local minimum in $\mathring{B}_r(x)$. Since $f - tq - h$ vanishes on $S_r(x)$, $f - tq - h \geq 0$ on

$B_r(x)$. In particular

$$f(x) \geq h(x) = \int (f(y) - tq(y))\sigma_r(x, dy) = \int f(y)\sigma_r(x, dy) - tr^2/2d.$$

Similarly $f(x) \leq \int f(y)\sigma_r(x, dy) - sr^2/2d$ and hence

$$s \leq \frac{2d}{r^2}\left[\int f(y)\sigma_r(x, dy) - f(x)\right] \leq t, \qquad 0 < r \leq \delta.$$

Thus $\tilde{\Delta} f(x) = \Delta f(x)$ for $x \in D$ as desired.

Theorem 6.2. *Let f be continuous on D. Then f is harmonic on D if and only if $\tilde{\Delta} f = 0$ on D.*

Proof. If f is harmonic on D, then f is twice continuously differentiable on D and hence $\tilde{\Delta} f = \Delta f = 0$ on D by Theorem 6.1. Suppose conversely that $\tilde{\Delta} f = 0$ on D. Choose $x \in D$ and define q on D by $q(y) = \|y - x\|^2$. Then $\tilde{\Delta} q = \Delta q = 2d$ on D. Let $r > 0$ be such that $B_r(x) \subset D$. Choose $\varepsilon > 0$ and let h be the harmonic function on $\mathring{B}_r(x)$ having boundary function $f - \varepsilon q$ on $S_r(x)$. Then $\tilde{\Delta}(f - \varepsilon q - h) < 0$ on $\mathring{B}_r(x)$ and hence $f - \varepsilon q - h$ cannot have a local minimum on $\mathring{B}_r(x)$. Consequently $f - \varepsilon q - h \geq 0$ on $\mathring{B}_r(x)$ and in particular

$$f(x) \geq h(x) = \int (f(y) - \varepsilon q(y))\sigma_r(x, dy) = \int f(y)\sigma_r(x, dy) - \varepsilon r^2.$$

Since ε can be made arbitrarily small $f(x) \geq \int f(y)\sigma_r(x, dy)$. Similarly the opposite inequality holds and hence $f(x) = \int f(y)\sigma_r(x, dy)$. Thus f is harmonic on D.

Let $r > 0$. By Proposition 2.1.1, $\int_{B_r} \|y\|^{-d}\, dy = \infty$ but $\int_{B_r} \|y\|^{\alpha - d}\, dy < \infty$ for $\alpha > 0$. Thus integrals of the form $\int_{B_r} \|y\|^{-d} |f(y) - f(0)|\, dy$ are infinite for some continuous functions f but are necessarily finite if f satisfies a Hölder condition on B_r. This motivates the following definition. A function f on D is said to be Hölder continuous if for every compact subset C of D there are positive constants α and M such that $|f(y) - f(x)| \leq M\|y - x\|^\alpha$ for $x, y \in C$.

Poisson's equation on $\mathbb{R}^d$ will now be considered. It is convenient to consider $d \geq 3$, $d = 2$ and $d = 1$ separately.

Theorem 6.3. *Suppose $d \geq 3$, $D = \mathbb{R}^d$, ρ is bounded on compacts and $\int (\|y\| + 1)^{2-d} |\rho(y)|\, dy < \infty$. Then $g\rho$ is continuously differentiable. If ρ is continuous, then $\tilde{\Delta} g\rho = -2\rho$. If ρ is Hölder continuous, then $g\rho$ is twice continuously differentiable and $\Delta g\rho = -2\rho$. Let ρ be bounded, continuous, and integrable. Then f is a continuous solution to $\tilde{\Delta} f = -2\rho$ which is bounded below or above if and only if $f = g\rho + \alpha$ for some $\alpha \in \mathbb{R}$.*

Proof. Let D_0 be an open subset of $\mathbb{R}^d$ and set $\rho_0 = \rho I_{D_0^c}$. It is easily seen that $g|\rho_0|$ is locally integrable on D_0 and that $g|\rho_0| = H_{D_0} g|\rho_0|$ on D_0. Thus by Theorem 3.7, $g\rho_0 = H_{D_0} g\rho_0$ is harmonic and hence infinitely differentiable on D_0. Consequently to prove that $g\rho$ is continuously differentiable on D_0 it suffices to prove that $g(\rho I_{D_0})$ is continuously differentiable. In other words, in proving that $g\rho$ is continuously differentiable it can be assumed that ρ is a bounded function which vanishes outside some compact set. For such a ρ it is easily seen that

$$\int \frac{y-x}{\|y-x\|^d}\rho(y)\,dy$$

is continuous in x and it now follows from Fubini's theorem that

$$\frac{d}{dx}\int \frac{\rho(y)}{\|y-x\|^{d-2}}\,dy = (d-2)\int \frac{(y-x)}{\|y-x\|^d}\rho(y)\,dy.$$

Thus the first conclusion of the theorem is valid.

Suppose now that ρ is continuous on $\mathbb{R}^d$. Choose $x \in \mathbb{R}^d$ and $r > 0$. Set $\tau_r = \tau_{S_r}(x)$. Then $E_x\tau_r = r^2/d$ by Proposition 2.2.21. By (1) in the introduction to Chapter 3

$$g\rho(x) = E_x\int_0^{\tau_r}\rho(X(t))\,dt + \int g\rho(y)\sigma_r(x,dy).$$

Consequently

$$\frac{2d}{r^2}\left[\int g\rho(y)\sigma_r(x,dy) - g\rho(x)\right] = -\frac{2}{E_x(\tau_r)}E_x\left[\int_0^{\tau_r}(\rho(X(t)) - \rho(x))\,dt\right] - 2\rho(x).$$

It follows by letting $r \to 0$ that $\tilde{\Delta} g\rho = -2\rho$. Thus the second conclusion of the theorem is valid.

Suppose ρ is Hölder continuous. Let $0 < r < s$. It is possible to express ρ in the form $\rho_1 + \rho_2$ where ρ_1 is Hölder continuous and vanishes outside B_s and ρ_2 vanishes on B_r. Then $g\rho_2$ is harmonic on $\mathring{B}_r$. Thus to prove that $g\rho$ is twice continuously differentiable on $\mathring{B}_r$ and $\Delta g\rho = -2\rho$ on $\mathring{B}_r$ it suffices to prove the same result with ρ replaced by ρ_1. In other words, in varifying the third conclusion of the theorem it can be assumed without loss of generality that ρ has compact support. Let $r > 0$ be such that ρ vanishes outside B_r. By Proposition 3.1.7

(1) $$g\rho(x) = \rho(x_0)\left(\frac{r^2}{d-2} - \frac{\|x\|^2}{d}\right) + \int_{B_r} g(x,y)(\rho(y) - \rho(x_0))\,dy, \qquad x_0, x \in \mathring{B}_r.$$

It follows easily as in the first paragraph of this proof that

$$\frac{d}{dx} g\rho(x) = \frac{-2\rho(x_0)}{d} x + \int_{B_r} \frac{d}{dx} g(x, y)(\rho(y) - \rho(x_0))\,dy, \qquad x_0, x \in \mathring{B}_r.$$

It will now be shown that

$$\text{(2)} \quad \frac{d^2}{dx^2} g\rho(x)\Big|_{x=x_0}$$

$$= \frac{-2\rho(x_0)}{d} I + \int_{Br} \frac{d^2}{dx^2} g(x, y)\Big|_{x=x_0} (\rho(y) - \rho(x_0))\,dy, \qquad x_0 \in B_r.$$

Here I is the $d \times d$ identity matrix and the second derivative is considered as the $d \times d$ matrix of second order partial derivatives. Choose $x_0 \in \mathring{B}_r$. It sufficies to show that

$$\frac{\partial}{\partial x_i} \int_{B_r} \frac{(y_j - x_j)}{\|y - x\|^d} (\rho(y) - \rho(x_0))\,dy\Big|_{x=x_0}$$

$$= \int_{B_r} \frac{\partial}{\partial x_i} \left(\frac{y_j - x_j}{\|y - x\|^d}\right)\Big|_{x=x_0} (\rho(y) - \rho(x_0))\,dy$$

for $1 \le i, j \le d$. Suppose for simplicity that $i \ne j$ (a similar proof works for $i = j$). It must be shown that

$$\frac{\partial}{\partial x_i} \int_{B_r(-x_0)} \frac{y_j}{\|y - x\|^d} (\rho(x_0 + y) - \rho(x_0))\,dy\Big|_{x=0}$$

$$= d \int_{B_r(-x_0)} \frac{y_i y_j}{\|y\|^{d+2}} (\rho(x_0 + y) - \rho(x_0))\,dy$$

or equivalently that

$$\text{(3)} \quad \lim_{h\to 0} \int_{B_r(-x_0)} \frac{1}{h}\left(\frac{1}{\|y - hu_i\|^d} - \frac{1}{\|y\|^d}\right) y_j(\rho(x_0 + y) - \rho(x_0))\,dy$$

$$= d \int_{B_r(-x_0)} \frac{y_i y_j}{\|y\|^{d+2}} (\rho(x_0 + y) - \rho(x_0))\,dy,$$

where $u_i = (0, \ldots, 1, \ldots, 0)$ has a 1 in the ith position and zeros elsewhere; note that the right-hand side of (3) is finite by the Hölder continuity of ρ. It is straightforward to verify (3) by integrating separately over $\|y\| \le h/2$, $h/2 < \|y\| \le 2h$, $2h < \|y\| \le s$ and $y \in B_r(-x_0)\backslash B_s$, where s is small, and using the Hölder continuity of ρ. The details are left to the interested reader. Therefore (2) is valid. It follows by decomposing the integral as in the proof of

(3) that the right-hand side of (2) depends continuously on x_0. Consequently $g\rho$ is twice continuously differentiable. The formula $\Delta g\rho = -2\rho$ now follows either from (2) or from the formula $\tilde{\Delta} g\rho = -2\rho$ together with Theorem 6.1. Thus the third conclusion of the theorem is valid.

Suppose ρ is bounded, continuous, and integrable. Then $g\rho$ is bounded and continuous and $\tilde{\Delta} g\rho = -2\rho$. Let f be a continuous solution to $\tilde{\Delta} f = -2\rho$ which is bounded below or above. Then $f - g\rho$ is harmonic on $\mathbb{R}^d$ and bounded below or above. Consequently $f = g\rho + \alpha$ for some $\alpha \in \mathbb{R}$ by Theorem 3.4. Thus the last conclusion of the theorem is valid and the proof of the theorem is complete.

Theorem 6.4. *Suppose $d = 2$, $D = \mathbb{R}^2$, ρ is bounded on compacts and $\int \log(\|y\| + 1)|\rho(y)|\, dy < \infty$. Then $k\rho$ is continuously differentiable. If ρ is continuous, then $\tilde{\Delta} k\rho = -2\rho$. If ρ is Hölder continuous, then $k\rho$ is twice continuously differentiable and $\Delta k\rho = -2\rho$. Let ρ be continuous and have compact support. Then f is a continuous solution to $\tilde{\Delta} f = -2\rho$ such that $f(x)/\log(\|x\| + 2)$, $x \in \mathbb{R}^2$, is bounded below or above if and only if $f = k\rho + \alpha$ for some $\alpha \in \mathbb{R}$.*

Proof. Only the last conclusion will be verified, since the proofs of the other conclusions of the theorem are similar to the proofs of the corresponding conclusions of Theorem 6.3.

Let ρ be continuous and have compact support. Then $\tilde{\Delta} k\rho = -2\rho$ by the second conclusion of the theorem. Set $\beta = \int \rho(y)\, dy$. By (3.4.2)

$$k\rho(x) - \beta k(x) = \int (k(x, y) - k(x))\rho(y)\, dy \to 0 \qquad \text{as } \|x\| \to \infty.$$

Thus $k\rho(x)/\log(\|x\| + 2)$, $x \in \mathbb{R}^2$, is bounded. Suppose f is a continuous solution to $\tilde{\Delta} f = -2\rho$ such that $f(x)/\log(\|x\| + 2)$, $x \in \mathbb{R}^2$, is bounded (say) below. Set $u = f - k\rho$. It must be shown that u is constant on $\mathbb{R}^2$. Now u is continuous and $\tilde{\Delta} u = 0$, so u is harmonic by Theorem 6.2. Also $u(x)/\log(\|x\| + 2)$, $x \in \mathbb{R}^2$, is bounded below. Thus there is an $M > 0$ such that $u(x) \geq -M \cdot \log\|x\|$ for $\|x\| \geq 2$. Let v be the conjugate harmonic function to u on $\mathbb{R}^2$ as defined in texts on complex variable theory and set $i = \sqrt{-1}$. Now $\mathbb{R}^2$ can be identified with the complex plane $\mathbb{C}$ in the usual manner. Then $F = \exp(-(u + iv)/2M)$ is analytic on $\mathbb{C}$ and $|F(z)| \leq |z|^{1/2}$ for $|z| \geq 2$. It follows easily from Cauchy's integral formula that $F'(z) = 0$ for $z \in \mathbb{C}$ and hence that F is constant on $\mathbb{C}$. Thus $|F| = \exp(-u/2M)$ is constant on $\mathbb{C}$ and hence u is constant on $\mathbb{R}^2$ as desired (see the discussion following Theorem 6.7.10 for an alternative proof). This completes the proof of the last conclusion of the theorem.

Proposition 6.5. *Suppose $d = 1$, $D = \mathbb{R}$, ρ is bounded on compacts and $\int |y| |\rho(y)| \, dy < \infty$. Then $k\rho$ is continuously differentiable and if ρ is continuous, then $(k\rho)'' = -2\rho$.*

Proof. This result follows easily from the formula $k(x) = -|x|$. The details are left to the reader.

In the next theorem Poisson's equation on a Greenian open set is considered. Observe that (4) below holds either if $d \geq 3$ and $\int_D (\|y\| + 1)^{2-d} |\rho(y)| \, dy < \infty$ or (by (5.1)) if ρ is integrable on D.

Theorem 6.6. *Suppose D is Greenian, ρ is bounded on compact subsets of D, and*

$$\int_{D \setminus B_r(x)} G_D(x, y) |\rho(y)| \, dy < \infty, \qquad x \in D \quad \text{and} \quad r > 0. \tag{4}$$

Then $G_D\rho$ is continuously differentiable. If ρ is continuous, then $\tilde{\Delta} G_D \rho = -2\rho$ on D. If ρ is Hölder continuous, then $G_D\rho$ is twice continuously differentiable and $\Delta G_D \rho = -2\rho$ on D.

Proof. It is easily seen that if $\rho = 0$ on an open subset D_0 of D, then $G_D\rho = H_{D_0} G_D \rho$ on D_0 and hence $G_D\rho$ is harmonic on D_0. It follows from this observation that, without loss of generality, ρ can be assumed to vanish outside some compact subset of D. Suppose $d \geq 3$. Then $g\rho = G_D\rho + H_D g\rho$ on D by the fundamental identity for Newtonian potentials. Now $H_D g\rho$ is harmonic on D, so the desired result follows from Theorem 6.3. Suppose next that $d \leq 2$. Then

$$k\rho = G_D\rho + H_D k\rho - W_{D^c} \int \rho(y) \, dy \qquad \text{on } D$$

by the fundamental identities for logarithmic and linear potentials. Now $H_D k\rho - W_{D^c} \int \rho(y) \, dy$ is harmonic on D, so the desired result follows from Theorem 6.4 and Proposition 6.5. This completes the proof of the theorem.

Theorem 6.7. *Suppose D is Greenian and that ρ is bounded and integrable on D. Then $G_D\rho$ has boundary value zero on $\partial D \cap (D^c)^r$.*

Proof. Choose $b \in \partial D \cap (D^c)^r$. Let C be a compact neighborhood of b and let ρ_1 and ρ_2 denote, respectively, the functions ρI_{C^c} and ρI_C on D. Let U be an open neighborhood of b whose closure is contained in the interior of C. Then $\sup[G_D(x, y) : x \in U \text{ and } y \notin C] < \infty$ by (5.1). Since $G_D(\cdot, y)$ has boundary value zero at b for $y \in D$, as was pointed out in Section 5, it follows from the dominated convergence theorem that $G_D\rho_1$ has boundary value zero at b. By the strong Markov property

$$\begin{aligned} G_D(\cdot, C) &= E_{\cdot} \int_0^{T_D} I_C(X(t))\,dt \\ &\le E_{\cdot}(T_D; T_D \le 1) + P_{\cdot}(T_D > 1)(1 + \sup_y G_D(y, C)) \end{aligned}$$

on D. Since $\sup_y G_D(y, C) < \infty$ by Proposition 2.2.7 and $\lim_{x \to b} P_x(T_D \ge t) = 0$ for all $t > 0$ by Proposition 2.3.5, $G_D(\cdot, C)$ has boundary value zero at b. Consequently $G_D\rho_2$ has boundary value zero at b. Therefore $G_D\rho$ has boundary value zero at b and hence $G_D\rho$ has boundary value zero on $\partial D \cap (D^c)^r$ as desired.

Poisson's equation with specified boundary values will now be considered. Suppose D is Greenian and that ρ is bounded, continuous, and integrable. Observe that if $d \ge 3$ or ρ vanishes outside a bounded subset of D, then $G_D\rho$ is bounded on D.

Theorem 6.8. *Suppose D is Greenian, ρ is bounded, continuous, and integrable on D and that $G_D\rho$ is bounded on D. Let φ be a bounded and essentially continuous function on ∂D. Then f is a bounded continuous solution to $\tilde{\Delta} f = -2\rho$ on D having boundary value $\varphi(b)$ at b for every $b \in \partial D \cap (D^c)^r$ at which φ is defined and continuous if and only if $f = G_D\rho + H_D\varphi + \alpha P_{\cdot}(T_D = \infty)$ on D for some $\alpha \in \mathbb{R}$.*

Proof. This result follows easily from Theorem 2.10 and various results obtained in this section. The details are left to the reader.

As an illustration of the results of this section the function $E_{\cdot}T_D = G_D 1$ on D will be computed when D is the annulus $\{x \in \mathbb{R}^d : r_1 < \|x\| < r_2\}$, where $0 < r_1 < r_2 < \infty$. For $x \in D$ set $p_i(x) = P_x(\|X(T_D)\| = r_i)$. Then $p_1(x) + p_2(x) = 1$. Also

$$\begin{aligned} p_2(x) &= \frac{\|x\| - r_1}{r_2 - r_1} && \text{if } d = 1, \\ &= \frac{\log\|x\| - \log r_1}{\log r_2 - \log r_1} && \text{if } d = 2, \\ &= \frac{\|x\|^{2-d} - r_1^{2-d}}{r_2^{2-d} - r_1^{2-d}} && \text{if } d \ge 3, \end{aligned}$$

by Propositions 2.2.20, 3.4.8, and 3.1.5. By Theorem 6.7, $G_D 1$ has boundary value zero on ∂D. By Theorem 6.6, $G_D 1$ is twice continuously differentiable on D and $\Delta G_D 1 = -2$ on D. Set $q(y) = \|y\|^2/d$ for $y \in \bar{D}$. Then $\Delta q = 2$ on D and hence $\Delta(G_D 1 + q) = 0$ on D. Consequently $G_D 1 + q$ is harmonic on D. Now $q = r_i^2/d$ on S_{r_i}, so by Theorem 2.4

$$G_D 1(x) + q(x) = \frac{p_1(x) r_1^2 + p_2(x) r_2^2}{d}, \qquad x \in D.$$

Therefore

$$E_x T_D = \frac{p_1(x) r_1^2 + p_2(x) r_2^2 - \|x\|^2}{d}, \qquad x \in D. \tag{5}$$

It is easily seen that (5) reduces to the formula $E_x T_D = (x - r_1)(r_2 - x)$, $r_1 < x < r_2$, given in Proposition 2.2.20 when $d = 1$.

7. Eigenfunction Expansion

Throughout this section D is a nonempty open subset of $\mathbb{R}^d$ having finite Lebesgue measure $|D|$. An eigenfunction expansion for $Q_D(t, x, y)$ will be obtained which extends the second formula for $Q_D(t, x, y)$ in Proposition 2.8.2. The asymptotic distribution of the eigenvalues and the asymptotic distribution of $P_x(T_D > t, X(T_D) \in A)$ will also be obtained.

Recall that for $t > 0$, $Q_D(t, x, y)$ is a nonnegative continuous symmetric function of x and y, which is positive if D is connected. Observe that

$$\int Q_D^2(t, x, y)\, dy = Q_D(2t, x, x) \le p(2t, 0), \qquad x \in D, \tag{1}$$

and hence that

$$\iint Q_D^2(t, x, y)\, dx\, dy = \int Q_D(2t, x, x)\, dx \le |D| p(2t, 0). \tag{2}$$

Let $\mathscr{L}^2 = \mathscr{L}^2(D)$ denote the Hilbert space of functions f on D such that $\|f\|^2 = \int f^2(x)\, dx < \infty$. For $t > 0$ let Q_D^t denote the operator on $\mathscr{L}^2$ defined by $Q_D^t f(x) = \int Q_D(t, x, y) f(y)\, dy$, $x \in D$. It follows from Schwarz's inequality, symmetry and Fubini's theorem that

$$\|Q_D^t f\|^2 = \int (Q_D^t f(x))^2\, dx \le \int \left(\int Q_D(t, x, y) f^2(y)\, dy \right) dx \le \int f^2(y)\, dy = \|f\|^2.$$

Thus Q_D^t is a bounded linear operator on $\mathscr{L}^2$ of norm at most one. It follows easily from Fubini's theorem and the boundedness and symmetry of $Q_D(t, x, y)$ that Q_D^t is a symmetric operator. Since $Q_D(t, x, y)$ is square integrable over

$(x, y) \in D \times D$, Q_D^t is completely continuous (see page 179 of Riesz and Sz.-Nagy [1]). It follows from the semigroup property of $Q_D(t, x, y)$ that the operators Q_D^t on $\mathscr{L}^2$ satisfy the semigroup property $Q_D^{s+t} = Q_D^s Q_D^t$ for $s, t > 0$.

It will now be shown that Q_D^t is positive definite. Note first that if f is a continuous function with compact support in D, then $\lim_{t \to 0} Q_D^t f = f$. Since the collection of all such functions is dense in $\mathscr{L}^2$ it follows that $\lim_{t \to 0} Q_D^t f = f$ for all $f \in \mathscr{L}^2$. Suppose now that $f \in \mathscr{L}^2$ and that $Q_D^t f = 0$ for some $t > 0$. Then $\|Q_D^{t/2} f\|^2 = \int f(x) Q_D^t f(x)\, dx = 0$ and hence $Q_D^{t/2} f = 0$. By induction $Q_D^{t/2^n} f = 0$ for all $n \geq 1$ and hence $f = \lim_n Q_D^{t/2^n} f = 0$. Thus $Q_D^t f$ is positive definite for all $t > 0$.

If μ is a real number and φ is a nonzero element of $\mathscr{L}^2$ such that $Q_D^t \varphi = \mu\varphi$, then φ is said to be an *eigenfunction* of Q_D^t corresponding to the *eigenvalue* μ. It follows from the boundedness and continuity of $Q_D(t, x, y)$ that the eigenfunctions can be taken to be bounded continuous functions on D. Since Q_D^t is positive definite and symmetric all its eigenvalues are positive and eigenfunctions corresponding to distinct eigenvalues are orthogonal.

Let μ be an eigenvalue of Q_D^1 and let φ be a corresponding eigenfunction. Set $\lambda = -\log \mu$. Then for $t > 0$, φ is an eigenfunction of Q_D^t corresponding to the eigenvalue $e^{-\lambda t}$. For observe that $0 = (Q_D^1 - \mu)\varphi = (Q_D^{1/2} + \mu^{1/2})(Q_D^{1/2} - \mu^{1/2})\varphi$. Set $\psi = (Q_D^{1/2} - \mu^{1/2})\varphi$. Then $(Q_D^{1/2} + \mu^{1/2})\psi = 0$, so

$$0 = \|(Q_D^{1/2} + \mu^{1/2})\psi\|^2 = \|Q_D^{1/2}\psi\|^2 + \mu\|\psi\|^2 + 2\mu^{1/2}\|Q_D^{1/4}\psi\|^2.$$

Consequently $\psi = 0$ and hence $Q_D^{1/2}\varphi = \mu^{1/2}\varphi$. By induction

$$Q_D^t \varphi = e^{-\lambda t} \varphi \tag{3}$$

holds for $t > 0$ of the form 2^{-n}, $n \geq 1$. By the semigroup property (3) holds for t of the form $m2^{-n}$ for $m, n \geq 1$. Thus by continuity (3) holds for all $t > 0$ as desired.

Since Q_D^1 is a completely continuous positive definite symmetric operator there is a sequence $\{\mu_n\}$ of eigenvalues of Q_D^1 and a corresponding sequence $\{\varphi_n\}$ of eigenfunctions such that the μ_n's are positive and nonincreasing in n and $\{\varphi_n\}$ is a complete orthonormal system (see page 234 of Riesz and Sz.-Nagy [1]). Also $\mu_1 \leq 1$ since Q_D^1 has norm at most one. Set $\lambda_n = -\log \mu_n$. Then λ_n is nonnegative and nondecreasing in n. The sequence $\{e^{-\lambda_n t}\}$ represents all the eigenvalues of Q_D^t with the proper multiplicities..

Now $Q_D(t, x, \cdot) \in \mathscr{L}^2$ by (1). It follows by expanding this function in terms of $\{\varphi_n\}$ that

$$\sum_n e^{-2\lambda_n t} \varphi_n^2(x) = \sum_n \left(\int Q_D(t, x, y) \varphi_n(y)\, dy \right)^2 = \int Q_D^2(t, x, y)\, dy = Q_D(2t, x, x)$$

and hence that

(4) $$\sum_n e^{-\lambda_n t}\varphi_n^2(x) = Q_D(t, x, x), \qquad x \in D.$$

Consequently

(5) $$\sum_n e^{-\lambda_n t} = \int Q_D(t, x, x)\, dx.$$

Since $|D| > 0$ and $Q_D(t, x, x) \le p(t, 0)$, which approaches zero as $t \to \infty$, it follows from (5) that $\lambda_1 > 0$. By (4), for $\varepsilon > 0$

(6) $$\varphi_n^2(x) \le e^{\lambda_n \varepsilon} p(\varepsilon, 0), \qquad x \in D.$$

Observe that $\{\varphi_m \varphi_n\}$ is a complete orthonormal system in $\mathscr{L}^2(D \times D)$. By (2), $Q_D(t, x, y)$ can be expanded in terms of this system. It is easily seen that

(7) $$Q_D(t, x, y) = \sum_n e^{-\lambda_n t}\varphi_n(x)\varphi_n(y)$$

in the sense of convergence in $\mathscr{L}^2(D \times D)$. It follows from (6) and Schwarz's inequality that for $0 < \varepsilon < t$

(8) $$\begin{aligned}\left(\sum_{n_0}^{\infty} e^{-\lambda_n t}|\varphi_n(x)\varphi_n(y)|\right)^2 &\le \sum_{n_0}^{\infty} e^{-\lambda_n t}\varphi_n^2(x) \sum_{n_0}^{\infty} e^{-\lambda_n t}\varphi_n^2(y) \\ &\le p^2(\varepsilon, 0)\left(\sum_{n_0}^{\infty} e^{-\lambda_n(t-\varepsilon)}\right)^2,\end{aligned}$$

which by (5) approaches zero as $n_0 \to \infty$. Thus the series in (7) converges absolutely and uniformly on $D \times D$.

It will now be shown that

(9) $$E_x T_D = \int_D G_D(x, y)\, dy \le 2 + \frac{(|D|p(1, 0))^2}{\lambda_1} < \infty, \qquad x \in D.$$

To verify (9) observe first that

$$e^{-\lambda_n}|\varphi_n(x)| \le \int Q(1, x, y)|\varphi_n(y)|\, dy \le p(1, 0)\int |\varphi_n(y)|\, dy.$$

By Schwarz's inequality

$$\left(\int |\varphi_n(y)|\, dy\right)^2 \le |D| \int \varphi_n^2(y)\, dy = |D|.$$

By (7) and the last two observations

$$\int Q_D(t, x, y)\, dy \le |D|p(1, 0)\sum_n e^{-\lambda_n(t-1)}, \qquad t > 1.$$

Consequently

$$\begin{aligned}\int_D G_D(x,y)\,dy &\leq 2 + \int_2^\infty \left(\int Q_D(t,x,y)\,dt\right)dy \\ &\leq 2 + |D|p(1,0)\sum_n \frac{e^{-\lambda_n}}{\lambda_n} \\ &\leq 2 + \frac{|D|p(1,0)}{\lambda_1}\sum_n e^{-\lambda_n}.\end{aligned}$$

Equation (9) now follows easily from (5).

Observe that for $x \in D$

$$\begin{aligned}\int G_D(x,y)\varphi_n(y)\,dy &= \int_0^\infty \left(\int Q_D(t,x,y)\varphi_n(y)\,dy\right)dt \\ &= \int_0^\infty e^{-\lambda_n t}\varphi_n(x)\,dt \\ &= \lambda_n^{-1}\varphi_n(x).\end{aligned}$$

Thus φ_n is an eigenfunction of G_D corresponding to the eigenvalue λ_n^{-1}. Suppose conversely that φ is an eigenfunction of G_D corresponding to an eigenvalue μ. Let $n \geq 1$. Then by (9)

$$\begin{aligned}&\iint G_D(x,y)|\varphi(x)|\,|\varphi_n(y)|\,dx\,dy \\ &\qquad \leq (\sup_y |\varphi_n(y)|)\left(\sup_x \int G_D(x,y)\,dy\right)\int |\varphi(x)|\,dx < \infty,\end{aligned}$$

so by Fubini's theorem

$$(\lambda_n^{-1} - \mu)\int \varphi(x)\varphi_n(x)\,dx = \int \varphi(x)G_D\varphi_n(x)\,dx - \int G_D\varphi(x)\varphi_n(x)\,dx = 0.$$

Since $\{\varphi_n\}$ is complete there is a n_0 such that $\int \varphi(x)\varphi_{n_0}(x)\,dx \neq 0$. Thus $\mu = \lambda_{n_0}^{-1}$. If $\lambda_n \neq \lambda_{n_0}$, then $(\lambda_n^{-1} - \mu) \neq 0$ and hence $\int \varphi(x)\varphi_n(x)\,dx = 0$. Consequently φ is a linear combination of the eigenfunctions φ_n corresponding to the eigenvalue λ_{n_0}.

Since $G_D\varphi_n = \lambda_n^{-1}\varphi_n$ it follows from Theorem 6.6 that φ_n is continuously differentiable and hence Hölder continuous on D. Thus by another application of Theorem 6.6, φ_n is twice continuously differentiable on D and

$$\tfrac{1}{2}\Delta\varphi_n = \tfrac{1}{2}\lambda_n \Delta G_D\varphi_n = -\lambda_n\varphi_n \qquad \text{on } D.$$

Now φ_n is bounded and $Q_D^t\varphi_n = e^{-\lambda_n t}\varphi_n$ for $t > 0$ so by Theorem 2.4.3, φ_n has boundary value zero on $\partial D \cap (D^c)^r$.

A function φ on D is said to be an eigenfunction of $\Delta/2$ corresponding to the eigenvalue λ if φ is bounded, twice continuously differentiable, and not identically zero on D, φ has boundary value zero on $\partial D \cap (D^c)^r$, and $\frac{1}{2}\Delta\varphi = \lambda\varphi$ on D. By the results of the previous paragraph φ_n is an eigenfunction of $\Delta/2$ corresponding to the eigenvalue $-\lambda_n$. Suppose conversely that φ is an eigenfunction of $\Delta/2$ corresponding to the eigenvalue $-\lambda$. Now $P_x(T_D < \infty) = 1$ for $x \in D$ by (9) so D^c is recurrent. It follows from Theorem 2.10 that $\lambda \neq 0$. Note that $\Delta G_D(\lambda\varphi) = -2\lambda\varphi = \Delta\varphi$ by Theorem 6.6. Thus $G_D(\lambda\varphi) - \varphi$ is harmonic on D. By Theorem 6.7 it has boundary value zero on $\partial D \cap (D^c)^r$. Consequently $G_D\lambda\varphi - \varphi = 0$ on D by Theorem 2.10. Thus φ is an eigenfunction of G_D corresponding to the eigenvalue λ^{-1}. By the argument two paragraphs above, $\lambda = \lambda_{n_0}$ for some positive integer n_0 and φ is a linear combination of the eigenfunctions φ_n corresponding to the eigenvalue λ_{n_0}.

Suppose, for example that $d = 1$ and $D = (a, b)$ is a bounded open interval. A function φ on (a, b) is an eigenfunction of $\Delta/2$ corresponding to the eigenvalue $-\lambda < 0$ if φ is twice continuously differentiable on (a, b), $\varphi'' = -2\lambda\varphi$ on (a, b), and φ has boundary value zero at a and b. The general solution to $\varphi'' = -2\lambda\varphi$ on (a, b) is given by

$$\varphi(x) = A\sin(\sqrt{2\lambda}(x-a)) + B\cos(\sqrt{2\lambda}(x-a)), \qquad a < x < b.$$

This solution has boundary value zero at a if and only if $B = 0$. Suppose $B = 0$ but φ is not identically zero. Then $A \neq 0$. The solution has boundary value zero at b if and only if $\lambda = \lambda_n = n^2\pi^2/2(b-a)^2$ for some positive integer n. The solution φ_n corresponding to the eigenvalue $-\lambda_n$ is normalized (i.e., $\|\varphi_n\| = 1$) if and only if $A = \sqrt{2/(b-a)}$. Therefore the eigenvalues and corresponding normalized eigenfunctions are given by $-\lambda_n = n^2\pi^2/2(b-a)^2$ and

$$\varphi_n(x) = \sqrt{2/(b-a)}\sin\left(n\pi\frac{x-a}{b-a}\right), \qquad a < x < b.$$

Consequently

$$\begin{aligned} Q_D(t, x, y) &= \sum_n e^{-\lambda_n t}\varphi_n(x)\varphi_n(y) \\ &= \frac{2}{b-a}\sum_n \exp\left[-\frac{n^2\pi^2 t}{2(b-a)^2}\right]\sin\left(n\pi\frac{x-a}{b-a}\right)\sin\left(n\pi\frac{y-a}{b-n}\right) \end{aligned}$$

for $t > 0$ and $a < x, y < b$. This agrees with the second formula for $Q_D(t, x, y)$ given in Proposition 2.8.2.

A formula due to Weyl for the asymptotic distribution of $\{\lambda_n\}$ will now be obtained. A corresponding result for $\{\varphi_n(x)\}$ due to Carleman will also be obtained. The connection with Brownian motion was discovered by Kac [1].

Theorem 7.1. *Let $x \in D$. Then*

$$\lim_{\lambda \to \infty} \lambda^{-d/2} \sum_{\lambda_n \le \lambda} \varphi_n^2(x) = \frac{1}{(2\pi)^{d/2}\Gamma(d/2+1)}.$$

Also

$$\lim_{\lambda \to \infty} \lambda^{-d/2} \sum_{\lambda_n \le \lambda} 1 = \frac{|D|}{(2\pi)^{d/2}\Gamma(d/2+1)}.$$

Proof. Choose $x \in D$. By Proposition 2.4.1

$$\lim_{t \to 0} \frac{Q_D(t,x,x)}{(2\pi t)^{-d/2}} = \lim_{t \to 0} \frac{Q_D(t,x,x)}{p(t,0)} = 1.$$

The first conclusion now follows from (4) and Karamata's Tauberian theorem (Theorem 2, page 445 of Feller [1]). The second conclusion follows from (5) and a similar argument.

Theorem 7.2. *Let $m \ge 1$ be such that $\lambda_1 = \cdots = \lambda_m < \lambda_{m+1}$ and let $A \subset D$. Then*

(10) $$\lim_{t \to \infty} e^{\lambda_1 t} P_x(T_D > t, X(t) \in A) = \int_A \left(\sum_{n=1}^{m} \varphi_n(x)\varphi_n(y) \right) dy, \qquad x \in D.$$

In particular

(11) $$\lim_{t \to \infty} e^{\lambda_1 t} P_x(T_D > t) = \int_D \left(\sum_{n=1}^{m} \varphi_n(x)\varphi_n(y) \right) dy, \qquad x \in D.$$

The right-hand side of (11) *is positive for some $x \in D$ and if D is connected it is positive for all $x \in D$.*

Proof. Since D has finite Lebesgue measure, so does A. By (5), (7), and (8)

$$\begin{aligned}\lim_{t \to \infty} e^{\lambda_1 t} P_x(T_D > t, X(t) \in A) &= \lim_{t \to \infty} e^{\lambda_1 t} \int_A Q_D(t,x,y)\,dy \\ &= \lim_{t \to \infty} \int_A \left[\sum_{n=1}^{m} \varphi_j(x)\varphi_j(y) + \sum_{n=m+1}^{\infty} e^{-(\lambda_n - \lambda_1)t} \varphi_n(x)\varphi_n(y) \right] dy \\ &= \int_A \left(\sum_{n=1}^{m} \varphi_n(x)\varphi_n(y) \right) dy.\end{aligned}$$

Thus (10) and (11) hold. Since each φ_n is continuous it follows from (10) that

(12) $$\sum_{n=1}^{m} \varphi_n(x)\varphi_n(y) \ge 0, \qquad x, y \in D.$$

If the left-hand side of (12) were identically zero, then $\sum_1^m \varphi_n^2(x)$ would be identically zero, which would contradict the fact that each φ_n has unit norm. Thus the left-hand side of (12) is somewhere positive and hence the right-hand side of (11) is somewhere positive.

Suppose D is connected. Let $f(x) = \sum_1^m \varphi_n(x) \int_D \varphi_n(y)\,dy$ denote the right-hand side of (11) for $x \in D$. Then f is a nonnegative continuous function on D which is not identically zero. It remains to show that f is positive on D. Choose $x \in D$ and let $t > 0$. Then $Q_D(t, x, \cdot) > 0$ on D and hence

$$0 < Q_D^t f(x) = \sum_1^m e^{-\lambda_1 t} \varphi_n(x) \left(\int_D \varphi_n(y)\,dy \right) = e^{-\lambda_1 t} f(x)$$

as desired.

Chapter 5

Superharmonic and Excessive Functions

Throughout this chapter D is a nonempty open subset of $\mathbb{R}^d$. Superharmonic and excessive functions on D are defined and studied. In particular it is shown that a nonnegative function on D is superharmonic if and only if it is excessive. This important result is the key to the probabilistic treatment of superharmonic functions. The connection between Brownian motion and superharmonic functions was established by Doob [1]. Excessive functions were defined in the context of a Markov process and their fundamental properties were developed by Hunt [2] (see Blumenthal and Getoor [1]).

1. Properties of Superharmonic and Excessive Functions

A function f on D is said to be *superharmonic* on D if f is lower semicontinuous on D, f is not identically infinite on any component of D, and for every $x \in D$ there is a $\delta > 0$ such that $B_\delta(x) \subset D$ and $f(x) \geq \int f(y)\sigma_r(x, dy)$ for $0 < r \leq \delta$. If f is superharmonic on D, then $\underline{\lim}_{y \to x} f(y) = f(x)$ for $x \in D$. A function is superharmonic on D if and only if it is superharmonic on each component of D. If f_1 and f_2 are superharmonic on D, so are $f_1 \wedge f_2$ and $c_1 f_1 + c_2 f_2$ for $c_1, c_2 \geq 0$.

A function f on D is said to be *subharmonic* on D if $-f$ is superharmonic on D. By Proposition 4.6.2 a function f is harmonic on D if and only if it is both subharmonic and superharmonic on D.

Proposition 1.1 *Suppose f_n is superharmonic on D for $n \geq 1$, $f_n \uparrow f$, and f is not identically infinite on any component of D. Then f is superharmonic on D.*

Proof. This result follows easily from the definition of superharmonicity and the monotone convergence theorem.

A function ψ on an interval $I \subset \mathbb{R}$ is said to be *concave* on I if ψ is finite on I and

$$\psi(\lambda x_1 + (1 - \lambda)x_2) \geq \lambda\psi(x_1) + (1 - \lambda)\psi(x_2), \qquad x_1, x_2 \in I \quad \text{and} \quad 0 < \lambda < 1.$$

Such a function is necessarily continuous on $\mathring{I}$. Suppose ψ is concave on an interval I which is unbounded above. Then ψ is monotone sufficiently far to the right, so it can be extended to a function on $I \cup \{\infty\}$ by setting $\psi(\infty) = \lim_{x \to \infty} \psi(x)$.

Proposition 1.2. *Let f be superharmonic on D. Let ψ be concave on an interval I and suppose that either I contains the range of f or I is unbounded from above and $I \cup \{\infty\}$ contains the range of f. If f is harmonic on D or ψ is nondecreasing on I, then $\psi \circ f$ is superharmonic on D.*

The proof, which depends on Jensen's inequality, is left to the reader.

It will now be shown that a superharmonic function on a connected open set having an interior minimum is necessarily constant.

Proposition 1.3. *Let D be connected and let f be a superharmonic function on D such that $f \geq f(x_0)$ on D for some $x_0 \in D$. Then f is constant on D.*

Proof. Set $A = \{x \in D : f(x) = f(x_0)\}$. Choose $x \in A$. If r is positive and sufficiently small, then $f(x) \geq \int f(y)\sigma_r(x, dy)$ and hence $f = f(x_0)$ almost surely with respect to $\sigma_r(x, \cdot)$ on $S_r(x)$. By lower semicontinuity $f = f(x_0)$ on $S_r(x)$. Consequently A is open. It follows from lower semicontinuity that $D \backslash A$ is open. Thus $A = D$ by connectivity and hence f is constant on D as desired.

Let f be defined on D and bounded below on compact subsets of D. The *lower regularization* $\underline{f}$ of f is the function defined on $\bar{D}$ by

$$\begin{aligned} \underline{f}(x) &= \varliminf_{y \to x} f(y), && x \in \partial D, \\ &= f(x) \wedge \varliminf_{y \to x} f(y), && x \in D. \end{aligned}$$

Note that $\underline{f} \le f$ on D and that $\underline{f}$ is lower semicontinuous on D. The function f is lower semicontinuous on D if and only if $\underline{f} = f$ on D. In particular if f is superharmonic on D, then $\underline{f} = f$ on D. If D is unbounded, set

$$\underline{f}(\infty) = \varliminf_{r \to \infty} [f(y) : y \in D \text{ and } \|y\| \ge r].$$

Proposition 1.4. *Suppose D is bounded. If f is superharmonic on D, then $\inf_{x \in \partial D} \underline{f}(x) = \inf_{x \in D} f(x)$.*

Proof. Choose $x_0 \in \bar{D}$ such that $\underline{f}(x_0) = \inf_{x \in D} f(x)$. If $x_0 \in \partial D$, the desired result holds. Otherwise f is constant on the component of D containing x_0 by Proposition 1.3, so the desired result again holds.

Proposition 1.5. *Suppose D is bounded and let f be superharmonic on D. If h is harmonic on D and continuous on $\bar{D}$ and if $\underline{f} \ge h$ on ∂D, then $f \ge h$ on D.*

Proof. Since $f - h$ is superharmonic on D and $\underline{f - h} = \underline{f} - h$, the desired result follows from Proposition 1.4.

An increasing limit of lower semicontinuous functions on D is lower semicontinuous on D. A converse to this result will now be given.

Proposition 1.6. *Let f be lower semicontinuous on D and let B be a compact subset of D such that f is not identically infinite on B. Define f_n on B by $f_n(x) = \inf_{z \in B}(f(z) + n\|x - z\|)$. Then f_n is bounded and continuous on B and $f_n \uparrow f$ on B.*

Proof. This result follows easily from the definition of lower semicontinuity. The details are left to the reader.

Proposition 1.5 will now be used to give an alternative characterization of superharmonic functions.

Proposition 1.7. *Let f be a lower semicontinuous function on D which is not identically infinite on any component of D. Then f is superharmonic on D if and only if it satisfies the following property: if D_0 is a relatively compact open subset of D, h is harmonic on D_0 and continuous on $\bar{D}_0$, and $\underline{f} \ge h$ on ∂D_0, then $f \ge h$ on D_0.*

Proof. It follows easily from Proposition 1.5 that superharmonic functions satisfy the indicated property. Suppose conversely that the indicated property holds. Let $x \in D$ and let $r > 0$ be such that $B_r(x) \subset D$. By Proposition 1.6 there is an increasing sequence of bounded continuous functions

f_n on $B_r(x)$ such that $f_n \uparrow f$ on $B_r(x)$. Set $h_n = H_{\mathring{B}_r(x)} f_n$ on $\mathring{B}_r(x)$ and $h_n = f_n$ on $S_r(x)$. Then h_n is harmonic on $\mathring{B}_r(x)$ and continuous on $B_r(x)$, so $f \geq h_n$ on $\mathring{B}_r(x)$. In particular, $f(x) \geq h_n(x) = H_{\mathring{B}_r(x)} f_n(x)$, so by the monotone convergence theorem $f(x) \geq H_{\mathring{B}_r(x)} f(x) = \int f(y)\sigma_r(x, dy)$. Thus f is superharmonic on D as desired.

It follows from Proposition 1.7 that a superharmonic function on an open subset D of the real line $\mathbb{R}$ is concave on each of the component intervals of D and therefore continuous on D.

Proposition 1.8. *Let f be superharmonic on D and let D_0 be a relatively compact open subset of D. Then $f \geq H_{D_0} f$ on D_0.*

Proof. By Proposition 4.2.5 there exist relatively compact regular open subsets D_n of D_0 such that $D_n \uparrow D_0$. By Proposition 1.6 there exist bounded continuous functions f_m on $\bar{D}_0$ such that $f_m \uparrow f$ on $\bar{D}_0$. Now $H_{D_n} f_m$ is harmonic on D_n and by Proposition 4.2.1 this function extends to a continuous function on $\bar{D}_n$ which equals f_m on ∂D_n. Consequently $f \geq H_{D_n} f_m$ on D_n by Proposition 1.7. Thus $f \geq H_{D_n} f$ on D_n by the monotone convergence theorem. In other words, $f \geq E_{\cdot} f(X(T_{D_n}))$ on D_n. Now $P_{\cdot}(T_{D_n} \uparrow T_{D_0}) = 1$ on D_0 by Proposition 2.3.8, so

$$\underline{\lim}_n f(X(T_{D_n})) \geq f(X(T_{D_0})) \quad \text{a.s. } (P_x), \qquad x \in D_0.$$

Since f is bounded below on $\bar{D}_0$ it now follows from Fatou's lemma that $f \geq E_{\cdot} f(X(T_{D_0})) = H_{D_0} f$ on D_0.

Proposition 1.9. *Let f be superharmonic and bounded below on D. If D^c is recurrent, then $f \geq H_D \underline{f}$ on D. If D^c is transient, then $\underline{f}(\infty) < \infty$ and $f \geq H_D \underline{f} + P_{\cdot}(T_D = \infty)\underline{f}(\infty)$ on D.*

Proof. Let D_n be relatively compact open subsets of D such that $D_n \uparrow D$. Then $f \geq H_{D_n} f$ on D_n by Proposition 1.8. Choose $x \in D$. Then $P_x(T_{D_n} \uparrow T_D) = 1$ by Proposition 2.3.8. Thus by Fatou's lemma

$$f(x) \geq E_x\left(\underline{\lim}_n f(X(T_{D_n}))\right) \geq H_D \underline{f}(x) + E_x\left(\underline{\lim}_n f(X(T_{D_n})); T_D = \infty\right).$$

The desired conclusion therefore holds if D^c is recurrent. Suppose now that D^c is transient and $d \geq 3$. Then $T_{D_n} \to \infty$ and hence $X(T_{D_n}) \to \infty$ a.s. (P_x) on $\{T_D = \infty\}$. Consequently

$$E_x\left(\underline{\lim}_n f(X(T_{D_n})); T_D = \infty\right) \geq P_x(T_D = \infty)\underline{f}(\infty),$$

so $f \geq H_D \underline{f} + P_{\cdot}(T_D = \infty)\underline{f}(\infty)$ on D. By Proposition 4.2.9, $\underline{f}(\infty) < \infty$. Suppose finally that D^c is transient and $d \leq 2$. Then D^c is polar by Proposition 2.2.10. The desired conclusion now simplifies to $f \geq \underline{f}(\infty)$ on D. To verify this inequality set $D_n = \{x \in D: \|x\| < n\}$. Then $f \geq H_{D_n} f = E_{\cdot} f(X(T_{D_n}))$ on D_n by what has already been shown. Since $\|X(T_{D_n})\| = n$ a.s. (P_x) for $x \in D_n$, it follows from Fatou's lemma that $f \geq \underline{f}(\infty)$ on D as desired. This completes the proof of the proposition.

Proposition 1.10. *Let f be superharmonic on D. Then f is locally integrable on D.*

Proof. Note first that if $x \in D$, $f(x) < \infty$, $\delta > 0$ and $B_\delta(x) \subset D$, then f is integrable on $B_\delta(x)$. For f is bounded below on $B_\delta(x)$ and $f(x) \geq \int f(y)\sigma_r(x, dy)$, $0 < r \leq \delta$, so the desired result follows from Proposition 2.1.1.

In proving that f is locally integrable on D it can be assumed without loss of generality that D is connected. Let D_0 be the set of points $x \in D$ such that f is integrable on some neighborhood of x. Since f is not identically infinite on D it follows from the first paragraph of this proof that D_0 is nonempty. Clearly D_0 is open. It is also true that D_0 is closed relative to D. For let $x \in \bar{D}_0 \cap D$ and let $\delta > 0$ be such that $B_{2\delta}(x) \subset D$. There is a $y \in D$ such that $\|y - x\| \leq \delta/2$ and $f(y) < \infty$. Then $B_\delta(y) \subset D$, so it follows from the first paragraph of the proof that f is integrable on $B_\delta(y)$ and hence also on $B_{\delta/2}(x)$. Consequently $x \in D_0$. Thus D_0 is a nonempty open relatively closed subset of the connected set D and hence $D_0 = D$. Therefore f is locally integrable on D as desired.

It follows from the last proposition that if f is superharmonic on D it cannot be identically infinite on any nonempty open subset of D, so it is superharmonic on every nonempty open subset of D. Conversely if f is superharmonic on some open neighborhood of x for every $x \in D$ it is superharmonic on D. Thus the property of being superharmonic is a purely local property.

Recall the definition of the generalized Laplacian $\tilde{\Delta}$ given in Section 4.6.

Proposition 1.11. *Suppose f is lower semicontinuous on D and not identically infinite on any component of D and that $\tilde{\Delta} f$ exists on D. Then f is superharmonic on D if and only if $\tilde{\Delta} f \leq 0$ on D.*

Proof. Clearly if f is superharmonic on D, then $\tilde{\Delta} f \leq 0$ on D. Suppose conversely that $\tilde{\Delta} f \leq 0$ on D. Define q on D by $q(x) = \|x\|^2$. Then $\tilde{\Delta} q = \Delta q = 2d$ on D. Set $f_n = f - q/n$ on D for $n \geq 1$. Then $\tilde{\Delta} f_n = \tilde{\Delta} f - \tilde{\Delta} q/n < 0$ on D. Thus for each $x \in D$ there is a $\delta > 0$ such that $B_\delta(x) \subset D$ and $f_n(x) \geq$

$\int f_n(y)\sigma_r(x, dy)$ for $0 < r \le \delta$. Consequently f_n is superharmonic on D. Since $f_n \uparrow f$ on D, f is superharmonic on D by Proposition 1.1. This completes the proof of the proposition.

Proposition 1.12. *Let f be twice continuously differentiable on D. Then f is superharmonic on D if and only if $\Delta f \le 0$ on D.*

Proof. Since $\tilde{\Delta} f = \Delta f$ on D by Theorem 4.6.1 the desired result follows from Proposition 1.11.

Although superharmonic functions can be discontinuous (see Theorem 2.7 below) the next result shows that they can be approximated pointwise by infinitely differentiable superharmonic functions.

Proposition 1.13. *Let f be superharmonic on D and let D_n be relatively compact open subsets of D such that $D_n \uparrow D$. Then there exist bounded infinitely differentiable superharmonic functions f_n on D_n such that $f_n \ge f_m$ on D_m for $n \ge m$ and $\lim_n f_n = f$ on D. If f is nonnegative, each f_n can be made nonnegative.*

Proof. Let ψ denote a nonnegative infinitely differentiable function on $[0, \infty)$ such that $\psi = 0$ on $[1, \infty)$ and $\int \psi(\|y\|^2)\,dy = 1$. Choose $\varepsilon_n > 0$ such that $\varepsilon_n \downarrow 0$ and $\varepsilon_n < \inf_{x \in D_n} d(x, D^c)$. Define f_n on D_n by

$$f_n(x) = \int f(x + \varepsilon_n y)\psi(\|y\|^2)\,dy = \varepsilon_n^{-d} \int f(y)\psi(\varepsilon_n^{-2}\|y - x\|^2)\,dy.$$

If f is nonnegative on D, f_n is nonnegative on D_n. Since f is locally integrable on D, f_n is bounded and infinitely differentiable on D_n. For $x \in D_n$ and $0 < r < \varepsilon_n$

$$\begin{aligned}\int f_n(z)\sigma_r(x, dz) &= \int \psi(\|y\|^2)\left(\int f(z + \varepsilon_n y)\sigma_r(x, dz)\right)dy\\ &= \int \psi(\|y\|^2)\left(\int f(z)\sigma_r(x + \varepsilon_n y, dz)\right)dy\\ &\le \int \psi(\|y\|^2) f(x + \varepsilon_n y)\,dy = f_n(x).\end{aligned}$$

Thus f_n is superharmonic on D_n. Set $s_d = 2\pi^{d/2}/\Gamma(d/2)$. By Proposition 2.1.1

$$\begin{aligned}f_n(x) &= s_d \int_0^1 \psi(r^2)\left(\int f(x + \varepsilon_n y)\sigma_r(dy)\right) r^{d-1}\,dr\\ &= s_d \int_0^1 \psi(r^2)\left(\int f(z)\sigma_{r\varepsilon_n}(x, dz)\right) r^{d-1}\,dr.\end{aligned}$$

Thus

$$f_n(x) \le f(x) s_d \int_0^1 \psi(r^2) r^{d-1}\, dr = f(x) \int \psi(\|y\|^2)\, dy = f(x).$$

Let $n \ge m$ and choose $x \in D_m$. If $\varepsilon_m = \varepsilon_n$, then $f_m(x) = f_n(x)$. Suppose $\varepsilon_m > \varepsilon_n$. By Proposition 2.2.21 and Theorem 2.6.7

$$\sigma_{r\varepsilon_m}(x, \cdot) = \int \sigma_{r\varepsilon_n}(x, dy) H_{\mathring{B}_{r\varepsilon_m}(x)}(y, \cdot).$$

Thus by Proposition 1.8

$$f_m(x) = s_d \int_0^1 \psi(r^2) \left(\int \sigma_{r\varepsilon_n}(x, dy) \int H_{\mathring{B}_{r\varepsilon_m}(x)}(y, dz) f(z) \right) r^{d-1}\, dr$$

$$\le s_d \int_0^1 \psi(r^2) \left(\int f(y) \sigma_{r\varepsilon_n}(x, dy) \right) r^{d-1}\, dr = f_n(x).$$

Hence $f_m \le f_n \le f$ on D_m. By the lower semicontinuity of f, $\underline{\lim}_n f_n \ge f$ on D. Therefore $\lim_n f_n = f$ on D, which completes the proof of the proposition.

Theorem 1.14. *Let f_1 and f_2 be superharmonic on D. If $f_1 \le f_2$ a.e. on D, then $f_1 \le f_2$ on D and if $f_1 = f_2$ a.e. on D, then $f_1 = f_2$ on D.*

Proof. Let D_n be relatively compact open subsets of D such that $D_n \uparrow D$. Let f_{n1} be the approximation to f_1 on D_n determined in the proof of Proposition 1.13 and let f_{n2} be the corresponding approximation to f_2. Then $f_{n1} \le f_{n2}$ on D_n and hence $f_1 = \lim_n f_{n1} \le \lim_n f_{n2} = f_2$ on D. If $f_1 = f_2$ a.e. on D, then $f_1 \le f_2$ and $f_2 \le f_1$ on D and hence $f_1 = f_2$ on D.

A generalization of Theorem 4.2.18 will now be considered. Let D_0 be an open subset of D such that $D \backslash D_0$ is polar and let f be superharmonic on D_0. Consider the problem of extending f to a superharmonic function on D. Since $D \backslash D_0$ is polar it has Lebesgue measure zero, so by Theorem 1.14 there is at most one such extension. If there is an extension it must be bounded below on compacts. Thus a necessary condition for an extension to exist is that f be bounded below on $D_0 \cap C$ for every compact subset C of D. The next result shows that this condition is also sufficient. The necessity of the original condition that $D \backslash D_0$ be polar will be considered in Theorem 4.11.

Theorem 1.15. *Let D_0 be an open subset of D such that $D \backslash D_0$ is polar. Let f be superharmonic on D_0 and bounded below on $C \cap D_0$ for every compact subset C of D. Then $\underline{f}$ is the unique superharmonic function on D which agrees with f on D_0.*

Proof. Now $\underline{f}$ is lower semicontinuous on D and $\underline{f} = f$ on D_0. Since f is locally integrable on D_0 it is finite almost everywhere on D_0 and hence $\underline{f}$ is finite almost everywhere on D. Thus $\underline{f}$ is not identically infinite on any component of D. The characterization of superharmonic functions given in Proposition 1.7 will now be used to verify that $\underline{f}$ is superharmonic on D.

Let D_1 be a relatively compact open subset of D. Let h be harmonic on D_1 and continuous on $\bar{D}_1$ and such that $\underline{f} \geq h$ on ∂D_1. It must be shown that $\underline{f} \geq h$ on D_1. To do so it suffices to show that $f \geq h$ on $D_2 = D_1 \cap D_0$. Now f is superharmonic and bounded below on D_2. Thus $f \geq H_{D_2}\underline{f}$ on D_2 by Proposition 1.9. Note that $H_{D_2}(x, \cdot) = H_{D_1}(x, \cdot)$ for $x \in D_2$ since $D_1 \backslash D_2$ is polar. Thus $f \geq H_{D_1}\underline{f} \geq H_{D_1}h$ on D_2. Since $h = H_{D_1}h$ on D_1 by Proposition 4.2.6, $f \geq h$ on D_2 as desired. This completes the proof of the theorem.

A function f on D is said to be *superinvariant* on D if $f \geq 0$ on D, f is not identically infinite on any component of D and $Q_D^t f \leq f$ on D for $t > 0$. It follows easily from the semigroup property of Q_D^t, $t \geq 0$, that if f is superinvariant on D, then $Q_D^t f \uparrow$ as $t \downarrow$ and hence that $\lim_{t \to 0} Q_D^t f$ exists. A function f on D is called *excessive* on D if f is superinvariant on D and $\lim_{t \to 0} Q_D^t f = f$ on D.

Theorem 1.16. *Let f be superinvariant on D. Then $Q_D^t f$ is excessive on D for $t > 0$ and $Q_D^t f \uparrow \underline{f}$ as $t \downarrow 0$. The function $\underline{f}$ is excessive on D and $\underline{f} = f$ a.e. on D. Finally, f is excessive on D if and only if it is lower semicontinuous on D.*

Proof. Now $Q_D^t f \leq f$ (on D) and $Q_D^s Q_D^t f = Q_D^t Q_D^s f \leq Q_D^t f$ for $s > 0$ so $Q_D^t f$ is superinvariant for $t > 0$. Also $Q_D(t, \cdot, y)$ is continuous on D for $t > 0$ and $y \in D$ by Theorem 2.4.3, so by Fatou's lemma $Q_D^t f = \int Q_D(t, \cdot, y) f(y)\, dy$ is lower semicontinuous on D.

It will now be shown that $Q_D^t f \uparrow \underline{f}$ as $t \downarrow 0$. There is a nonnegative function f_1 on D such that $Q_D^t f \uparrow f_1$ as $t \downarrow 0$. Since f_1 is an increasing limit of lower semicontinuous functions, it is lower semicontinuous. Clearly $f_1 \leq f$ and hence $f_1 \leq \underline{f}$. It follows easily from Fatou's lemma and continuity of paths that

$$f_1 = \lim_{t \to 0} Q_D^t f = \lim_{t \to 0} E_{\cdot}(f(X(t)); T_D > t) \geq \underline{f} P_{\cdot}(T_D > 0) = \underline{f}.$$

Thus $f_1 = \underline{f}$ and hence $Q_D^t f \uparrow \underline{f}$ as $t \downarrow 0$, which was to be shown.

By the monotone convergence theorem and the semigroup property of Q_D^t, $t > 0$,

$$Q_D^t \underline{f} = Q_D^t\left(\lim_{s \to 0} Q_D^s f\right) = \lim_{s \to 0} Q_D^s(Q_D^t f) \leq \lim_{s \to 0} Q_D^s f = \underline{f} \qquad \text{for } t > 0.$$

Thus $\underline{f}$ is superinvariant. Observe that $Q_D^t \underline{f} \geq Q_D^t Q_D^t f = Q_D^{2t} f$ for $t > 0$ and hence that $\lim_{t \to 0} Q_D^t \underline{f} \geq \underline{f}$. Thus $\lim_{t \to 0} Q_D^t \underline{f} = \underline{f}$ and hence $\underline{f}$ is excessive. Now f is excessive if and only if $\lim_{t \to 0} Q_D^t f = f$ and hence if and only if $\underline{f} = f$. Thus f is excessive if and only if it is lower semicontinuous. Since for $t > 0$, $Q_D^t f$ is superinvariant and lower semicontinuous, it is excessive.

It remains to show that in general $\underline{f} = f$ a.e. on D. Without loss of generality it can be assumed that D is connected. Choose $\lambda > 0$ and set $G_D^\lambda = \int_0^\infty e^{-\lambda t} Q_D^t \, dt$. Then $G_D^\lambda \underline{f} \leq G_D^\lambda f \leq \lambda^{-1} f$. On the other hand $Q_D^t \underline{f} \geq Q_D^t Q_D^s f$ for $s, t > 0$ and hence

$$G_D^\lambda \underline{f} = \int_0^\infty e^{-\lambda t} Q_D^t \underline{f} \, dt \geq \int_0^\infty e^{-\lambda t} Q_D^{s+t} f \, dt = e^{\lambda s} \int_s^\infty e^{-\lambda t} Q_D^t f \, dt.$$

It follows by letting $s \to 0$ that $G_D^\lambda \underline{f} \geq G_D^\lambda f$ and hence that $G_D^\lambda \underline{f} = G_D^\lambda f \leq \lambda^{-1} f$. Choose $x_0 \in D$ such that $f(x_0) < \infty$. Then $G_D^\lambda f(x_0) = G_D^\lambda \underline{f}(x_0) < \infty$ and hence $\int G_D^\lambda(x_0, y)(f(y) - \underline{f}(y)) \, dy = 0$. Since $f - \underline{f} \geq 0$ and $G_D^\lambda(x_0, \cdot) > 0$ on D by Theorem 2.4.3, $\underline{f} = f$ a.e. on D as desired. This completes the proof of the theorem.

Theorem 1.17. *Let f be excessive on D and let τ be a stopping time. Then $f \geq E_{\cdot}(f(X(\tau)); \tau < T_D)$ on D.*

Proof. The proof uses a reduction to discrete time and an argument similar to that used in proving the optional sampling theorem for submartingales (see the proof of Theorem 9.3.4 of Chung [1]).

Choose $\varepsilon > 0$ and let τ_ε be the stopping time defined by $\tau_\varepsilon = \infty$ if $\tau = \infty$ and $\tau_\varepsilon = m\varepsilon$ if $m\varepsilon - \varepsilon \leq \tau < m\varepsilon$ for the positive integer m. Now $f \geq Q_D^\varepsilon f$ on D so by the Markov property

$$\begin{aligned} &E_{\cdot}(f(X(m\varepsilon - \varepsilon)); \tau_\varepsilon > m\varepsilon - \varepsilon, T_D > m\varepsilon - \varepsilon) \\ &\quad \geq E_{\cdot}(Q_D^\varepsilon f(X(m\varepsilon - \varepsilon); \tau_\varepsilon > m\varepsilon - \varepsilon, T_D > m\varepsilon - \varepsilon) \\ &\quad = E_{\cdot}(f(X(m\varepsilon)); \tau_\varepsilon > m\varepsilon - \varepsilon, T_D > m\varepsilon - \varepsilon, T_D \circ \theta_{m\varepsilon - \varepsilon} > \varepsilon) \\ &\quad = E_{\cdot}(f(X(m\varepsilon)); \tau_\varepsilon > m\varepsilon - \varepsilon, T_D > m\varepsilon). \end{aligned}$$

Consequently

$$\begin{aligned} &E_{\cdot}(f(X(m\varepsilon - \varepsilon)); \tau_\varepsilon > m\varepsilon - \varepsilon, T_D > m\varepsilon - \varepsilon) \\ &\quad \geq E_{\cdot}(f(X(\tau_\varepsilon); \tau_\varepsilon = m\varepsilon, T_D > \tau_\varepsilon) + E_{\cdot}(f(X(m\varepsilon)); \tau_\varepsilon > m\varepsilon, T_D > m\varepsilon). \end{aligned}$$

It follows by summing both sides of this inequality on m that

$$E_{\cdot}(f(X(\tau_\varepsilon)); \tau_\varepsilon < T_D) \leq f \qquad \text{on } D.$$

Now $\tau = \lim_n \tau_{1/n}$ so $\underline{\lim}_n f(X(\tau_{1/n})) \geq f(X(\tau))$ on $\{\tau < T_D\}$ by continuity of paths and the lower semicontinuity of f. Therefore by Fatou's lemma

$$E_x(f(X(\tau));\tau < T_D) \le \varliminf_n E_.(f(X(\tau_{1/n}));\tau_{1/n} < T_D) \le f \ \le \text{ on } D$$

as desired.

The next result is required for the proof of the theorem which follows it.

Proposition 1.18. *Let f be bounded on $\mathbb{R}^d$ and twice continuously differentiable on D. Then*

$$\lim_{t\to 0} \frac{p^t f - f}{t} = \tfrac{1}{2}\Delta f \text{ uniformly on compact subsets of } D.$$

Proof. Choose $x \in D$. Let $f'(x) \in \mathbb{R}^d$ be the first derivative (gradient) of f at x and let $f''(x)$ denote the second derivative of f at x, that is, the $d \times d$ matrix of elements $\partial^2 f(x)/\partial x_i \partial x_j$ where $x = (x_1, \ldots, x_d)$. Then $\Delta f(x) = \operatorname{trace} f''(x)$. Let $t > 0$. Note that if X has density $p(t,\cdot)$ it has mean zero and covariance matrix t times the identity matrix. Consequently

$$\frac{1}{t}\int p(t,y) y \cdot f''(x) y \, dy = \Delta f(x), \qquad t > 0.$$

Therefore

$$\frac{p^t f(x) - f(x)}{t} = \tfrac{1}{2}\Delta f(x) + \frac{1}{t}\int p(t,y)[f(x+y) - f(x) - y\cdot f'(x) - \tfrac{1}{2} y\cdot f''(x) y]\, dy.$$

Thus to complete the proof of the proposition it suffices to show that if C is a compact subset of D, then

$$\text{(1)} \qquad \lim_{t\to 0}\frac{1}{t}\int p(t,y)[f(x+y) - f(x) - y\cdot f'(x) - \tfrac{1}{2} y\cdot f''(x) y]\, dy = 0$$

uniformly for $x \in C$.

Choose $\varepsilon > 0$. There is a $\delta > 0$ such that

$$|f(x+y) - f(x) - y\cdot f'(x) - \tfrac{1}{2} y\cdot f''(x) y| \le \varepsilon \|y\|^2, \qquad x \in C \quad \text{and} \quad \|y\| \le \delta.$$

Since $\int \|y\|^2 p(t,y)\,dy = td$, it now follows that

$$\text{(2)} \qquad \frac{1}{t}\int_{\|y\|\le\delta} p(t,y)|f(x,y) - f(x) - y\cdot f'(x) - \tfrac{1}{2} y\cdot f''(x) y|\, dy \le \varepsilon d, \qquad x \in C \quad \text{and} \quad t > 0.$$

There is a positive constant M such that

$$|f(x+y) - f(x) - y \cdot f'(x) - \tfrac{1}{2} y \cdot f''(x) y| \le M \|y\|^2, \qquad x \in C \quad \text{and} \quad \|y\| > \delta.$$

Since

$$\lim_{t \to 0} \frac{1}{t} \int_{\|y\| > \delta} p(t, y) \|y\|^2 \, dy = \lim_{t \to 0} \int_{\|y\| > t^{-1/2}\delta} p(1, y) \|y\|^2 \, dy = 0,$$

it follows that

$$\text{(3)} \quad \lim_{t \to 0} \frac{1}{t} \int_{\|y\| > \delta} p(t, y) |f(x+y) - f(x) - y \cdot f'(x) - \tfrac{1}{2} y \cdot f''(x) y| \, dy = 0$$

uniformly for $x \in C$. Now (1) follows from (2) and (3), so the proof of the proposition is complete.

The important connection between superharmonic and excessive functions will now be obtained.

Theorem 1.19. *Let f be nonnegative on D. Then f is superharmonic on D if and only if it is excessive on D.*

Proof. Suppose f is excessive on D. Choose $x \in D$ and $r > 0$ such that $B_r(x) \subset D$. By Theorem 1.17 applied to $\tau = \tau_{S_r(x)}$, $f(x) \ge \int \sigma_r(x, dy) f(y)$. Now f is lower semicontinuous by Theorem 1.16 and by definition f is not identically infinite on any component of D, so f is superharmonic on D.

Suppose conversely that f is superharmonic on D. Then f is lower semicontinuous on D. Thus by Theorem 1.16, to verify that f is excessive on D it suffices to show that this function is superinvariant on D.

Let D_n be relatively compact open subsets of D such that $D_n \uparrow D$. By Proposition 1.13 there exist bounded, nonnegative, infinitely differentiable superharmonic functions f_n on D_n such that $f_n \ge f_m$ on D_m for $n \ge m$ and $\lim_n f_n = f$ on D. Let $t > 0$, let m and n be positive integers with $n \ge m$ and let $h > 0$. Then on D_m

$$\begin{aligned} \frac{1}{h} \int_0^h Q^s_{D_m} f_n \, ds - \frac{1}{h} \int_t^{t+h} Q^s_{D_m} f_n \, ds &= \int_0^t Q^s_{D_m} (h^{-1}(f_n - Q^h_{D_m} f_n)) \, ds \\ &\ge \int_0^t Q^s_{D_m} (h^{-1}(f_n - p^h f_n)) \, ds. \end{aligned}$$

Now $Q^s_{D_m} f_n$ is continuous in s by Theorem 2.4.3. It follows by letting $h \to 0$ and using Propositions 1.12 and 1.18 that $f_n - Q^t_{D_m} f_n \ge 0$ on D_m. Consequently

$f \geq Q_{D_m}^t f$ on D_m. Since $P_.(T_{D_m} \uparrow T_D) = 1$ on D by Proposition 2.3.8, it follows from Fatou's lemma and the lower semicontinuity of f that $f \geq Q_D^t f$ on D for $t > 0$ and hence that f is superinvariant on D. This completes the proof of the theorem.

By Theorem 1.19 the terms "excessive" and "nonnegative superharmonic" are interchangeable. In particular Theorem 1.17 applies to nonnegative superharmonic functions and the various results proven earlier for superharmonic functions are valid for excessive functions.

Suppose D^c is polar. Then there are no nonconstant bounded harmonic functions on D. This is obvious if $d = 1$, in which case D^c is empty and follows from Theorem 4.2.17 if $d \geq 2$. The next result shows in particular that if $d \leq 2$ there are no nonconstant harmonic functions on D which are bounded below on D. The assumption that $d \leq 2$ is necessary. For if $d \geq 3$, then $\|x\|^{2-d}$ defines a nonconstant positive harmonic function on $\mathbb{R}^d \setminus \{0\}$.

Theorem 1.20. *Suppose $d \leq 2$ and D^c is polar and let f be superharmonic and bounded below on D. Then f is constant on D.*

Proof. Choose $x_0 \in D$ and $c < f(x_0)$. Since $\underline{\lim}_{y \to x_0} f(y) = f(x_0)$ there is an $r > 0$ such that $B_r(x_0) \subset D$ and $f \geq c$ on $B_r(x_0)$. Now $P_.(\tau_{B_r(x_0)} < \infty) = 1$ on D by Proposition 2.2.10, so $f \geq E_. f(X(\tau_{B_r(x_0)})) \geq c$ on D by Theorem 1.17. Consequently $f \geq f(x_0)$ on D. Since x_0 is an arbitrary point of D, f is constant on D as desired.

Proposition 1.21. *Let $B \in \mathscr{B}$. Then $P_.(\tau_B < T_D)$ is superharmonic on D.*

Proof. Set $f = P_.(\tau_B < T_D)$ and let $t > 0$. It follows easily from the Markov property that $Q_D^t f = P_.(t + \tau_B \circ \theta_t < T_D)$. Note that $t + \tau_B \circ \theta_t = \inf[s > t : X(s) \in B] \downarrow \tau_B$ as $t \downarrow 0$. Consequently $Q_D^t f \leq f$ and $\lim_{t \to 0} Q_D^t f = f$, so $f = P_.(\tau_B < T_D)$ is excessive and hence superharmonic on D.

Let f be nonnegative on D, let D_0 be an open subset of D, and set $h = E_.(f(X(T_{D_0})); T_{D_0} < T_D)$ on D_0. It follows easily from the strong Markov property that if D_1 is a relatively compact open subset of D_0, then $h = H_{D_1} h$ on D_1. Thus by Theorem 4.3.7, either h is harmonic on D_1 or h is identically infinite on D_1. Consequently if h is locally integrable on D_0 it is harmonic on D_0. Suppose now that f is a nonnegative superharmonic function on D and hence locally integrable on D by Proposition 1.10. Since $h \leq f$ on D_0 by Theorem 1.17, h is locally integrable and hence harmonic on D. The following result is therefore valid.

Proposition 1.22. *Let f be a nonnegative superharmonic function on D and let D_0 be an open subset of D. Then $E_.(f(X(T_{D_0})); T_{D_0} < T_D)$ is harmonic on D_0.*

Let f be superharmonic on D. A function h on D is said to be a *harmonic minorant* of f on D if h is harmonic on D and $h \le f$ on D. If h is a harmonic minorant of f on D and $h \ge h_1$ whenever h_1 is a harmonic minorant of f on D, then h is called the *greatest harmonic minorant* of f on D. Suppose f has a harmonic minorant h on an open subset D_0 of D. Then $f - h$ is excessive on D_0. In particular if D_0 is a relatively compact open subset of D, then f is bounded below on D_0 and hence $f + M$ is excessive on D_0 for some real number M. These observations can be used to obtain properties of superharmonic functions which are not necessarily nonnegative from those of excessive functions.

Theorem 1.23. *Let f be a superharmonic function on D which has a harmonic minorant on D. Then f has a greatest harmonic minorant h on D. Let D_n be relatively compact open subsets of D such that $D_n \uparrow D$. Then $H_{D_n} f$ is a harmonic minorant of f on D_n, $H_{D_n} f \le H_{D_m} f$ on D_m for $n \ge m$ and $\lim_n H_{D_n} f = h$ on D.*

Proof. Let h' be a harmonic minorant of f on D. Since $H_{D_n} h' = h'$ on D_n by Proposition 4.2.6, it suffices to verify that the theorem holds when f is replaced by $f - h'$. In other words, without loss of generality it can be assumed that f is nonnegative on D.

Now $H_{D_n} f$ is a harmonic minorant of f on D_n by Proposition 1.22 and Theorem 1.17. Let $n \ge m$. By Theorems 2.6.5, 2.6.7, and 1.17

$$H_{D_n} f = \int_{D_n} H_{D_m}(\cdot, dy) H_{D_n} f(y) + \int_{\partial D_n} H_{D_m}(\cdot, dy) f(y) \le H_{D_m} f \qquad \text{on } D_m.$$

Thus $h = \lim_n H_{D_n} f$ exists and is finite on D. Clearly $h \le f$ on D and by Harnack's theorem (Theorem 4.3.6) h is harmonic on D. Consequently h is a harmonic minorant of f on D.

Let h' be any harmonic minorant of f on D. Then $H_{D_n} f \ge H_{D_n} h' = h'$ on D_n by Proposition 4.2.6, so $h \ge h'$ on D. Therefore h is the greatest harmonic minorant of f on D. This completes the proof of the theorem.

Let $b \in \partial D$. A function f is called a *barrier* at b if there is an open neighborhood V of b such that f is a positive superharmonic function on $U = V \cap D$ and $f(x) \to 0$ as $x \to b$ within D.

Theorem 1.24. *If $b \in \partial D$, then $b \in (D^c)^r$ if and only if there is a barrier at b. In particular D is regular if and only if there is a barrier at b for each $b \in \partial D$.*

Proof. It follows immediately from Theorem 4.2.15 that if $b \in \partial D \cap (D^c)^r$, there is a barrier at b. Suppose now that $b \in \partial D \backslash (D^c)^r$. Let V be an open neighborhood of b and let f be a positive superharmonic function defined on $U = V \cap D$. Then $b \in \partial U \backslash (U^c)^r$, so after replacing D by U if necessary it can be assumed that f is a positive superharmonic function on D. By Proposition 4.2.14 there is a compact subset B of D such that

(4) $$\overline{\lim_{\substack{x \to b \\ x \in D}}} P_x(\tau_B < T_D) > 0.$$

Since f is a positive lower semicontinuous function on D there is a $\delta > 0$ such that $f \geq \delta$ on B. By Theorem 1.17, $f \geq \delta P_{\cdot}(\tau_B < T_D)$ on D. It now follows from (4) that f is not a barrier at b. This completes the proof of the theorem.

Theorem 1.25. *Let D be a connected Greenian open set, let $b \in \partial D$ and let $x \in D$. Then $b \in (D^c)^r$ if and only if $G_D(x, y) \to 0$ as $y \to b$ within D.*

Proof. The necessity of the indicated condition was pointed out in Section 4.5. Suppose conversely that $G_D(x, y) \to 0$ as $y \to b$ within D. Now $G_D(x, \cdot)$ is harmonic on $D \backslash \{x\}$ by Proposition 4.5.4 and $G_D(x, \cdot) > 0$ on D so $G_D(x, \cdot)$ is a barrier at b. Consequently $b \in (D^c)^r$ by Theorem 1.24.

2. Superharmonic Functions and Polar Sets

Throughout this section $d \geq 2$ (the corresponding results for $d = 1$ are all trivial). An analytic concept, thinness, developed by Brelot [2], [4], [6] is introduced and it is shown that a point $x \in \mathbb{R}^d$ is irregular for a set $B \in \mathscr{B}$ if and only if B is thin at x.

Recall that if f is superharmonic in a neighborhood of x, then $\underline{\lim}_{y \to x} f(y) = f(x)$. A subset B of $\mathbb{R}^d$ is said to be *thin* at $x \in \mathbb{R}^d$ if $x \notin \overline{B \backslash \{x\}}$ or if $x \in \overline{B \backslash \{x\}}$ and there is a superharmonic function f on some open neighborhood of x such that

(1) $$\underline{\lim_{\substack{y \to x \\ y \in B}}} f(y) > f(x).$$

Clearly if $x \in \mathring{B}$, then B is not thin at x. Thinness is a local property. That is, if U is an open neighborhood of x, then B is thin at x if and only if $B \cap U$ is thin at x. To show that B is not thin at x it suffices to show that if D_0 is an open neighborhood of x and f is excessive on D_0, then f fails to satisfy (1).

Proposition 2.1. *Let A and B be subsets of $\mathbb{R}^d$ each of which is thin at $x \in \mathbb{R}^d$. Then $A \cup B$ is thin at x.*

Proof. Suppose $x \in \overline{(A \setminus \{x\})} \cap \overline{(B \setminus \{x\})}$ (the conclusion clearly holds otherwise). Then there is a neighborhood U of x and there are subharmonic functions f_1 and f_2 on U such that

$$\varliminf_{\substack{y \to x \\ y \in A}} f_1(y) > f_1(x) \quad \text{and} \quad \varliminf_{\substack{y \to x \\ y \in B}} f_2(y) > f_2(x).$$

By the lower semicontinuity of f_1 and f_2

$$\varliminf_{\substack{y \to x \\ y \in A \cup B}} (f_1(y) + f_2(y)) > (f_1(x) + f_2(x)),$$

so $A \cup B$ is thin at x.

Proposition 2.2. *Let $B \in \mathscr{B}$ and $x \in B^r$. Then B is not thin at x.*

Proof. Let D_0 be an open neighborhood of x and let f be excessive on D_0. Choose $r > 0$ such that $B_r(x) \subset D_0$ and set $A = (B \setminus \{x\}) \cap B_r(x)$. Then $A \in \mathscr{B}$ and $x \in A^r$. Let B_n be compact sets such that $B_n \uparrow A$. Then $\tau_{B_n} \downarrow \tau_A$ and hence $P_x(\tau_{B_n} \downarrow 0) = 1$. Consequently $\lim_n P_x(\tau_{B_n} < T_{D_0}) = 1$. By Theorem 1.17

$$f(x) \geq E_x(f(X(\tau_{B_n}); \tau_{B_n} < T_{D_0}) \geq P_x(\tau_{B_n} < T_{D_0}) \inf_{y \in A} f(y).$$

Therefore $f(x) \geq \inf_{y \in A} f(y)$. It follows by letting $r \to 0$ that

$$f(x) \geq \varliminf_{\substack{y \to x \\ y \in B}} f(y)$$

and hence that (1) fails to hold. Thus B is not thin at x.

Proposition 2.2 is restated below together with its converse as Theorem 2.10.

Proposition 2.3. *Let f be superharmonic and bounded below on D and set $D_n = \{x \in D : f(x) > n\}$. Then $P_.(\tau_{D_n} < T_D$ for all $n \geq 1) = 0$ on D.*

Proof. Without loss of generality it can be assumed that f is nonnegative and hence excessive on D. Now D_n is open since f is lower semicontinuous on D. According to Proposition 2.2, D_n is not thin at any $y \in D_n^r \cap D$. Consequently

$$f(y) = \varliminf_{\substack{z \to y \\ z \in D_n}} f(z) \geq n, \qquad y \in D_n^r \cap D.$$

Choose $x \in D$ such that $f(x) < \infty$. Observe that $X(\tau_{D_n}) \in D_n^r \cap D$ a.s. (P_x) on $\{\tau_{D_n} < T_D\}$ by Theorem 2.6.5. Thus $f(X(\tau_{D_n})) \geq n$ a.s. (P_x) on $\{\tau_{D_n} < T_D\}$.

Theorem 1.17 now implies that

$$\infty > f(x) \geq E_x(f(X(\tau_{D_n})); \tau_{D_n} < T_D) \geq nP_x(\tau_{D_n} < T_D).$$

Therefore $\lim_n P_x(\tau_{D_n} < T_D) = 0$ and hence $P_x(\tau_{D_n} < T_D \text{ for all } n \geq 1) = 0$. Set $\Lambda = \{\tau_{D_n} < T_D \text{ for all } n \geq 1\}$. Then $P_.(\Lambda) = 0$ on $\{x \subset D : f(x) < \infty\}$ so $P_.(\Lambda) = 0$ a.e. on D by Proposition 1.10. For $\varepsilon > 0$ set $\Lambda^\varepsilon = \{\varepsilon + \tau_{D_n} \circ \theta_\varepsilon < T_D$ for all $n \geq 1\}$. By the Markov property $P_.(\Lambda^\varepsilon) = E_.(P_{X(\varepsilon)}(\Lambda);\ T_D > \varepsilon) = 0$ on D. Now $\Lambda = \bigcup_m \Lambda^{1/m}$, so $P_.(\Lambda) = 0$ on D as desired.

It follows from Proposition 2.3 that if f is superharmonic on D and $\{x \in D: f(x) = \infty\} \in \mathscr{B}$, then this set is polar. In general the following result is valid.

Theorem 2.4. *Let f be superharmonic on D. Then every compact subset of $\{x \in D: f(x) = \infty\}$ is polar.*

Proof. Without loss of generality it can be assumed that f is bounded below on D. Set $D_n = \{x \in D: f(x) > n\}$. Let B be a compact subset of $\{x \in D: f(x) = \infty\}$ and let $x \in B$. Then $\{X(0) = x, \tau_B = 0\} \subset \{\tau_{D_n} < T_D \text{ for all } n \geq 1\}$, so $P_x(\tau_B = 0) = 0$ by Proposition 2.3. Thus B^r is empty, so B is polar by Theorem 2.6.4.

Proposition 2.5. *Let $B \in \mathscr{B}$ be polar. Then there is a Greenian open set D containing B.*

Proof. Since B is polar it has Lebesgue measure zero. Thus there is an open set D containing B such that D^c has positive Lebesgue measure. Then D^c is nonpolar so D is Greenian by Proposition 4.5.1.

Proposition 2.6. *Let B be a compact polar subset of a Greenian open set D and let B_n be compact subsets of D such that $B_n \downarrow B$. Then $P_.(\tau_{B_n} < T_D) \downarrow 0$ on $D \backslash B$.*

Proof. Since B is polar, $P_.(\tau_B = \infty) = 1$ on D. Now $P_.(\tau_{B_n} \uparrow \tau_B) = 1$ on $D \backslash B$ by Proposition 2.3.8, so $P_.(\tau_{B_n} \uparrow \infty) = 1$ on $D \backslash B$. Suppose $d = 2$. Since D is Greenian, D^c is nonpolar by Proposition 4.5.1, so D^c is recurrent by Proposition 2.2.10. Thus $P_.(T_D < \infty) = 1$ on D and hence $P_.(\tau_{B_n} < T_D) \downarrow 0$ on $D \backslash B$. Suppose instead that $d \geq 3$, in which case the Brownian motion process is transient. Then $P_.(\tau_{B_n} < \infty \text{ for all } n) = 0$ on $D \backslash B$ and hence $P_.(\tau_{B_n} < T_D) \downarrow 0$ on $D \backslash B$. This completes the proof of the proposition.

Theorem 2.7. *Suppose D is Greenian. Let $B \in \mathscr{B}$ be a polar subset of D and let $x \in D \backslash B$. Then there is a nonnegative superharmonic function f on D such that $f = \infty$ on B and $f(x) < \infty$.*

Proof. Without loss of generality it can be assumed that D is connected. Suppose first that B is compact. Choose $c > 0$ such that $\{y : d(y, B) \le c\} \subset D$. Set $B_n = \{y : d(y, B) \le c/n\}$. Then $B_n^r \supset B$ and hence $P_\cdot(\tau_{B_n} < T_D) = 1$ on B. By Proposition 2.6, $\lim_n P_x(\tau_{B_n} < T_D) = 0$. Thus there is a strictly increasing sequence $\{m_n\}$ of positive integers such that $\sum_n P_x(\tau_n < T_D) < \infty$, where τ_n is the hitting time of B_{m_n}. Set $f = \sum_n P_\cdot(\tau_n < T_D)$. Then $f \ge 0$ on D, $f = \infty$ on B, and $f(x) < \infty$. It follows from Propositions 1.1 and 1.21 that f is superharmonic on D.

Consider now the general case. Let B_n be compact polar subsets of B such that $B_n \uparrow B$. Let f_n be a nonnegative superharmonic function on D such that $f_n = \infty$ on B_n and $f_n(x) < \infty$. Choose $a_n > 0$ such that $\sum_n a_n f_n(x) < \infty$ and set $f = \sum_n a_n f_n$. Then f has the desired properties.

The term "polar" and its definition in terms of the analytic criterion for polarity in the next result are due to Brelot [3].

Theorem 2.8. *Suppose D is Greenian and let $B \in \mathscr{B}$ be a subset of D. Then B is polar if and only if there is a superharmonic function on D which is identically infinite on B.*

Proof. This result follows from Theorems 2.4 and 2.7

Theorem 2.9. *Suppose D is Greenian, $B \in \mathscr{B}$ is a polar subset of ∂D and $x \in D$. Then there is a nonnegative superharmonic function f on D such that $\underline{f} = \infty$ on B and $f(x) < \infty$.*

Proof. If $d \ge 3$, the result follows immediately from Theorem 2.7 since $\mathbb{R}^d$ is Greenian. Suppose $d = 2$. Then D^c is nonpolar by Proposition 4.5.1 and hence recurrent by Proposition 2.2.10. Choose $x_0 \in D$. Then $H_D(x_0, \cdot)$ is a probability measure which by Theorem 2.6.5 is concentrated on $\partial D \cap (D^c)^r$ and does not charge polar sets. Since one-point sets are polar, there exist two distinct points $b_1, b_2 \in \partial D \cap (D^c)^r$. Set $r = \|b_1 - b_2\|/3$ and $D_i = D \cup (\mathbb{R}^2 \backslash B_r(b_i))$ for $i = 1, 2$. Now b_i is regular for D_i^c, so D_i^c is nonpolar and hence D_1 and D_2 are both Greenian. By Theorem 2.7 there is a nonnegative superharmonic function f_i on D_i such that $f_i = \infty$ on $D_i \cap B$ and $f_i(x) < \infty$. Then $f = f_1 + f_2$ satisfies the conclusion of the theorem.

Let $B \in \mathscr{B}$. The next result, obtained by Doob [1], yields analytic criteria for B to be polar and for a given point $x \in \mathbb{R}^d$ to be irregular for B.

Theorem 2.10. *Let $B \in \mathscr{B}$. Then $x \in \mathbb{R}^d$ is irregular for B if and only if B is thin at x. The set B is polar if and only if B is thin everywhere.*

Proof. If $x \in B^r$, then, by Proposition 2.2, B is not thin at x. If $x \notin \bar{B}$, then x is irregular for B and B is thin at x. Suppose $x \in \bar{B}\backslash B^r$. Then $P_x(\tau_B > c) > 0$ for some $c > 0$. There is an $r > 0$ such that $P_x(T_{\mathring{B}_r(x)} > c) \le P_x(\|X(c) - x\| < r) < P_x(\tau_B > c)$ and hence $P_x(\tau_B < T_{\mathring{B}_r(x)}) < 1$. Set $D_0 = \mathring{B}_r(x)$ and $f_1 = P_.(\tau_B < T_{D_0})$ on D_0. Then f_1 is excessive on D_0 by Proposition 1.21, $f_1 = 1$ on $B^r \cap D_0$ and $f_1(x) < 1$. Since $B\backslash B^r$ is polar, Theorem 2.7 implies that there is a superharmonic function f_2 on D_0 such that $f_2 = \infty$ on $[D_0 \cap (B\backslash B^r)]\backslash\{x\}$ and $f_2(x) < \infty$. Set $f = f_1 + f_2$ on D_0. If $x \in \overline{B^r}$, then

$$\varliminf_{\substack{y \to x \\ y \in B^r}} f(y) = 1 + \varliminf_{\substack{y \to x \\ y \in B^r}} f_2(y) \ge 1 + f_2(x) > f_1(x) + f_2(x) = f(x).$$

If $x \in \overline{B\backslash B^r}$, then

$$\varliminf_{\substack{y \to x \\ y \in B\backslash B^r}} f(y) = \infty > f(x).$$

Consequently f satisfies (1) and hence B is thin at x. This completes the proof of the first conclusion of the theorem.

By Theorem 2.6.4, B is polar if and only if B^r is empty, so the second conclusion of the theorem follows from the first conclusion.

Suppose $B \in \mathscr{B}$ is polar. Then B is thin everywhere by Theorem 2.10, so B^c is nowhere thin by Proposition 2.1. See the proof of Theorem 6.3.7 below for an application of this fact.

Let $B \in \mathscr{B}$ and $x \in \mathbb{R}^d$. Wiener's test (Theorem 3.3.2 for $d \ge 3$ and Theorem 6.7.35 below for $d = 2$) is a necessary and sufficient condition for x to be regular for B. By Theorem 2.10 it is also necessary and sufficient for B not to be thin at x; this result is due to Brelot [4], [6]. In particular Proposition 3.3.5 yields a necessary and sufficient condition for a thorn not to be thin at its vertex. By Theorems 2.10 and 4.2.2 a point $b \in \partial D$ is regular for the Dirichlet problem on D if and only if D^c is not thin at x; this result is due to Brelot [2].

3. Resolutivity

The PWB method due to Perron [1], Wiener [4], and Brelot [1], [4] can be used to solve Dirichlet and generalized Dirichlet problems (see Brelot [8] or Helms [1]). Since these problems have already been solved in Section 4.2 by a different method, it is unnecessary to give an independent solution by means of the PWB method. However, resolutivity, a key concept of the method, is itself interesting and will be treated here. The connection between resolutivity and solutions to Dirichlet and generalized Dirichlet problems is indicated briefly at the end of the section.

Throughout this section it is assumed that D is a nonempty connected open set. This assumption involves no loss of generality since resolutivity relative to an arbitrary open set can be treated by considering each component separately.

Let φ be defined on ∂D. The upper class $\mathscr{U}_\varphi$ for φ consists of the function on D which is identically equal to ∞ and all superharmonic functions f on D satisfying the following properties: f is bounded below on D; $f \geq \varphi$ on ∂D; and, if D^c is transient, $\underline{f}(\infty) \geq 0$. The lower class $\mathscr{L}_\varphi$ for φ is defined by $\mathscr{L}_\varphi = \{-f : f \in \mathscr{U}_{-\varphi}\}$. Set $\bar{H}_D\varphi = \inf[f : f \in \mathscr{U}_\varphi]$ and $\underline{H}_D\varphi = \sup[f : f \in \mathscr{L}_\varphi] = -\bar{H}_D(-\varphi)$. If $\bar{H}_D\varphi = \underline{H}_D\varphi$, then $\bar{H}_D c\varphi = \underline{H}_D c\varphi = c\bar{H}_D\varphi$ for $c \in \mathbb{R}$. If $\varphi_1 \leq \varphi_2$, then $\bar{H}_D\varphi_1 \leq \bar{H}_D\varphi_2$ and $\underline{H}_D\varphi_1 \leq \underline{H}_D\varphi_2$. If $\bar{H}_D\varphi_1 + \bar{H}_D\varphi_2$ is well defined, then $\bar{H}_D(\varphi_1 + \varphi_2) \leq \bar{H}_D\varphi_1 + \bar{H}_D\varphi_2$. The proofs of these results are left to the reader.

Suppose $H_D\varphi$ is well defined. Then $\bar{H}_D\varphi \geq H_D\varphi$ by Proposition 1.9, so to prove that $\bar{H}_D\varphi = H_D\varphi$ it must be shown that $\bar{H}_D\varphi \leq H_D\varphi$.

Proposition 3.1. *If φ is bounded below or above on ∂D, then $\bar{H}_D\varphi = \underline{H}_D\varphi = H_D\varphi$.*

Proof. Throughout this proof x is an arbitrary point in D and ε is an arbitrary positive number.

Consider first an indicator function $\varphi = I_B$ where $B \subset \partial D$. By Theorems 2.9 and 2.6.3 there is a nonnegative superharmonic function f_0 on D such that $\underline{f_0} = \infty$ on $\partial D \setminus (D^c)^r$ and $f_0(x) < \infty$. Let A be a relatively compact open subset of ∂D (depending on x and ε) such that $B \subset A$ and $H_D(x, A) \leq H_D(x, B) + \varepsilon$. Set $f = H_D(\cdot, A) + \varepsilon f_0$. Then $f \in \mathscr{U}_{I_B}$ and hence

$$\bar{H}_D\varphi(x) = \bar{H}_D I_B(x) \leq f(x) \leq H_D(x, B) + \varepsilon(1 + f_0(x)) = H_D\varphi(x) + \varepsilon(1 + f_0(x)).$$

Therefore $\bar{H}_D\varphi = H_D\varphi$. Let A now be a compact subset of B such that $H_D(x, A) \geq H_D(x, B) - \varepsilon$. Set $f = -H_D(\cdot, A) + \varepsilon f_0$. Note that $\underline{f}(\infty) \geq 0$ if D^c is transient (if D^c is polar this is obvious; otherwise $d \geq 3$ by Proposition 2.2.10, so the desired result follows from Proposition 2.2.11). Thus $f \in \mathscr{U}_{-I_B}$ and hence

$$\begin{aligned}\bar{H}_D(-\varphi)(x) = \bar{H}_D(-I_B)(x) \leq f(x) &\leq -H_D(x, B) + \varepsilon(1 + f_0(x)) \\ &= H_D(-\varphi)(x) + \varepsilon(1 + f_0(x)).\end{aligned}$$

Therefore $\bar{H}_D(-\varphi) = H_D(-\varphi)$ or equivalently $\underline{H}_D\varphi = H_D\varphi$. Consequently $\bar{H}_D\varphi = \underline{H}_D\varphi = H_D\varphi$.

Consider next a finite linear combination $\varphi = \sum_i c_i I_{A_i}$ of indicator functions. Now

$$\bar{H}_D\varphi \le \sum_i \bar{H}_D(c_i I_{A_i}) = \sum_i c_i \bar{H}_D I_{A_i} = \sum_i c_i H_D I_{A_i} = H_D\left(\sum_i c_i I_{A_i}\right) = H_D\varphi$$

and hence $\bar{H}_D\varphi = H_D\varphi$. Since $-\varphi$ is of the same form $\underline{H}_D\varphi = -\bar{H}_D(-\varphi) = -H_D(-\varphi) = H_D\varphi$ and hence $\bar{H}_D\varphi = \underline{H}_D\varphi = H_D\varphi$.

Suppose next that φ is bounded below. There exist functions φ_n on ∂D, each of which is a finite linear combination of indicator functions, such that $\varphi_n \uparrow \varphi$ on D. Choose $f_1 \in \mathscr{U}_{\varphi_1}$ such that $f_1(x) \le \bar{H}_D\varphi_1(x) + 2^{-1}\varepsilon = H_D\varphi_1(x) + 2^{-1}\varepsilon$ and for $n \ge 2$ choose $f_n \in \mathscr{U}_{\varphi_n - \varphi_{n-1}}$ such that

$$f_n(x) \le \bar{H}_D(\varphi_n - \varphi_{n-1})(x) + 2^{-n}\varepsilon = H_D(\varphi_n - \varphi_{n-1})(x) + 2^{-n}\varepsilon.$$

Set $f = \sum_{n\ge 1} f_n$. Then $f \in \mathscr{U}_\varphi$ and hence

$$\bar{H}_D\varphi(x) \le f(x) \le H_D\varphi_1(x) + \sum_{n\ge 2} H_D(\varphi_n - \varphi_{n-1})(x) + \varepsilon = H_D\varphi(x) + \varepsilon.$$

Therefore $\bar{H}_D\varphi = H_D\varphi$. Now choose $f_1 \in \mathscr{U}_{-\varphi_1}$ such that $f_1(x) \le \bar{H}_D(-\varphi_1)(x) + 2^{-1}\varepsilon$ and for $n \ge 2$ let $f'_n \in \mathscr{U}_{(\varphi_{n-1} - \varphi_n)}$ be such that $f'_n(x) \le \bar{H}_D(\varphi_{n-1} - \varphi_n)(x) + 2^{-n}\varepsilon$. Set $f_n = f_1 + f'_2 + \cdots + f'_n$. Then $f_n \in \mathscr{U}_{-\varphi}$ and hence

$$\bar{H}_D(-\varphi)(x) \le f_n(x) \le H_D(-\varphi_n)(x) + \varepsilon.$$

It follows from the monotone convergence theorem that $\bar{H}_D(-\varphi)(x) \le H_D(-\varphi)(x) + \varepsilon$. Therefore $\bar{H}_D(-\varphi) = H_D(-\varphi)$ or equivalently $\underline{H}_D\varphi = H_D\varphi$. Consequently $\bar{H}_D\varphi = \underline{H}_D\varphi = H_D\varphi$.

Suppose finally that φ is bounded above. Then $-\varphi$ is bounded below and hence $\bar{H}_D(-\varphi) = \underline{H}_D(-\varphi) = H_D(-\varphi)$. Since $\bar{H}_D\varphi = -\underline{H}_D(-\varphi)$ and $\underline{H}_D\varphi = -\bar{H}_D(-\varphi)$, $\bar{H}_D\varphi = \underline{H}_D\varphi = H_D\varphi$. This completes the proof of the proposition.

By Theorem 4.3.7, either $H_D\varphi^+ < \infty$ on D or $H_D\varphi^+ = \infty$ on D. Similarly either $H_D\varphi^- < \infty$ on D or $H_D\varphi^- = \infty$ on D.

Proposition 3.2. *Let φ be defined on ∂D. Then $\bar{H}_D\varphi = H_D\varphi$ if $H_D\varphi^+ < \infty$ and $\bar{H}_D\varphi = \infty$ if $H_D\varphi^+ = \infty$. Also $\underline{H}_D\varphi = H_D\varphi$ if $H_D\varphi^- < \infty$ and $\underline{H}_D\varphi = -\infty$ if $H_D\varphi^- = \infty$.*

Proof. Suppose first that $H_D\varphi^+ < \infty$. Set $\varphi_n = \varphi \vee (-n)$. Then $\bar{H}_D\varphi_n = H_D\varphi_n$ by Proposition 3.1. Now $\bar{H}_D\varphi \le \bar{H}_D\varphi_n = H_D\varphi_n \downarrow H_D\varphi$ by the monotone convergence theorem, so $\bar{H}_D\varphi \le H_D\varphi$. Therefore $\bar{H}_D\varphi = H_D\varphi$. Suppose instead that $H_D\varphi^+ = \infty$. Choose $f \in \mathscr{U}_\varphi$. Then $f \ge -M$ for some positive number M by the definition of $\mathscr{U}_\varphi$, so $f + M \in \mathscr{U}_{\varphi^+}$. By Proposition 3.1,

$f + M \geq \bar{H}_D(\varphi^+) = \infty$ and hence $f = \infty$. Therefore $\bar{H}_D\varphi = \infty$. Since $\underline{H}_D\varphi = -\bar{H}_D(-\varphi)$ the second conclusion of the proposition follows from the first conclusion.

A function φ on ∂D is said to be *resolutive* if $\bar{H}_D\varphi = \underline{H}_D\varphi$ and $\bar{H}_D\varphi$ is harmonic. The next two results are due to Brelot [1], [5].

Theorem 3.3. *Let φ be defined on ∂D. Then φ is resolutive if and only if $H_D|\varphi| < \infty$, in which case $\bar{H}_D\varphi = \underline{H}_D\varphi = H_D\varphi$.*

Proof. This result follows from Propositions 3.2 and 4.3.7.

Let φ be defined on ∂D and let $\alpha \in \mathbb{R}$. The upper class $\mathscr{U}_{\varphi,\alpha}$ for (φ, α) consists of the function on D which is identically equal to ∞ and all superharmonic functions f on D satisfying the following properties: f is bounded below on D; $\underline{f} \geq \varphi$ on ∂D; and, if D^c is transient, $\underline{f}(\infty) \geq \alpha$. The lower class $\mathscr{L}_{\varphi,\alpha}$ for (φ, α) is defined by $\mathscr{L}_{\varphi,\alpha} = \{-f : f \in \mathscr{U}_{-\varphi,-\alpha}\}$. Set $\bar{H}_D(\varphi, \alpha) = \inf[f : f \in \mathscr{U}_{\varphi,\alpha}]$ and $\underline{H}_D(\varphi, \alpha) = \sup[f : f \in \mathscr{L}_{\varphi,\alpha}] = -\bar{H}_D(-\varphi, -\alpha)$. The pair (φ, α) is said to be resolutive if $\bar{H}_D(\varphi, \alpha) = \underline{H}_D(\varphi, \alpha)$ and $\bar{H}_D(\varphi, \alpha)$ is harmonic. Clearly $\bar{H}_D(\varphi, \alpha) = \alpha + \bar{H}_D(\varphi - \alpha)$, $H_D(\varphi, \alpha) = \alpha + \underline{H}_D(\varphi - \alpha)$, and $\alpha + H_D(\varphi - \alpha) = \alpha + H_D\varphi - \alpha P_\cdot(T_D < \infty) = H_D\varphi + \alpha P_\cdot(T_D = \infty)$ if $H_D|\varphi| < \infty$. Thus Theorem 3.3 yields the following more general result.

Theorem 3.4. *Let φ be defined on ∂D and let $\alpha \in \mathbb{R}$. Then (φ, α) is resolutive if and only if $H_D|\varphi| < \infty$, in which case*

$$\bar{H}_D(\varphi, \alpha) = \underline{H}_D(\varphi, \alpha) = H_D\varphi + \alpha P_\cdot(T_D = \infty).$$

Suppose φ is a bounded and essentially continuous function on ∂D and let $\alpha \in \mathbb{R}$. Then $\bar{H}_D(\varphi, \alpha) = H_D\varphi + \alpha P_\cdot(T_D = \infty)$ by Theorem 3.4, so the results in Section 4.2 yield various senses in which $\bar{H}_D(\varphi, \alpha)$ solves a Dirichlet or generalized Dirichlet problem.

4. Behavior along Brownian Motion Paths

Let f be superharmonic on D and let $x \in D$. In this section the function $f(X(t))$, $0 \leq t < T_D$, is studied. In particular it is shown that a.s. (P_x) this function is continuous and, if f is bounded below on D, it has a finite limit at T_D. These results were obtained by Doob [1].

First an important technique involving time-reversal will be considered. For $0 < r < s$ let $\mathscr{C}([r, s])$ denotes the collection of continuous functions $x(\cdot)$

from $[r,s]$ to $\mathbb{R}^d$. Let $\mathscr{G}_{rs}$ denote the σ-field on $\mathscr{C}([r,s])$ generated by the functions f_t defined by $f_t(x(\cdot)) = x(t)$ as t ranges over $[r,s]$. For $A \in \mathscr{G}_{rs}$ set $A_{rs} = \{x(r+s-\cdot):x(\cdot) \in A\}$. Also set $\mathfrak{F}_{rs} = \sigma(X(t), r \le t \le s)$. Then $\Lambda \in \mathfrak{F}_{rs}$ if only if there is an $A \in \mathscr{G}_{rs}$ such that

$$\Lambda = \{\omega\text{:the restriction of } X(\cdot,\omega) \text{ to } [r,s] \text{ is in } A\},$$

in which case set

$$\Lambda_{rs} = \{\omega\text{:the restriction of } X(\cdot,\omega) \text{ to } [r,s] \text{ is in } A_{rs}\}.$$

Then $\Lambda_{rs} \in \mathfrak{F}_{rs}$. The next result is a modification of Theorem 4.12 of Dynkin [1].

Theorem 4.1. *Let* $0 < r < s$, $\Lambda \in \mathfrak{F}_{rs}$, *and* $x \in \mathbb{R}^d$. *Then*

$$P_x(\Lambda_{rs}) = E_x\left(\frac{p(r,x,X(s))}{p(r,x,X(r))} I_\Lambda\right). \tag{1}$$

In particular $P_x(\Lambda_{rs}) = 0$ *if and only if* $P_x(\Lambda) = 0$.

Proof. It suffices to verify (1) for sets Λ of the form $\Lambda = \{X(t_i) \in B_i$ for $0 \le i \le n\}$, where n is a positive integer, $r = t_0 < \cdots < t_n = s$, and $B_0, \ldots, B_n$ are subsets of $\mathbb{R}^d$; in this case $\Lambda_{rs} = \{X(r+s-t_i) \in B_i$ for $0 \le i \le n\}$. Now

$$
\begin{aligned}
&P_x(\Lambda_{rs})\\
&= \int \cdots \int p(r,x,y_0) I_{B_n}(y_0) \prod_{i=1}^{n} p(t_{n-i+1} - t_{n-i}, y_{i-1}, y_i) I_{B_{n-i}}(y_i)\, dy_0 \cdots dy_n\\
&= \int \cdots \int p(r,x,y_n) I_{B_n}(y_n) \prod_{i=1}^{n} p(t_{n-i+1} - t_{n-i}, y_{n-i+1}, y_{n-i}) I_{B_{n-i}}(y_{n-i})\, dy_0 \cdots dy_n\\
&= \int \cdots \int p(r,x,y_n) I_{B_n}(y_n) \prod_{i=0}^{n-1} p(t_{i+1} - t_i, y_{i+1}, y_i) I_{B_i}(y_i)\, dy_0 \cdots dy_n\\
&= \int \cdots \int p(r,x,y_n) I_{B_n}(y_n) \prod_{i=0}^{n-1} p(t_{i+1} - t_i, y_i, y_{i+1}) I_{B_i}(y_i)\, dy_0 \cdots dy_n\\
&= \int \cdots \int \frac{p(r,x,y_n)}{p(r,x,y_0)} p(r,x,y_0) I_{B_0}(y_0) \prod_{i=1}^{n} p(t_i - t_{i-1}, y_{i-1}, y_i) I_{B_i}(y_i)\, dy_0 \cdots dy_n\\
&= E_x\left(\frac{p(r,x,X(s))}{p(r,x,X(r))} I_\Lambda\right)
\end{aligned}
$$

as desired.

Let $x \in D$. Then $X(T_D) \in (D^c)^r$ a.s. (P_x) on $\{T_D < \infty\}$ by Theorem 2.6.5. Time-reversal will now be used to show that $X(T_D) \in D^r$ a.s. (P_x) on $\{T_D < \infty\}$ and hence that $\tau_D \circ \theta_{T_D} = 0$ a.s. (P_x) on $\{T_D < \infty\}$. This implies that Brownian motion starting at $x \in D$ will almost surely reenter D immediately after first leaving this set. It also implies that $H_D(x, \cdot)$ is concentrated on $\partial D \cap D^r \cap (D^c)^r$.

Theorem 4.2. *Let $x \in D$. Then $X(T_D) \in D^r$ a.s. (P_x) on $\{T_D < \infty\}$.*

Proof. By the strong Markov property

(2) $\quad P_x(T_D < \infty, X(T_D) \notin D^r) = P_x(T_D < \infty, X(T_D) \notin D^r, \tau_D \circ \theta_{T_D} > 0).$

Let Q^+ denote the set of positive rationals. Then

$$\begin{aligned}\{X(0) = x, T_D < \infty, X(T_D) \notin D^r, \tau_D \circ \theta_{T_D} > 0\} \\ \subset \bigcup_{\substack{r,s \in Q^+ \\ r<s}} \{r < r + T_D \circ \theta_r < s, X(r + T_D \circ \theta_r) \notin D^r, X(t) \notin D \\ \text{for } r + T_D \circ \theta_r < t < s\}.\end{aligned}$$

Let $0 < r < s$. Set

$$\begin{aligned}\Lambda = \{r < r + T_D \circ \theta_r < s, X(r + T_D \circ \theta_r) \notin D^r, X(t) \notin D \\ \text{for } r + T_D \circ \theta_r < t < s\}.\end{aligned}$$

By (2) it suffices to show that $P_x(\Lambda) = 0$ or equivalently by Theorem 4.1 that $P_x(\Lambda_{rs}) = 0$. Now

$$\begin{aligned}\Lambda_{rs} &= \{r < r + \tau_D \circ \theta_r < s, X(r + \tau_D \circ \theta_r) \notin D^r, X(t) \in D \text{ for } r + \tau_D \circ \theta_r < t < s\} \\ &\subset \{r < r + \tau_D \circ \theta_r < s, X(r + \tau_D \circ \theta_r) \notin D^r\}.\end{aligned}$$

It follows from Theorem 2.6.5 and the strong Markov property that $P_x(\Lambda_{rs}) = 0$. This completes the proof of the theorem.

Proposition 4.3. *Let f be a finite-valued superharmonic function on D and let $\varepsilon > 0$ and $x \in D$. Set $\tau = \inf[t \geq 0: X(t) \in D^c$ or $|f(X(t)) - f(X(0))| > \varepsilon]$. Then τ is a stopping time, $|f(X(\tau)) - f(X(0))| \geq \varepsilon$ a.s. (P_x) on $\{\tau < T_D\}$ and $P_x(\tau = 0) = 0$.*

Proof. Observe first that $\{x \in D: f(x) \leq b\}$ is a relatively closed subset of D for $b \in \mathbb{R}$. Consequently $\{x \in D: f(x) < a\} \in \mathscr{B}$ for $a \in \mathbb{R}$. Choose a, $b \in \mathbb{R}$ such that $a < b$. Set $B_{ab} = \{x \in D: f(x) < a$ or $f(x) > b\}$. Then $B_{ab} \in \mathscr{B}$. It follows from Proposition 2.2 and the lower semicontinuity of f that if $x \in B_{ab}^r \cap D$, then $f(x) \leq a$ or $f(x) \geq b$.

Let Q denote the set of rational numbers. Choose $t > 0$. Then $\{\tau < t, X(0) \in D^c\} = \{X(0) \in D^c\} \in \mathfrak{F}_t$ and

$$\{\tau < t, X(0) \in D\} = \{X(0) \in D\}$$

$$\cap \left(\bigcup_{\substack{a,b \in Q \\ a<b}} \{a < f(X(0)) < b, \tau_{B_{a-\varepsilon,b+\varepsilon}} \wedge T_D < t\} \right) \in \mathfrak{F}_t.$$

Thus $\{\tau < t\} \in \mathfrak{F}_t$ for all $t > 0$ and hence τ is a stopping time. Set $B = B_{f(x)-\varepsilon, f(x)+\varepsilon}$. Then $B \in \mathscr{B}$ and $|f(y) - f(x)| \geq \varepsilon$ for $y \in B^r \cap D$. Now $\tau = \tau_B$ a.s. (P_x) on $\{\tau < T_D\}$ and hence $X(\tau) \in B^r \cap D$ a.s. (P_x) on $\{\tau < T_D\}$ by Theorem 2.6.5. Consequently $|f(X(\tau)) - f(X(0))| \geq \varepsilon$ a.s. (P_x) on $\{\tau < T_D\}$ and therefore $P_x(\tau = 0) = 0$. This completes the proof of the proposition.

Proposition 4.4. *Let f be bounded and superharmonic on D and let $\varepsilon > 0$. Set*

$$\tau_1 = \inf[t \geq 0 : X(t) \in D^c \text{ or } |f(X(t)) - f(X(0))| > \varepsilon]$$

and let τ_n, $n \geq 2$, be defined successively by $\tau_n = \tau_{n-1} + \tau_1 \circ \theta_{\tau_{n-1}}) \leq T_D$ if $\tau_{n-1} < T_D$ and $\tau_n = T_D$ if $\tau_{n-1} = T_D$. Then $P_x(\tau_n = T_D$ for n sufficiently large$) = 1$ for $x \in D$.

Proof. Without loss of generality it can be assumed that f is nonnegative. By Proposition 4.3 together with Propositions 1.4.2 and 1.4.6, each τ_n is a stopping time. Define the bounded random variable Y_n by $Y_n = f(X(\tau_n))$ if $\tau_n < T_D$ and $Y_n = 0$ if $\tau_n = T_D$. Then Y_n is $\mathfrak{F}_{\tau_n+}$-measurable by Proposition 1.4.5. Note that $\mathfrak{F}_{\tau_n+}\uparrow$ by Proposition 1.4.3. Choose $x \in D$ and $\Lambda \in \mathfrak{F}_{\tau_n+}$. By Theorem 1.17 and the strong Markov property

$$\begin{aligned} E_x(Y_{n+1}; \Lambda) &= E_x(f(X(\tau_{n+1})); \Lambda, \tau_{n+1} < T_D) \\ &= E_x(E_{X(\tau_n)}(f(X(\tau_1)); \tau_1 < T_D); \Lambda, \tau_n < T_D) \\ &\leq E_x(f(X(\tau_n)); \Lambda, \tau_n < T_D) \\ &= E_x(Y_n; \Lambda). \end{aligned}$$

Consequently $E_x(Y_{n+1} | \mathfrak{F}_{\tau_n+}) \leq Y_n$ a.s. (P_x). Therefore by the supermartingale convergence theorem (see page 335 of Chung [1]), $\lim_n Y_n$ exists a.s. (P_x). It follows from Proposition 4.3 that $|f(X(\tau_{n+1})) - f(X(\tau_n))| \geq \varepsilon$ a.s. (P_x) on $\{\tau_{n+1} < T_D\}$. Now if $\tau_n < T_D$ and $|f(X(\tau_{n+1})) - f(X(\tau_n))| \geq \varepsilon$ for all n, Y_n fails to have a limit as $n \to \infty$. Thus $P_x(\tau_n = T_D$ for n sufficiently large$) = 1$, which completes the proof of the proposition.

In the following theorem and its applications it is convenient to interpret a.s. (P_x) as meaning that the exceptional set is contained in an event Λ such that $P_x(\Lambda) = 0$.

Theorem 4.5. *Let f be superharmonic on D. Then the function $f(X(\cdot))$ on $[0, T_D)$ is finite-valued and continuous a.s. (P_x) for each $x \in D$. If f is bounded below on D, then $\lim_{t \uparrow T_D} f(X(t))$ exists and is finite a.s. (P_x) for each $x \in D$ and $E_{\cdot}(\lim_{t \uparrow T_D} f(X(t)))$ is a harmonic minorant of f on D.*

Proof. Without loss of generality it can be assumed that f is a nonnegative superharmonic function (for the first conclusion of the theorem approximate D by relatively compact open sets). Suppose f is bounded and for $\varepsilon > 0$ let the stopping times τ_n in Proposition 4.4. be written as τ_n^ε to indicate the dependence on ε. Set

$$\Lambda_1 = \bigcap_{m \geq 1} \{\tau_n^{1/m} = T_D \text{ for } n \text{ sufficiently large}\}.$$

Then $P_{\cdot}(\Lambda_1) = 1$ on D. Set

$$\Lambda = \{f(X(\cdot)) \text{ is right-continuous on } [0, T_D) \\ \text{and has left-hand limits on } (0, T_D]\}.$$

It is left to the reader to check that $\Lambda = \Lambda_1$. Thus $P_{\cdot}(\Lambda) = 0$ on D. In other words $f(X(\cdot))$ is right-continuous on $[0, T_D)$ and has left-hand limits on $(0, T_D]$ a.s. (P_x) for each $x \in D$.

Let f now be an arbitrary nonnegative superharmonic function on D. Then $1 - e^{-f}$ is a bounded nonnegative superharmonic function on D by Proposition 1.2. Thus by Proposition 2.3 and the first paragraph of this proof, $f(X(\cdot))$ is finite and right-continuous on $[0, T_D)$ and has finite left-hand limits on $(0, T_D]$ a.s. (P_x) for each $x \in D$. Let D_n be relatively compact open subsets of D such that $D_n \uparrow D$. By Theorem 1.17, Proposition 2.3.8, and Fatou's lemma $f \geq E_{\cdot}(\lim_n f(X(T_{D_n}))) = E_{\cdot}(\lim_{t \uparrow T_D} f(X(t)))$ on D. Thus $E_{\cdot}(\lim_{t \uparrow T_D} f(X(t)))$ is locally integrable on D. It now follows easily from the strong Markov property that this function satisfies the definition of harmonicity on D and hence is a harmonic minorant of f on D.

Let $x \in D$. It remains to show that $f(X(\cdot))$ is left-continuous on $(0, T_D)$ a.s. (P_x). It suffices to show that, given $s > 0$, $f(X(\cdot))$ is left-continuous on $(0, s)$ a.s. (P_x) on $\{T_D \geq s\}$. To prove this result it is enough to show that, for $0 < r < s$, $f(X(\cdot))$ is left-continuous on (r, s) a.s. (P_x) on $\{T_D \circ \theta_r \geq s - r\}$. Set

$$\Lambda = \{T_D \circ \theta_r \geq s - r \text{ and } f(X(\cdot)) \text{ is left-continuous on } (r, s)\}.$$

Then

$$\Lambda_{rs} = \{T_D \circ \theta_r \geq s - r \text{ and } f(X(\cdot)) \text{ is right-continuous on } (r, s)\}.$$

The desired result now follows from Theorem 4.1 and the second paragraph of the present proof. This completes the proof of the theorem.

Proposition 4.6. *Let φ be a function on ∂D such that $H_D|\varphi| < \infty$ on D and let $x \in D$. Then $\lim_{t \uparrow T_D} H_D\varphi(X(t)) = \varphi(X(T_D))$ a.s. (P_x) on $\{T_D < \infty\}$ and $\lim_{t \uparrow \infty} H_D\varphi(X(t)) = 0$ a.s. (P_x) on $\{T_D = \infty\}$.*

Proof. Let D_n be relatively compact subsets of D such that $D_n \uparrow D$. Set $T_n = T_{D_n}$ and let $\mathfrak{F}$ denote the σ-field generated by $\bigcup_n \mathfrak{F}_{T_n}$. Then T_n is $\mathfrak{F}_{T_n}$-measurable and $T_n \uparrow T_D$ on $\{X(0) \in D\}$. Thus T_D is $\mathfrak{F}$-measurable on $\{X(0) \in D\}$. Let Y denote the random variable defined by $Y = \varphi(X(T_D))$ on $\{X(0) \in D, T_D < \infty\}$ and $Y = 0$ otherwise. Then Y is $\mathfrak{F}$-measurable and $E_{.}Y = H_D\varphi$ on D. It follows from the strong Markov property that for $x \in D_n$

$$E_x(Y \mid \mathfrak{F}_{T_n}) = E_{X(T_n)}Y = H_D\varphi(X(T_n)) \text{ a.s. } (P_x).$$

Choose $x \in D$. Then by the martingale convergence theorem (page 340 of Chung [1])

$$\lim_n H_D\varphi(X(T_n)) = E(Y \mid \mathfrak{F}) = Y \text{ a.s. } (P_x).$$

Now $H_D\varphi^+$ and $H_D\varphi^-$ are nonnegative harmonic functions on D by Theorem 4.3.7 and $H_D\varphi = H_D\varphi^+ - H_D\varphi^-$, so $\lim_{t \uparrow T_D} H_D\varphi(X(t))$ exists a.s. (P_x) by Theorem 4.5. Since $T_n \uparrow T_D$ a.s. (P_x) it follows that $\lim_{t \uparrow T_D} H_D\varphi(X(t)) = Y$ a.s. (P_x), which is equivalent to the conclusion of the proposition.

Proposition 4.7. *Let $x \in D$. Then $\lim_{t \uparrow T_D} P_{X(t)}(T_D = \infty) = 0$ a.s. (P_x) on $\{T_D < \infty\}$ and $\lim_{t \uparrow \infty} P_{X(t)}(T_D = \infty) = 1$ a.s. (P_x) on $\{T_D = \infty\}$.*

Proof. This result follows from Proposition 4.6 with $\varphi = 1$ on ∂D.

Proposition 4.8. *Let f be superharmonic and bounded below on D. Then there is an $\alpha \in \mathbb{R}$ such that $\lim_{t \to \infty} f(X(t)) = \alpha$ a.s. (P_x) on $\{T_D = \infty\}$ for $x \in D$.*

Proof. Without loss of generality it can be assumed that f is nonnegative and hence excessive on D. Suppose first that f is bounded on D. Then $Q_D^t f \downarrow h$ as $t \uparrow \infty$, where h is bounded on D and $Q_D^t h = \lim_{s \to \infty} Q_D^t Q_D^s f = h$ on D for all $t > 0$ by the dominated convergence theorem. By Proposition 4.2.7 $h = \alpha P_{.}(T_D = \infty)$ on D for some nonnegative constant α. Thus

$$\alpha P_{.}(T_D = \infty) = \lim_{t \to \infty} Q_D^t f = \lim_{t \to \infty} E_{.}(f(X(t)); T_D > t).$$

Now $\lim_{t \to \infty} f(X(t))$ exists a.s. (P_x) on $\{T_D = \infty\}$ by Theorem 4.5, so by the dominated convergence theorem $\alpha P_{.}(T_D = \infty) = E_{.}(\lim_{t \to \infty} f(X(t));$

$T_D = \infty$). Let $x \in D$ and choose $r > 0$ and $\Lambda \in \mathfrak{F}_r$. It follows from the Markov property that for $s \geq r$

$$E_x\left(\lim_{t\to\infty} f(X(t));\ T_D = \infty,\ \Lambda\right) = E_x\left(E_{X(s)}\left(\lim_{t\to\infty} f(X(t));\ T_D = \infty\right);\ T_D > s,\ \Lambda\right)$$

$$= \alpha E_x(P_{X(s)}(T_D = \infty);\ T_D > s,\ \Lambda).$$

Hence by Proposition 4.7

$$E_x\left(\lim_{t\to\infty} f(X(t)); T_D = \infty, \Lambda\right) = \alpha P_x(T_D = \infty, \Lambda). \tag{3}$$

Choose $\Lambda \in \mathfrak{F}_\infty$. Then for every $\varepsilon > 0$ there is an $r > 0$ and a set $\Lambda_1 \in \mathfrak{F}_r$ such that $P_x(\Lambda \backslash \Lambda_1) \leq \varepsilon$ and $P_x(\Lambda_1 \backslash \Lambda) \leq \varepsilon$. Consequently (3) holds for all $\Lambda \in \mathfrak{F}_\infty$. Therefore $\lim_{t\to\infty} f(X(t)) = \alpha$ a.s. (P_x) on $\{T_D = \infty\}$.

Suppose now that f is an arbitrary nonnegative superharmonic function on D. Then $f_1 = 1 - e^{-f}$ is a bounded nonnegative superharmonic function on D by Proposition 1.2. Thus by what has already been shown there is a constant $\beta \in [0, 1]$ such that $\lim_{t\to\infty} f_1(X(t)) = \beta$ a.s. (P_x) on $\{T_D = \infty\}$ for each $x \in D$. Set $\alpha = -\log(1 - \beta)$. Then $\lim_{t\to\infty} f(X(t)) = \alpha$ a.s. (P_x) on $\{T_D = \infty\}$ for each $x \in D$. It follows from Theorem 4.5 that α is finite. This completes the proof of the proposition.

Let φ be a function on ∂D and let $\alpha \in \mathbb{R}$. A solution to the *stochastic Dirichlet problem* for (φ, α) is a harmonic function f on D such that for $x \in D$

$$\lim_{t\uparrow T_D} f(X(t)) = \varphi(X(T_D)) \qquad \text{a.s. } (P_x) \quad \text{on} \quad \{T_D < \infty\}$$

$$= \alpha \qquad \text{a.s. } (P_x) \quad \text{on} \quad \{T_D = \infty\}.$$

The stochastic version of the Dirichlet problem was first formulated and solved by Doob [1].

Theorem 4.9. *Let φ be a function on ∂D such that $H_D|\varphi| < \infty$ on D and let $\alpha \in \mathbb{R}$. Then $H_D\varphi + \alpha P_{\cdot}(T_D = \infty)$ is a solution to the stochastic Dirichlet problem for (φ, α). If φ is bounded on ∂D, then $H_D\varphi + \alpha P_{\cdot}(T_D = \infty)$ is the unique bounded solution to the stochastic Dirichlet problem for (φ, α).*

Proof. It follows from Propositions 4.6 and 4.7 that $H_D\varphi + \alpha P_{\cdot}(T_D = \infty)$ is a solution to the stochastic Dirichlet problem for (φ, α). Suppose φ is a bounded function and let f be a bounded solution to the stochastic Dirichlet problem for (φ, α). Then $H_D\varphi + \alpha P_{\cdot}(T_D = \infty) \leq f$ by Theorem 4.5. Similarly $H_D(-\varphi) - \alpha P_{\cdot}(T_D = \infty) \leq -f$ and hence $H_D\varphi + \alpha P_{\cdot}(T_D = \infty) = f$. This completes the proof of the theorem

Suppose $d \geq 2$ and $r > 0$. Let f be a nonnegative harmonic function on $\mathring{B}_r$, let μ_f be as in Section 4.4 and write $\mu_f = \mu_f^a + \mu_f^s$, where μ_f^a is absolutely continuous with respect to σ_r and μ_f^s is singular with respect to σ_r. Let φ_f be the function defined on S_r by $\varphi_f = d\mu_f^a/d\sigma_r$.

Proposition 4.10. *Suppose $d \geq 2$ and $r > 0$ and let f be a nonnegative harmonic function on $D = \mathring{B}_r$. Then $\lim_{t \uparrow T_D} f(X(t)) = \varphi_f(X(T_D))$ a.s. (P_x) for $x \in D$.*

Proof. Let f^a and f^s be the nonnegative harmonic functions on D corresponding to μ_f^a and μ_f^s, respectively. Then $f = f^a + f^s$ on D and and $f^a = H_D\varphi_f$ on D by Proposition 4.4.5. Let $x \in D$. By Proposition 4.6, $\lim_{t \uparrow T_D} f^a(X(t)) = \varphi_f(X(T_D))$ a.s. (P_x). Thus it suffices to show that $\lim_{t \uparrow T_D} f^s(X(t)) = 0$ a.s. (P_x). In other words it suffices to show that if f is a nonnegative harmonic function on D such that μ_f is singular with respect to σ_r, then $\lim_{t \uparrow T_D} f(X(t)) = 0$ a.s. (P_x).

To this end note that $f \wedge 1$ is nonnegative and superharmonic on D. By Theorems 1.23 and 4.5

$$f_1 = E_{\cdot}\left(\lim_{t \uparrow T_D} (f \wedge 1)(X(t))\right) = E_{\cdot}\left(\lim_{t \uparrow T_D} f(X(t)) \wedge 1\right)$$

is the greatest harmonic minorant of $f \wedge 1$ on D. In particular f_1 is harmonic on D and $f_1 \leq f \wedge 1$ on D. Thus $\mu_{f_1} \leq \mu_f$ and $\mu_{f_1} \leq \sigma_r$ by Proposition 4.4.4. Since μ_f is singular with respect to σ_r, $\mu_{f_1} = 0$ and hence $f_1 = 0$. Thus $(\lim_{t \uparrow T_D} f(X(t))) \wedge 1 = 0$ a.s. (P_x) and hence $\lim_{t \uparrow T_D} f(X(t)) = 0$ a.s. (P_x) as desired. This completes the proof of the proposition.

Let f be nonnegative and superharmonic on D. By Theorem 4.5, $f_1 = E_{\cdot}(\lim_{t \uparrow T_D} f(X(t)))$ is a harmonic minorant of f on D. Proposition 4.10 can be used to show that f_1 need not be the greatest harmonic minorant of f on D. The details are left to the reader.

Let D_0 be an open subset of D. In Theorem 4.2.18 and Theorem 1.15 conditions were given which guarantee that harmonic and superharmonic functions on D_0 can be extended, respectively, to harmonic and superharmonic functions on D. In both results it was assumed that $D \backslash D_0$ is polar. The necessity of this condition will now be considered. Suppose first that $d = 1$ and that D_0 is an open interval. Then a harmonic function on D_0 is necessarily linear and can therefore be extended to a harmonic function on $\mathbb{R}$. If $D_0 \neq \mathbb{R}$, then D_0 must have at least one finite end point. There is a nonnegative superharmonic (concave) function on D_0 whose first derivative becomes infinite at the finite end points of D_0. Such a function cannot be extended to a superharmonic function on any larger interval. The remaining cases are covered by the following result.

Theorem 4.11. *Let D be connected and let D_0 be a nonempty open subset of D such that $D \backslash D_0$ is nonpolar and D_0 is not a subinterval of $\mathbb{R}$. Then there is a bounded harmonic function on D_0 which cannot be extended to a superharmonic function on D.*

Proof. Suppose first that $d = 1$. Then D_0 must be the union of two or more disjoint open subintervals of $\mathbb{R}$. Let f be a harmonic function on D_0 which equals zero on one component of D_0 and one on the remainder of D_0. Then f is a bounded harmonic function on D_0 which cannot be extended to a superharmonic function on the interval D.

Suppose next that $d \geq 2$ and set $B = D \backslash D_0$. Now $P_{.}(\tau_B < T_D)$ is a nonnegative superharmonic function on D by Proposition 1.21. Also $P_{.}(\tau_B < T_D) = 1$ on $B \cap B^r$, which is nonempty since B is nonpolar and $B \backslash B^r$ is polar. Thus $P_{.}(\tau_B < T_D) > 0$ on D by Proposition 1.3.

Let $x \in D_0$ be fixed and let μ denote the measure $P_x(\tau_B < T_D, X(\tau_B) \in \cdot)$. Then $\mu(\mathbb{R}^d) = P_x(\tau_B < T_D) > 0$. Let $S \subset \partial D_0 \cap B$ denote the support of μ. Now μ is nonatomic since one-point sets are polar for $d \geq 2$, so there is a countable dense subset S_0 of S and necessarily $\mu(S_0) = 0$. Thus there is an open set U containing S_0 such that $\mu(S \backslash U) > 0$. Set $A_1 = S \cap U$ and $A_2 = \partial D_0 \backslash A_1$. Then $A_2 \cap S = S \backslash U$ and hence $\mu(A_2 \cap S) > 0$. If $y \in S$ and N is an open neighborhood of y, then $\mu(A_1 \cap N) > 0$ (for N contains a point $y_0 \in S_0$ and $U \cap N$ is an open neighborhood of y_0, so $\mu(A_1 \cap N) = \mu(S \cap U \cap N) = \mu(U \cap N) > 0$).

Set $\varphi = I_{A_2}$ on ∂D_0 and $f = H_{D_0}\varphi$ on D_0. Then f is a bounded harmonic function on D_0 and $\lim_{t \uparrow T_{D_0}} f(X(t)) = \varphi(X(T_{D_0}))$ a.s. (P_x) on $\{T_{D_0} < \infty\}$ by Proposition 4.6. In particular $\lim_{t \uparrow \tau_B} f(X(t)) = \varphi(X(\tau_B))$ a.s. (P_x) on $\{\tau_B < T_D\}$, since $T_{D_0} = \tau_B$ when $\tau_B < T_D$. Suppose f extends to a superharmonic function on D, still denoted by f. Then $\lim_{t \uparrow \tau_B} f(X(t)) = f(X(\tau_B))$ a.s. (P_x) on $\{\tau_B < T_D\}$ by Theorem 4.5. Consequently $f = \varphi$ a.e. (μ) on S. Thus there is a $y \in A_2 \cap S$ such that $f(y) = 1$. Also $f = 0$ a.e. (μ) on A_1, so by the last sentence of the previous paragraph there are points $y_n \in A_1$ such that $f(y_n) = 0$ and $y_n \to y$. Hence f is not lower semicontinuous on D, which contradicts the supposition that f is superharmonic on D. Therefore the bounded harmonic function f on D_0 has no superharmonic extension to D. This completes the proof of the theorem.

Chapter 6

Potential Theory

In the fist six sections of this chapter, D is a nonempty Greenian open subset of $\mathbb{R}^d$. The theory of Green potentials $G_D\mu$ is obtained by using probabilistic methods developed mainly by Hunt [2] (see Blumenthal and Getoor [1], especially Section 6 of Chapter III, Section 1 of Chapter V, and Sections 1, 2, and 4 of Chapter VI). If $d \geq 3$ then $\mathbb{R}^d$ is Greenian and $G_{\mathbb{R}^d}\mu = g\mu$, so the theory of Green potentials includes that of Newtonian potentials as a special case (some of the papers cited in the text treat only Newtonian potentials). An application of this theory to electrostatics is presented in Section 6. The theory of Green potentials is used in Section 7 to obtain a corresponding theory of logarithmic potentials. For the analytic approach to classical potential theory see Brelot [8], Helms [1], and Landkof [1]. For detailed discussions of the interesting historical development of the analytic approach see Brelot [7], [9] and the comments at the end of Helms [1] and Landkof [1]. For discussions of the development of probabilistic potential theory see the comments at the end of Dynkin [1], Blumenthal and Getoor [1], Kemeny, Snell and Knapp [1] and Revuz [1].

1. Green Potentials

Let μ be a measure on D (when convenient, μ can be thought of as a measure on $\mathbb{R}^d$ with $\mu(\mathbb{R}^d\backslash D) = 0$). The *Green potential* $G_D\mu$ of μ is defined by $G_D\mu = \int G_D(\cdot, y)\mu(dy)$. Let D_0 be a component of D. Then $G_D\mu = \int_{D_0} G_{D_0}(\cdot, y)\mu(dy)$

on D_0, where for each $x \in D_0$, $G_{D_0}(x, \cdot)$ is positive on D_0 and bounded away from zero on compact subsets of D_0. Thus $\mu(D_0) = 0$ if and only if $G_D\mu = 0$ on D_0. If $G_D\mu$ is not identically infinite on D_0, then μ assigns finite measure to each compact subset of D_0. Therefore if $G_D\mu$ is not identically infinite on any component of D and in particular if $G_D\mu$ is superharmonic on D, then μ is a Radon measure on D.

Let μ and ν be measures on D. It follows from Fubini's theorem and the symmetry of $G_D(x, y)$ in x and y that $\int G_D\mu \, d\nu = \int G_D\nu \, d\mu$, where $\int G_D\mu \, d\nu = \int G_D\mu(x)\nu(dx)$. In particular if f is nonnegative on D, then $\int G_D\mu(x)f(x)\, dx = \int G_D f \, d\mu$.

Set $Q_D^t\mu = \int Q_D(t, \cdot, y)\mu(dy)$ for $t > 0$. Then $G_D\mu = \int_0^\infty Q_D^t\mu \, dt$.

Theorem 1.1. *Let μ be a measure on D. Then $G_D\mu$ is either superharmonic on D or identically infinite on some component D_0 of D such that $\mu(D_0) = \infty$.*

Proof. Without loss of generality it can be assumed that D is connected. By Fatou's lemma, $G_D\mu = \int G_D(\cdot, y)\mu(dy)$ is lower semicontinuous on D. For $t > 0$

$$Q_D^t G_D\mu = Q_D^t \int_0^\infty Q_D^s\mu \, ds = \int_0^\infty Q_D^{s+t}\mu \, ds = \int_t^\infty Q_D^s\mu \, ds \le G_D\mu.$$

Thus by Theorem 5.1.16, $G_D\mu$ is either superharmonic on D or identically infinite on D. Suppose μ is a finite measure. If A is a compact subset of D, then $\int_A G_D\mu(x)\, dx = \int G_D I_A \, d\mu < \infty$ by Proposition 2.2.7. Thus $G_D\mu$ is locally integrable and hence superharmonic on D. This completes the proof of the theorem.

Let $\mathscr{M}_D$ denote the collection of measures μ on D such that $G_D\mu$ is superharmonic on D. For any measure μ on D, the following statements are equivalent: $\mu \in \mathscr{M}_D$; $G_D\mu$ is superharmonic on D; $G_D\mu$ is excessive on D; $G_D\mu$ is locally integrable on D; $G_D\mu$ is not identically infinite on any component of D. If $\mu \in \mathscr{M}_D$, then μ is a Radon measure. For $B \subset \mathbb{R}^d$ let $\mathscr{M}_D(B)$ denote the collection of measures $\mu \in \mathscr{M}_D$ which are concentrated on B.

Proposition 1.2. *Let $\mu \in \mathscr{M}_D$. Then the greatest harmonic minorant of $G_D\mu$ on D is zero.*

Proof. Let h be the greatest harmonic minorant of $G_D\mu$ on D, let C be a compact subset of D, and let D_n be relatively compact open subsets of D such that $C \subset D_1$ and $D_n \uparrow D$. It follows from (1) in the introduction to Chapter 3 that

$$G_D(x, y) = g_{D_n^c}(x, y) + \int H_{D_n}(x, dz)G_D(z, y), \qquad x, y \in D.$$

Consequently by the finiteness of $G_D(x, y)$ for $x \neq y$ and the symmetry of $G_D(x, y)$ and $g_{D_n^c}(x, y)$ in x and y

$$\int H_{D_n}(x, dz)G_D(z, y) = \int H_{D_n}(y, dz)G_D(z, x), \qquad x, y \in D.$$

Thus by the strong Markov property

$$\int_C H_{D_n}G_D\mu(x)\,dx = \int \mu(dy)H_{D_n}G_D I_C(y) = \int \mu(dy)\int_{T_{D_n}}^{T_D} I_C(X(t))\,dt.$$

Now $H_{D_1}G_D\mu$ is integrable on C since $H_{D_1}G_D\mu \leq G_D\mu$ on D_1 by Theorem 5.1.17; also $P_.(T_{D_n} \uparrow T_D) = 1$ on D by Proposition 2.3.8. Therefore by the dominated convergence theorem $\lim_n \int_C H_{D_n}G_D\mu(x)\,dx = 0$. Now $H_{D_n}G_D\mu \downarrow h$ on D by Theorem 5.1.23, so $\int_C h(x)\,dx = 0$. Since C is an arbitrary compact subset of D, $h = 0$ on D. This completes the proof of the proposition.

Suppose $d \geq 2$ and let $\mu \in \mathcal{M}_D$. One might expect that

$$\text{(1)} \qquad \lim_{x \to b} G_D\mu(x) = \lim_{x \to b} \int G_D(x, y)\mu(dy)$$
$$= \int \left(\lim_{x \to b} G_D(x, y)\right)\mu(dy) = 0, \qquad b \in \partial D \cap (D^c)^r,$$

but taking the limit under the integral sign is not necessarily valid (it is valid when $d = 1$). Indeed since $G_D(x, x) = \infty$ for $x \in D$ by Proposition 4.5.4 it is easy to construct examples in which (1) fails to hold. Let x_n, $n \geq 1$, be distinct points of D such that $\lim_n x_n = b \in \partial D \cap (D^c)^r$. Let μ be a finite measure on D which is concentrated on $\{x_n : n \geq 1\}$ and such that $\mu(\{x_n\}) > 0$ for $n \geq 1$. Then $\mu \in \mathcal{M}_D$ but $G_D\mu(x_n) = \infty$ for $n \geq 1$, so (1) fails to hold. Equation (1) does hold under some restrictions on μ. By Theorem 4.6.7 (1) holds if μ is a finite absolutely continuous measure having a bounded density. By (4.5.1), it also holds if μ is a finite measure such that $\mu(B_r(b)) = 0$ for some $r > 0$. In general it follows from Theorem 5.4.5 that $P_.(\lim_{t \uparrow T_D} G_D\mu(X(t))$ exists$) = 1$ on D and that $E_.(\lim_{t \uparrow T_D} G_D\mu(X(t)))$ is a harmonic minorant of $G_D\mu$ on D. Thus by Proposition 1.2

$$P_.\left(\lim_{t \uparrow T_D} G_D\mu(X(t)) = 0\right) = 1 \qquad \text{on } D,$$

which is suprisingly close to (1).

Theorem 1.3. *Let $\mu, \nu \in \mathcal{M}_D$ be such that $G_D\mu = G_D\nu$ a.e. on D. Then $\mu = \nu$.*

Proof. Without loss of generality it can be assumed that D is connected. Let $t > 0$. Note that $Q_D^t G_D \mu = Q_D^t G_D \nu$. If $x \in D$ and $G_D\mu(x) < \infty$, then

$$G_D\mu(x) - Q_D^t G_D \mu(x) = \int_0^t Q_D^s \mu(x)\, ds.$$

Since the same result holds with μ replaced by ν,

$$\int_0^t Q_D^s \mu\, ds = \int_0^t Q_D^s \nu\, ds \qquad \text{a.e. on } D. \tag{2}$$

Let f be a bounded nonnegative function on D which vanishes outside some compact subset of D and is such that $\int f(x)\, dx > 0$. Then $G_D f > 0$ on D by the positivity of G_D on $D \times D$, $G_D f$ is bounded on D by Proposition 2.2.7, and $G_D f$ is continuous on D by Theorem 4.6.6. Also $\int G_D f\, d\mu = \int G_D \mu(x) f(x)\, dx < \infty$ by Proposition 5.1.10 and similarly $\int G_D f\, d\nu < \infty$. Let φ be a continuous function on D having compact support and such that $0 \le \varphi \le 1$. By the symmetry of $Q_D(s, x, y)$ in x and y, $\int Q_D^s(\varphi G_D f)\, d\mu = \int Q_D^s \mu(x) \varphi(x) G_D f(x)\, dx$ and the same formula holds with μ replaced by ν. Thus by (2)

$$\int \left(\frac{1}{t} \int_0^t Q_D^s(\varphi G_D f)\, ds \right) d\mu = \int \left(\frac{1}{t} \int_0^t Q_D^s(\varphi G_D f)\, ds \right) d\nu. \tag{3}$$

Now $\varphi G_D f$ is a bounded continuous function on D, so

$$\lim_{t \to 0} \frac{1}{t} \int_0^t Q_D^s(\varphi G_D f)\, ds = \varphi G_D f \qquad \text{on } D. \tag{4}$$

Since

$$\frac{1}{t} \int_0^t Q_D^s(\varphi G_D f)\, ds \le \frac{1}{t} \int_0^t Q_D^s(G_D f)\, ds \le G_D f,$$

it follows from (3), (4), and the dominated convergence theorem that $\int \varphi G_D f\, d\mu = \int \varphi G_D f\, d\nu$. Since this is true for every such function φ, the two measures $G_D f\, d\mu$ and $G_D f\, d\nu$ are equal. Since $G_D f > 0$ on D, $\mu = \nu$ as desired.

Let μ be a measure on D and let $B \subset D$. The measure $\mu|_B$ defined on D by $\mu|_B(A) = \mu(A \cap B)$ is called the *restriction of* μ *to* B. It is sometimes convenient to view $\mu|_B$ as the measure on B given by $\mu|_B(A) = \mu(A)$ for $A \subset B$.

Let $B \in \mathscr{B}$. The *hitting distribution* $h_{B,D}(x, \cdot)$ of B for Brownian motion on D starting at $x \in D$ is the measure on D defined by

$$h_{B,D}(x, A) = P_x(\tau_B < T_D, X(\tau_B) \in A) = h_{B \cup D^c}(x, A), \qquad A \subset D.$$

Note that $h_{B,D}(x, \cdot) = h_{B \cap D, D}(x, \cdot) = h_{B \cup D^c}(x, \cdot)|_D$ for $x \in D$. If $x \in B^r \cap D$, then $h_{B,D}(x, \cdot) = \delta_x$. It follows from Theorem 2.6.5 that if $x \in D \backslash B^r$, then $h_{B,D}(x, \cdot)$ is concentrated on $\partial B \cap B^r$ and does not charge polar sets. It follows as in the proof of the corresponding result of Theorem 2.6.5 that if $x \in D \backslash B^r$ and B is closed, then $h_{B,D}(x, \cdot) = h_{\partial B,D}(x, \cdot)$. Observe that

$$h_{B \cup D^c}(x, D^c \backslash (D^c)^r) = 0, \qquad x \in D. \tag{5}$$

For if $x \in D$, then $\tau_{B \cup D^c} = \tau_{D^c}$ on $\{X(0) = x, X(\tau_{B \cup D^c}) \in D^c\}$ so by Theorem 2.6.5

$$X(\tau_{B \cup D^c}) = X(\tau_{D^c}) \in (D^c)^r \qquad \text{a.s. } (P_x) \text{ on } \{X(\tau_{B \cup D^c}) \in D^c\}.$$

It follows from (5) and Theorem 2.6.7 that if $A \in \mathscr{B}$ and $A \supset B$, then $h_{A,D} h_{B,D} = h_{B,D} h_{A,D} = h_{B,D}$.

Let $B \in \mathscr{B}$. Then

$$G_D(x, y) = g_{B \cup D^c}(x, y) + \int h_{B,D}(x, dz) G_D(z, y), \qquad x, y \in D. \tag{6}$$

This equation is called the *fundamental identity for Green potentials.* Note that the first term on the right-hand side of (6) vanishes if either $x \in B^r$ or $y \in B^r$. Equation (6) follows from Eq. (1) in the introduction to Chapter 3 together with (5) and the fact that $g_{D^c}(z, y) = 0$ for $z \in (D^c)^r$ and $y \in D$.

Theorem 1.4. *Let $\mu \in \mathscr{M}_D$ and let D_0 be an open subset of D. Then $G_D \mu$ is harmonic on D_0 if and only if $\mu(D_0) = 0$.*

Proof. Suppose first that $\mu(D_0) = 0$. Let D_1 be a relatively compact open subset of D_0. By the fundamental identity $G_D \mu = H_{D_1} G_D \mu$ on D_1, so $G_D \mu$ is harmonic on D_1 by Theorem 4.3.7. Consequently $G_D \mu$ is harmonic on D_0. Suppose next that $\mu(D_0) > 0$. Choose $x \in D_0$ such that x is in the support of μ and set $D_1 = \mathring{B}_r(x)$ for some $r > 0$ such that $B_r(x) \subset D_0$. Then $G_{D_1} \mu > 0$ on D_1. By the fundamental identity $G_D \mu \geq G_{D_1} \mu + H_{D_1} G_D \mu > H_{D_1} G_D \mu$ on D_1. Thus $G_D \mu$ is not harmonic on D_0.

Theorem 1.5. *Let $\mu \in \mathscr{M}_D$ and let D_0 be an open subset of D. Then there is a nonnegative harmonic function h on D_0 such that $G_D \mu = G_{D_0} \mu|_{D_0} + h$ on D_0.*

Proof. Now $G_D \mu = G_D \mu|_{D_0} + G_D \mu|_{D \backslash D_0}$ and $h_1 = G_D \mu|_{D \backslash D_0}$ is harmonic on D_0 by Theorem 1.4. By the fundamental identity

$$G_D \mu\Big|_{D_0} = G_{D_0} \mu\Big|_{D_0} + \int_D H_{D_0}(\cdot, dz) G_D \mu\Big|_{D_0}(z) \qquad \text{on } D_0.$$

Let h_2 denote the second term on the right-hand side of this equation. Then h_2 is harmonic on D_0 by Theorem 4.3.7. Consequently, $G_D \mu = G_{D_0} \mu|_{D_0} + h$ on D_0, where $h = h_1 + h_2$ is harmonic on D_0 as desired.

A generalization of Theorem 1.4 will now be obtained.

Theorem 1.6. *Let $\mu, \nu \in \mathscr{M}_D$ and let D_0 be an open subset of D. Then $G_D\nu = G_D\mu + h$ on D_0 for some harmonic function h on D_0 if and only if $\nu|_{D_0} = \mu|_{D_0}$.*

Proof. By Theorem 1.5, $G_D\mu = G_{D_0}\mu|_{D_0} + h_1$ on D_0 and $G_D\nu = G_{D_0}\nu|_{D_0} + h_2$ on D_0 where h_1 and h_2 are harmonic on D_0. Suppose $\nu|_{D_0} = \mu|_{D_0}$. Then

$$G_D\nu = G_{D_0}\mu|_{D_0} + h_2 = G_D\mu + h_2 - h_1 = G_D\mu + h \qquad \text{on } D_0,$$

where $h = h_2 - h_1$ is harmonic on D_0. Suppose conversely that $G_D\nu = G_D\mu + h$ on D_0 where h is harmonic on D_0. Then

$$G_{D_0}\nu\Big|_{D_0} = G_{D_0}\mu\Big|_{D_0} + h + h_1 - h_2 \qquad \text{on } D_0.$$

Now $h + h_1 - h_0$ is a harmonic minorant of $G_{D_0}\nu|_{D_0}$ on D_0, so $h + h_1 - h_2 \le 0$ on D_0 by Proposition 1.2. Similarly $-(h + h_1 - h_2)$ is a harmonic minorant of $G_D\mu|_{D_0}$ on D_0 and hence $h + h_1 - h_2 \ge 0$ on D_0. Consequently $h + h_1 - h_2 = 0$ on D_0 and hence $G_{D_0}\nu|_{D_0} = G_{D_0}\mu|_{D_0}$ on D_0. Thus $\nu|_{D_0} = \mu|_{D_0}$ by Theorem 1.3. This completes the proof of the theorem.

A necessary and sufficient condition for a measure on D to be in $\mathscr{M}_D$ will now be obtained.

Theorem 1.7. *Let μ be a measure on D. Then $\mu \in \mathscr{M}_D$ if and only if*

$$\int_{\|y-x\|>r} G_D(x, y)\mu(dy) < \infty, \qquad x \in D \text{ and } r > 0. \tag{7}$$

Proof. Suppose first that $\mu \in \mathscr{M}_D$. Choose $x \in D$ and $r > 0$. Then $G_D\mu|_{D\setminus B_r(x)}$ is harmonic on $\mathring{B}_r(x)$ by Theorem 1.4. In particular

$$\int_{\|y-x\|>r} G_D(x, y)\mu(dy) = G_D\mu\Big|_{D\setminus B_r(x)}(x) < \infty.$$

Therefore (7) holds. Suppose conversely that (7) holds. Choose $x \in D$ and $r > 0$ such that $B_r(x) \subset D$. Set $\mu_1 = \mu|_{B_r(x)}$ and $\mu_2 = \mu|_{D\setminus B_r(x)}$. By (7), μ_1 is a finite measure on D and hence $\mu_1 \in \mathscr{M}_D$. Thus $G_D\mu_1 < \infty$ a.e. on D. Choose $y \in \mathring{B}_r(x)$ such that $G_D\mu_1(y) < \infty$. There is an $s > 0$ such that $B_s(y) \subset B_r(x)$. Now

$$G_D\mu_2(y) = \int_{\|z-x\|>r} G_D(y, z)\mu(dz) \le \int_{\|z-y\|>s} G_D(y, z)\mu(dz) < \infty$$

by (7). Consequently $G_D\mu(y) < \infty$. This shows that $G_D\mu < \infty$ a.e. on $\mathring{B}_r(x)$ and therefore that $G_D\mu < \infty$ a.e. on D. Thus $\mu \in \mathscr{M}_D$, which completes the proof of the theorem.

The next theorem, called the *maximum principle* for Green potentials, was obtained by Maria [1] (and extended shortly thereafter to Riesz potentials by Frostman [1]).

Theorem 1.8. *Let μ be a measure on D and let B be a subset of D on which μ is concentrated. Then* $\sup_{x \in D} G_D\mu(x) = \sup_{x \in B} G_D\mu(x)$.

Proof. This result follows by the same argument used to prove the maximum principle for λ-potentials (Theorem 2.5.1). The details are left to the reader.

The next theorem, due to Evans [1] and Vasilesco [1], is called the *continuity principle* for Green potentials.

Theorem 1.9. *Let μ be a measure on D which is supported on a relatively closed subset B of D and such that the restriction of $G_D\mu$ to B is continuous on B. Then $G_D\mu$ is continuous on D.*

Proof. Since $G_D\mu$ is assumed to be finite on B, it follows from Theorem 1.1 that $\mu \in \mathscr{M}_D(B)$. By Theorem 1.4, $G_D\mu$ is harmonic and hence continuous on $D\backslash B$. Choose $x \in B$ and $\varepsilon > 0$. Since $G_D\mu(x) < \infty$, there is an $r > 0$ such that

$$G_D\mu\Big|_{B_r(x)}(x) = \int_{B_r(X)} G_D(x, y)\mu(dy) \le \varepsilon/2.$$

Observe that

$$G_D\mu = G_D\mu\Big|_{B_r(x)} + G_D\mu\Big|_{D\backslash B_r(x)} \qquad \text{on } D. \tag{8}$$

Now $G_D\mu|_{D\backslash B_r(x)}$ is harmonic and hence continuous on $\mathring{B}_r(x)$ by Theorem 1.4, so it follows from (8) and the hypothesis of the theorem that the restriction of $G_D\mu|_{B_r(x)}$ to B is continuous on $B \cap \mathring{B}_r(x)$. Thus by making the positive number r smaller if necessary it can be assumed that $G_D\mu|_{B_r(x)} \le \varepsilon$ on $B \cap B_r(x)$. Consequently by the maximum principle

$$G_D\mu\Big|_{B_r(x)} \le \varepsilon \qquad \text{on } D. \tag{9}$$

It follows from (8) and (9) and the continuity of $G_D\mu|_{D\ B_r(x)}$ on $\mathring{B}_r(x)$ that

$$\overline{\lim_{y \to x}}\, G_D\mu(y) \le G_D\mu\,|_{D\backslash B_r(x)}(x) + \varepsilon \le G_D\mu(x) + \varepsilon.$$

Therefore $\overline{\lim}_{y \to x} G_D\mu(y) \le G_D\mu(x)$ for $x \in B$. Since $G_D\mu$ is lower semicontinuous on D, it is continuous on B and hence on all of D. This completes the proof of the theorem. (For an alternative method of proof see the proof of Theorem 4.12 of Chapter VI of Blumenthal and Getoor [1].)

Theorem 1.10. *Let B be a compact subset of D and let* $\mu \in \mathscr{M}_D(B)$ *be such that* $G_D\mu < \infty$ *a.e.* (μ). *Then for every* $\varepsilon > 0$ *there is a compact subset C of B such that* $\mu(B \backslash C) \leq \varepsilon$ *and* $G_D\mu|_C$ *is continuous on D.*

Proof. By hypothesis $\mu(B) < \infty$. Choose $\varepsilon > 0$. By Lusin's theorem there is a compact subset C of B such that $\mu(B\backslash C) \leq \varepsilon$ and the restriction of $G_D\mu$ to C is continuous on C. Now $G_D\mu = G_D\mu|_C + G_D\mu|_{B\backslash C}$. Since $G_D\mu|_{B\backslash C}$ is lower semicontinuous on D, the restriction of $G_D\mu|_C$ to C is upper semicontinuous on C. Since $G_D\mu|_C$ is lower semicontinuous on D, the restriction of this function to C is continuous on C. By the continuity principle, $G_D\mu|_C$ is continuous on D.

2. Riesz Decomposition Theorem

In this section it is determined when a superharmonic function on D is a Green potential and when it can be written in the form $G_D\mu + h$ for some $\mu \in \mathscr{M}_D$ and harmonic function h on D. It will first be shown that every nonnegative superharmonic function on D is an increasing limit of continuous Green potentials.

Theorem 2.1. *Let f be a nonnegative superharmonic function on D. Then there exist measures* $\mu_n \in \mathscr{M}_D$ *such that* $G_D\mu_n$ *is continuous and* $G_D\mu_n \uparrow f$ *on D.*

Proof. Let C_n be compact subsets of D such that $C_n \uparrow D$. Then $nG_DI_{C_n}$ is bounded and excessive on D for each n and $nG_DI_{C_n} \uparrow \infty$ on D. Set $f_n = f \wedge nG_DI_{C_n}$. Then f_n is bounded and excessive (recall that the minimum of two superharmonic functions is superharmonic) and $f_n \uparrow f$. By the dominated convergence theorem, for each $n \geq 1$

$$Q_D^t f_n \leq nQ_D^t G_D I_{C_n} = n \int_{C_n} \left(\int_t^\infty Q_D(u, \cdot, y)\, du \right) dy \downarrow 0 \qquad \text{as } t \uparrow \infty.$$

Choose $s > 0$. It follows by letting $r \uparrow \infty$ in the identity

$$\left(\int_0^r Q_D^t \, dt \right) (f_n - Q_D^s f_n) = \int_0^s Q_D^t f_n \, dt - \int_r^{r+s} Q_D^t f_n \, dt$$

that

$$G_D\left(\frac{f_n - Q_D^s f_n}{s} \right) = \frac{1}{s} \int_0^s Q_D^t f_n \, dt \leq \frac{1}{s} \int_0^s Q_D^t f \, dt \leq f.$$

Set $\varphi_n = n(f_n - Q_D^{1/n} f_n)$. Then $G_D\varphi_n \le f$. Also $G_D\varphi_n = n\int_0^{1/n} Q_D^t f_n\,dt$ is nondecreasing in n since

$$n\int_0^{1/n} Q_D^t f_n\,dt \ge n\int_0^{1/n} Q_D^t f_m\,dt \ge m\int_0^{1/m} Q_D^t f_m\,dt, \qquad n \ge m.$$

Since $Q_D(t,x,y)$ is jointly continuous on $(0,\infty)\times D\times D$ by Theorem 2.4.3 and $Q_D(t,x,y) \le p(t,x,y)$ it follows from the boundedness of f_n that $G_D\varphi_n$ is continuous on D. Let m be a positive integer. Then

$$G_D\varphi_n = n\int_0^{1/n} Q_D^t f_n\,dt \ge n\int_0^{1/n} Q_D^t f_m\,dt, \qquad n \ge m.$$

Now $Q_D^t f_m \uparrow f_m$ as $t \downarrow 0$ since f_m is excessive, so $\underline{\lim}_n G_D\varphi_n \ge f$ and therefore $G_D\varphi_n \uparrow f$. Let μ_n be the measure on D given by $\mu_n(dx) = \varphi_n(x)\,dx$. Then μ_n satisfies the conclusion of the theorem.

Let $\mu_n \in \mathscr{M}_D$ be such that $\lim_n G_D\mu_n = f$ for some function f on D. Suppose also that there is a locally integrable function f_1 on D such that $G_D\mu_n \le f_1$ for all n (which condition is automatically satisfied if $G_D\mu_n \downarrow f$ or if $G_D\mu_n \uparrow f$ and f is locally integrable on D). Then f is locally integrable on D. For each $t > 0$, $Q_D^t G_D\mu_n \le G_D\mu_n$ and hence by Fatou's lemma $Q_D^t f \le f$. Consequently f is superinvariant. Thus by Theorem 5.1.16, $\underline{f}$ is a nonnegative superharmonic function on D and $\underline{f} = f$ a.e. on D. The following theorem is taken from Chapter VI of Blumenthal and Getoor [1].

Theorem 2.2. *Let $\mu_n \in \mathscr{M}_D$ be such that $\lim_n G_D\mu_n = f$ for some function f on D. Suppose also that there is a locally integrable function f_1 on D such that $G_D\mu_n \le f_1$ for all n. Then the following condition is necessary and sufficient for there to exist a measure $\mu \in \mathscr{M}_D$ such that $G_D\mu = \underline{f}$ and μ_n converges vaguely to μ: for every compact subset C of D and every $\varepsilon > 0$ there is a compact subset A of D such that $\int_{A^c} G_D I_C\,d\mu_n \le \varepsilon$ for all n.*

Proof. Suppose the indicated condition holds. It will first be shown that $\sup_n \mu_n(C) < \infty$ for every compact subset C of D. Without loss of generality it can be assumed that C lies in a single component D_0 of D. Choose $x_0 \in D_0$ such that $f(x_0) < \infty$. Then $\alpha = \min_{y\in C} G_D(x_0, y) > 0$ and $G_D\mu_n(x_0) \ge \alpha\mu_n(C)$. Consequently $\overline{\lim}_n \mu_n(C) \le \alpha^{-1} f(x_0) < \infty$ and hence $\sup_n \mu_n(C) < \infty$ as desired.

By the result of the first paragraph there is a Radon measure μ on D and a strictly increasing sequence $\{n_j\}$ of positive integers such that μ_{n_j} converges vaguely to μ. For each $x \in D$, $G_D(x,\cdot)$ is nonnegative and continuous in the extended sense on D and hence

$$f(x) = \lim_j \int G_D(x,y)\mu_{n_j}(dy) \ge \int G_D(x,y)\mu(dy) = G_D\mu(x).$$

Consequently $f \geq G_D\mu$ on D. Since f is locally integrable on D, so is $G_D\mu$. Therefore $\mu \in \mathcal{M}_D$

Let C be a compact subset of D and let $\varepsilon > 0$. Let A be a compact subset of D such that $\int_{A^c} G_D I_C \, d\mu_n \leq \varepsilon$ for all $n \geq 1$. Let φ be a continuous function on D having compact support and such that $0 \leq \varphi \leq 1$ and $\varphi = 1$ on A. Then $G_D I_C$ is continuous on D by Theorem 4.6.6 and hence $\varphi G_D I_C$ is continuous on D and has compact support. For each j

$$\int_C G_D\mu_{n_j}(x)\, dx = \int G_D I_C \, d\mu_{n_j} \leq \int \varphi G_D I_C \, d\mu_{n_j} + \varepsilon.$$

Thus by Fatou's lemma

$$\int_C f(x)\, dx \leq \int \varphi G_D I_C \, d\mu + \varepsilon \leq \int G_D I_C \, d\mu + \varepsilon.$$

Since ε can be made arbitrarily small, $\int_C f(x)\, dx \leq \int_C G_D\mu(x)\, dx$ for every compact subset C of D. But $G_D\mu \leq f$ on D, so $G_D\mu = f$ a.e. on D. By Theorem 5.1.16, $G_D\mu = \underline{f}$ on a.e. on D. Thus by Theorem 5.1.14, $G_D\mu = \underline{f}$ on D. It follows from Theorem 1.3 that the measure μ is independent of the particular choice of the sequence $\{n_j\}$, so μ_n converges vaguely to μ. This completes the proof of the sufficiency of the condition.

Suppose conversely that $\mu \in \mathcal{M}_D$, $G_D\mu = \underline{f}$ and μ_n converges vaguely to μ. Let C be a compact subset of D and let $\varepsilon > 0$. By Proposition 5.1.10, $\int G_D I_C \, d\mu < \infty$ and $\int G_D I_C \, d\mu_n < \infty$ for each n. There is a compact subset A_1 of D such that $\int_{A_1^c} G_D I_C \, d\mu \leq \varepsilon/2$. By the dominated convergence theorem

$$\lim_n \int G_D I_C \, d\mu_n = \lim_n \int_C G_D\mu_n(x)\, dx = \int_C G_D\mu(x)\, dx = \int G_D I_C \, d\mu.$$

Let A_2 be a compact subset of D containing A_1 in its interior. Since $G_D I_C$ is continuous on D it follows from the vague convergence of μ_n to μ that

$$\underline{\lim}_n \int_{A_2} G_D I_C \, d\mu_n \geq \int_{A_1} G_D I_C \, d\mu \geq \int G_D I_C \, d\mu - \varepsilon/2.$$

Consequently $\overline{\lim}_n \int_{A_2^c} G_D I_C \, d\mu_n \leq \varepsilon/2$. Hence there is a positive integer n_0 such that $\int_{A_2^c} G_D I_C \, d\mu_n \leq \varepsilon$ for all $n > n_0$. There is a compact subset A of D containing A_2 such that $\int_{A^c} G_D I_C \, d\mu_n \leq \varepsilon$ for $1 \leq n \leq n_0$. Therefore $\int_{A^c} G_D I_C \, d\mu_n \leq \varepsilon$ for all n. Thus the indicated condition holds. This completes the proof of the theorem.

Let $\mu_n \in \mathcal{M}_D$ for $n \geq 1$. The condition in Theorem 2.2 is clearly satisfied if the measures μ_n are all supported on a fixed compact subset of D. It is also satisfied if $G_D\mu_n \leq f_1$ for all n where f_1 is a nonnegative superharmonic function on D whose greatest harmonic minorant on D is zero. To see this let C be a compact subset of D and let $\varepsilon > 0$. By Theorem 5.1.23 and the dominated convergence theorem there is a relatively compact open subset D_0 of D such that $C \subset D_0$ and $\int_C H_{D_0} f_1(x) \leq \varepsilon$. Set $A = \bar{D}_0$. By the fundamental identity, for all $n \geq 1$

$$G_D \mu_n|_{A^c} = H_{D_0} G_D \mu_n|_{A^c} \leq H_{D_0} G_D \mu_n \leq H_{D_0} f_1$$

on D_0 and hence

$$\int G_D I_C \, d\mu_n|_{A^c} = \int_C dx \, G_D \mu_n|_{A^c}(x) \leq \int_C H_{D_0} f_1(x) \, dx \leq \varepsilon$$

for all n.

A necessary and sufficient condition for a nonnegative superharmonic function to be a Green potential will now be given.

Theorem 2.3. *Let f be a nonnegative superharmonic function on D. Then f is the Green potential of a measure on D if and only if the greatest harmonic minorant of f on D is zero.*

Proof. The necessity of the condition has already been shown in Proposition 1.2. Suppose conversely that the greatest harmonic minorant of f on D is zero. By Theorem 2.1 there exist measures $\mu_n \in \mathscr{M}_D$ such that $G_D \mu_n \uparrow f$ on D. By the argument following the proof of Theorem 2.2, the condition in that theorem is satisfied. Thus by Theorem 2.2 there is a measure $\mu \in \mathscr{M}_D$ such that $G_D \mu = f$ on D. This completes the proof of the theorem.

Theorem 2.4. *Let $\mu_n \in \mathscr{M}_D$ be such that $G_D \mu_n \leq G_D \nu$ for all n, where $\nu \in \mathscr{M}_D$, and $\lim_n G_D \mu_n = f$ for some function f on D. Then there is a measure $\mu \in \mathscr{M}_D$ such that $G_D \mu = \underline{f}$ and μ_n converges vaguely to μ.*

Proof. By Theorem 2.3 the greatest harmonic minorant of $G_D \nu$ on D is zero. By the argument following the proof of Theorem 2.2 the condition in that theorem is satisfied. The desired result now follows from Theorem 2.2.

Theorem 2.5. *Let f_1 and f_2 be nonnegative superharmonic functions on D. If f_2 is the Green potential of a measure on D and $f_1 \leq f_2$ on D, then f_1 is the Green potential of a measure on D.*

Proof. The greatest harmonic minorant of f_2 on D is zero by Theorem 2.3, so the greatest harmonic minorant of f_1 on D is zero. The desired conclusion follows by another application of Theorem 2.3.

The next result is due to Riesz [1].

Theorem 2.6. *Let f be superharmonic on D. Then $f = G_D\mu + h$ for some measure $\mu \in \mathcal{M}_D$ and harmonic function h on D if and only if f has a harmonic minorant on D, in which case h is the greatest harmonic minorant of f on D and μ is also uniquely determined by f.*

Proof. If $f = G_D\mu + h$ for some measure $\mu \in \mathcal{M}_D$ and harmonic function h on D, then h is a harmonic minorant of f on D. Suppose conversely that f has a harmonic minorant on D. Let h denote the greatest harmonic minorant of f on D. Then $f - h$ is a nonnegative superharmonic function on D having greatest harmonic minorant zero on D. Thus by Theorem 2.3 there is a measure $\mu \in \mathcal{M}_D$ such that $f - h = G_D\mu$ and hence $f = G_D\mu + h$ on D. Suppose also that $f = G_D\mu_1 + h_1$ on D where μ_1 is a measure on D and h_1 is harmonic on D. Then $G_D\mu_1 = G_D\mu + h - h_1$ on D, so $\mu_1 = \mu$ by Theorem 1.6. Thus $h_1 = h$, which completes the proof of the theorem.

Stronger versions of the next result have been obtained by Evans [2], Deny [1], and Choquet [3] (see the comments on page 271 of Helms [1]).

Theorem 2.7. *Let $B \in \mathcal{B}$ be a polar subset of D. Then there is a measure $\mu \in \mathcal{M}_D$ such that $G_D\mu = \infty$ on B.*

Proof. Suppose $d \geq 2$ (the result is vacuous if $d = 1$). By Theorem 5.2.7 there is a nonnegative superharmonic function f on D such that $f = \infty$ on B. By Theorem 2.6, $f = G_D\mu + h$ where $\mu \in \mathcal{M}_D$ and h is harmonic on D. In particular h is finite on B and hence $G_D\mu = \infty$ on B as desired.

Theorem 2.8. *Let μ be a measure on D and let B be a subset of D such that μ is concentrated on B and $G_D\mu < \infty$ on B. Then μ does not charge polar sets.*

Proof. Now $G_D\mu$ is superharmonic on D by Theorem 1.1. Thus to prove the desired result it suffices to show that if A is a compact polar subset of B and $G_D\mu \leq M < \infty$ on A, then $\mu(A) = 0$. Now $G_D\mu|_A \leq M$ on D by the maximum principle. By Theorem 2.7 there is a measure $\nu \in \mathcal{M}_D$ such that $G_D\nu = \infty$ on A. By Theorem 1.6 it can be assumed that ν is supported on a compact subset of D and hence that $\nu(D) < \infty$. Since

$$\int_A G_D\nu \, d\mu = \int G_D\mu\Big|_A d\nu \leq M\nu(D) < \infty,$$

$\mu(A) = 0$ as desired.

The next result, called the *Riesz decomposition theorem*, is due to Riesz [1]. The measure μ occurring in the statement of this theorem is necessarily a Radon measure.

Theorem 2.9. *Let U be an open subset of D and let f be superharmonic on U. Then there is a unique measure μ on D which is concentrated on U such that, given any relatively compact open subset D_0 of U, $f = G_D\mu|_{D_0} + h_0$ on D_0 where h_0 is harmonic on D_0.*

Proof. Let D_0 and D_1 be relatively compact open subsets of U and let V be a relatively compact open subset of U containing $\bar{D}_0$ and $\bar{D}_1$. By Theorem 2.6 there is a measure $\nu \in \mathscr{M}_V$ and a harmonic function h' on V such that $f = G_V\nu + h'$ on V. For $i = 0, 1$ set $\mu_i = \nu|_{D_i}$. By Theorem 1.5 there exist harmonic functions h_i' on D_i such that $f = G_{D_i}\mu_i + h_i'$ on D_i. Observe that $\mu_i \in \mathscr{M}_D$ since μ_i is a finite measure. Thus by another application of Theorem 1.5 there exist harmonic functions h_i on D_i such that $f = G_D\mu_i + h_i$ on D_i. By Theorem 1.6, $\mu_0|_{D_0 \cap D_1} = \mu_1|_{D_0 \cap D_1}$. The existence of a measure μ on D having the properties specified in the statement of the theorem follows easily from these observations. By Theorem 1.6, $\mu|_{D_0}$ is uniquely determined for every relatively compact subset D_0 of U, so μ is uniquely determined. This completes the proof of the theorem.

Recall that g denotes the Newtonian potential kernel for $d \geq 3$. The next is due to Bôcher [1].

Theorem 2.10. *Let $d \geq 3$ and $r > 0$ and let f be a harmonic function on $\mathring{B}_r \backslash \{0\}$ such that f/g is bounded above or below on $\mathring{B}_r \backslash \{0\}$. Then there is an $\alpha \in \mathbb{R}$ and a harmonic function h on $\mathring{B}_r$ such that $f = \alpha g + h$ on $\mathring{B}_r \backslash \{0\}$.*

Proof. Without loss of generality it can be assumed that f/g is bounded below on $\mathring{B}_r \backslash \{0\}$. Thus there is a number $M > 0$ such that $f_1 = f + Mg$ is a nonnegative harmonic function on $\mathring{B}_r \backslash \{0\}$. By Theorem 5.1.15, f_1 extends to a nonnegative superharmonic function on $\mathring{B}_r$. Let D_0 be a relatively compact open subset of $\mathring{B}_r$ containing the origin. It follows from the Riesz decomposition theorem (with $D = \mathbb{R}^d$ and $U = \mathring{B}_r$) that there is a finite measure μ on $\mathbb{R}^d$ which is concentrated on D_0 and a harmonic function h_1 on D_0 such that $f_1 = g\mu + h_1$ on D_0. Since f_1 is harmonic on $D_0 \backslash \{0\}$ it follows from Theorem 1.4 that $\mu(D_0 \backslash \{0\}) = 0$ and hence that μ is concentrated at the origin. Set $\alpha_1 = \mu(\{0\})$. Then $f_1 = \alpha_1 g + h_1$ on D_0. Set $h = f_1 - \alpha_1 g$ on $\mathring{B}_r$ (with $h(0) = h_1(0)$). Then $h = h_1$ on D_0, so h is harmonic on D_0. Since h is harmonic on $\mathring{B}_r \backslash \{0\}$, it is harmonic on $\mathring{B}_r$. Now $f_1 = \alpha_1 g + h$ on $\mathring{B}_r$ and hence $f = \alpha g + h$ on $\mathring{B}_r \backslash \{0\}$ where $\alpha = \alpha_1 - M$.

Theorem 2.11. *Let $d \geq 3$ and $r > 0$ and let f be a harmonic function on $(B_r)^c$ which is bounded above or below on $(B_r)^c$. Then f has a finite limit at infinity.*

Proof. Without loss of generality it can be assumed that $f \geq 0$ on $(B_r)^c$. Let * denote inversion relative to the sphere S_r. Then the Kelvin transformation f^* of f is the harmonic function on $\mathring{B}_r \backslash \{0\}$ defined by

$$f^*(x) = (r/\|x\|)^{d-2} f(r^2 x/\|x\|^2).$$

Note that $f^* \geq 0$ on $\mathring{B}_r \backslash \{0\}$. By Theorem 2.10 there is an $\alpha \in \mathbb{R}$ and a harmonic function h on $\mathring{B}_r$ such that

$$f^*(x) = \alpha(r/\|x\|)^{d-2} + h(x), \qquad x \in \mathring{B}_r \backslash \{0\},$$

or equivalently

$$f(y) = \alpha + (r/\|y\|)^{d-2} h(r^2 y/\|y\|^2), \qquad y \in (B_r)^c.$$

Thus $\lim_{\|y\| \to \infty} f(y) = \alpha$ as desired.

Let U be an open subset of D and let f be superharmonic on U. Let D_0 be a relatively compact open subset of U and let D_n be relatively compact open subsets of D_0 such that $D_n \uparrow D_0$. By Theorem 5.1.23, $h = \lim_n H_{D_n} f = \lim_n E_{.} f(X(T_{D_n}))$ is the greatest harmonic minorant of f on D_0. Since

$$P_{.}\left(\lim_n f(X(T_{D_n})) = f(X(T_{D_0})) \right) = 1 \qquad \text{on } D_0$$

by Theorem 5.4.5, it is reasonable to conjecture that $h = E_{.} f(X(T_{D_0})) = H_{D_0} f$ on D_0. But the next result, taken from Brelot [8], shows that this conjecture is not true for all such superharmonic functions f on U unless D_0 is regular.

Theorem 2.12. *Let U be an open subset of D, let f be superharmonic on U, and let μ be the measure corresponding to f in the Riesz decomposition theorem. Let D_0 be a relatively compact connected open subset of U. Then $H_{D_0} f$ is the greatest harmonic minorant of f on D_0 if and only if $\mu(\partial D_0 \backslash (D_0^c)^r) = 0$.*

Proof. By the Riesz decomposition theorem it can be assumed that $f = G_D \mu$ on U for some measure μ on D. Set $\mu_0 = \mu|_{D_0}$ and $\mu_1 = \mu|_{\partial D_0 \backslash (D_0^c)^r}$. By the fundamental identity

$$\begin{aligned} f &= G_{D_0}\mu_0 + \int g_{D_0^c}(\cdot, y)\mu_1(dy) + H_{D_0} G_D \mu \\ &= G_{D_0}\mu_0 + H_{D_0} f + \int g_{D_0^c}(\cdot, y)\mu_1(dy) \end{aligned}$$

on D_0. Now $H_{D_0} f + \int g_{D_0^c}(\cdot, y)\mu_1(dy)$ is harmonic on D_0 and zero is the greatest harmonic minorant of $G_{D_0}\mu_0$ on D_0. Thus $H_{D_0} f + \int g_{D_0^c}(\cdot, y)\mu_1(dy)$ is the greatest harmonic minorant of f on D_0. Consequently $H_{D_0} f$ is the greatest harmonic minorant of f on D_0 if and only if

(1) $$\int g_{D_0^c}(\cdot, y)\mu_1(dy) = 0 \qquad \text{on } D_0.$$

If $\mu(\partial D_0 \backslash (D_0^c)^r) = 0$, then μ_1 is the zero measure and hence $H_{D_0} f$ is the greatest harmonic minorant of f on D_0. Suppose conversely that $H_{D_0} f$ is the greatest harmonic minorant of f on D_0. It must be shown that $\mu(\partial D_0 \backslash (D_0^c)^r) = 0$ or equivalently that μ_1 is the zero measure. Since (1) holds and μ_1 is concentrated on $\partial D_0 \backslash (D_0^c)^r$, it suffices to show that

(2) $$g_{D_0^c}(\cdot, y) > 0 \qquad \text{on } D_0, \quad y \in \partial D_0 \backslash (D_0^c)^r.$$

Let $y \in \partial D \backslash (D_0^c)^r$. Choose $x \in D_0$. Then for $t > 0$

$$g_{D_0^c}(x, y) = g_{D_0^c}(y, x) \geq \int_{D_0} q_{D_0^c}(t, y, z) G_{D_0}(z, x)\, dz.$$

Now $G_{D_0}(\cdot, x) > 0$ on D_0 since D_0 is assumed to be connected, and (at least) for t sufficiently small, $\int q_{D_0^c}(t, y, z)\, dz = P_y(\tau_{D_0^c} > t) > 0$. Thus $g_{D_0^c}(x, y) > 0$. Consequently (2) holds, which completes the proof of the theorem.

3. Balayage Problem

Let $d \geq 3$ and let B be a fixed compact nonpolar subset of $\mathbb{R}^d$. In 1840 Gauss considered the following problem: given a finite measure μ which is concentrated on B^c, find a measure μ' which is supported on B and whose potential agrees with that of μ on B. This problem, called after work of Poincare in 1890 the balayage (sweeping-out) problem, arises naturally in electrostatics. Let B be a grounded conductor and let μ be a fixed distribution of positive charges on the exterior of the conductor. Then μ should induce a corresponding distribution $-\mu'$ of negative charges on the conductor such that the combined potential $g(\mu - \mu')$ vanishes on the conductor; that is, μ' should solve the balayage problem. Gauss thought he had solved the balayage problem by means of the Dirichlet principle but his reasoning was invalid. Indeed the balayage problem as defined above need not have a solution. Specifically let U denote the unbounded component of B^c. It will be shown in the next paragraph that *if the balayage problem is solvable for every such measure μ then U is regular.* It is easy to use Lebesgue thorns to give an example of a compact nonpolar set B such that $U = B^c$ is connected but not regular. For such a set B the balayage problem is not always solvable.

Suppose the balayage problem is solvable for every finite measure μ which is concentrated on B^c. Choose $r > 0$ such that B is contained in the open ball $\mathring{B}_r$. Let $\mu = \mu_{S_r}$ denote the equilibrium measure of S_r (see Proposition 3.1.9). Then $g\mu = 1$ on B_r and hence $g\mu = 1$ on B. Let μ' be a solution to the balayage problem. Then μ' is supported on B and $g\mu' = 1$ on B (so μ' is the equilibrium measure of B as defined in Section 3.1). Now $g\mu' \leq 1$ on $\mathbb{R}^d$ by the maximum principle, $g\mu'$ has limit zero at infinity, $g\mu'$ is superharmonic on $\mathbb{R}^d$, and $g\mu'$ is harmonic on B^c by Theorem 1.4. Set $h = g\mu'$ on the unbounded component U of B^c. Then h is harmonic on U, $0 \leq h \leq 1$ on U and h has limit zero at infinity; since $g\mu'$ is lower semicontinuous and equals 1 on B, h has boundary value 1 on ∂U. Now $0 < h < 1$ on U by Proposition 4.1.3. Consequently $1 - h$ is a positive harmonic function on U which has boundary value zero on ∂U. Therefore U is regular by Theorem 4.2.15.

The above discussion suggests that the balayage problem, like the Dirichlet problem, be generalized by requiring that $g\mu'$ equal $g\mu$ on B^r instead of on all of B. In this form it will be seen that there is always a solution to the problem. But if $B \backslash B^r$ is nonempty the solution is not unique. One method of guaranteeing uniqueness is to require that μ' be concentrated on B^r.

In general, then, let $B \in \mathscr{B}$ and $\mu \in \mathscr{M}_D$. The *balayage problem* is to find a measure $\mu' \in \mathscr{M}_D(B^r)$ such that $G_D\mu' = G_D\mu$ on $B^r \cap D$. It follows from the fundamental identity and symmetry, as usual by now, that

$$(1) \qquad \int h_{B,D}(x, dz)G_D(z, y) = \int h_{B,D}(y, dz)G_D(z, x), \qquad x, y \in D.$$

Set $\mu h_{B,D} = \int \mu(dy)h_{B,D}(y, \cdot)$. Then $\mu h_{B,D}$ is concentrated on B^r by Theorem 2.6.5 and $G_D(\mu h_{B,D}) = h_{B,D}G_D\mu \leq G_D\mu$ by (1) and the fundamental identity. Thus $\mu h_{B,D} \in \mathscr{M}_D(B^r)$ and $G_D(\mu h_{B,D}) = G_D\mu$ *on* $B^r \cap D$. Consequently $\mu h_{B,D}$ solves the balayage problem. Suppose ν also solves the balayage problem. By the fundamental identity $G_D\nu = h_{B,D}G_D\mu = G_D(\mu h_{B,D})$, so $\nu = \mu h_{B,D}$. Therefore $\mu h_{B,D}$ *is the unique solution to the balayage problem.* Note that $\mu h_{B,D}(D) \leq \mu(D)$; note also that $\mu h_{B,D} = \mu$ if and only if μ is concentrated on B^r.

The motivation for the condition that μ' be concentrated on B^r rather than on B will now be discussed. First of all if B is not closed it is required for existence of a solution, since if μ is concentrated on B^c, μ' is supported on ∂B. Secondly the condition is required for uniqueness. Consider again Newtonian potentials on $\mathbb{R}^d$, $d \geq 3$. Suppose B is compact and $B \backslash B^r$ is nonempty. Choose $b \in B \backslash B^r$ and set $\mu = \delta_b$. Then δ_b and $h_B(b, \cdot)$ are two distinct measures which are supported on B and whose potentials equal the potential of μ on B^r. Even if B is compact and μ is concentrated on B^c the same phenomenon can occur. Choose $B = \{0\} \cup S_1(0)$ and let μ be the uniform probability distribution $\sigma_2(0, \cdot)$ on $S_2(0)$. Then $g\mu$ is constant on $B_1(0)$ and hence on B^r. Thus one can find distinct measures of the form $\mu_1' = c\sigma_1(0, \cdot)$ and $\mu_2' = c\delta_0$ which are supported on B and whose potential agrees with that of μ on B^r. (Perhaps the reader can find a more convincing example.)

In general, if B is closed and $\mu \in \mathscr{M}_D$ is such that $\mu(B \backslash B^r) = 0$, there is a unique measure μ' on D such that μ' is supported on B, $G_D\mu' \leq G_D\mu$ on D

and $G_D\mu' = G_D\mu$ on B^r. This follows easily from Theorem 3.11 below (see the proof of Theorem 3.12).

Let f be a nonnegative superharmonic function on D and let $B \in \mathscr{B}$. If $f = G_D\mu$ for some measure $\mu \in \mathscr{M}_D$, then $h_{B,D}f$ is the Green potential of $\mu h_{B,D}$. Suppose conversely that $h_{B,D}f$ is the Green potential of a measure $v \in \mathscr{M}_D$. By Theorem 2.6.7, $h_{B,D}f = h_{B,D}h_{B,D}f = h_{B,D}G_Dv = G_D(vh_{B,D})$ and hence $v = vh_{B,D}$. Thus v is concentrated on B^r. Properties of $h_{B,D}$ in general will now be determined.

Theorem 3.1. *Let f be a nonnegative superharmonic function on D and let $B \in \mathscr{B}$. Then $h_{B,D}f$ is nonnegative and superharmonic and $h_{B,D}f \le f$ on D. If B is a relatively compact subset of D, then $h_{B,D}f$ is the Green potential of a measure $\mu \in \mathscr{M}_D(B^r)$. If $B_n \in \mathscr{B}$ for each n and $B_n \uparrow B$, then $h_{B_n,D}f \uparrow h_{B,D}f$. Finally*

$$h_{B,D}f = \sup[G_Dv : v \in \mathscr{M}_D(B^r) \text{ and } G_Dv \le f \text{ on } D].$$

Proof. By Theorem 2.1 there exist measures $v_n \in \mathscr{M}_D$ such that $G_Dv_n \uparrow f$. Set $\mu_n = v_nh_{B,D}$. Then $\mu_n \in \mathscr{M}_D(\bar{B})$ and $G_D\mu_n = h_{B,D}G_Dv_n \uparrow h_{B,D}f$. Now $h_{B,D}f \le f$ by Theorem 5.1.17. Thus $h_{B,D}f$ is superharmonic on D by Proposition 5.1.1. It is clearly nonnegative so the first conclusion of the theorem is valid. Suppose B is a relatively compact subset of D. Since the measures μ_n are all supported on the compact subset $\bar{B}$ of D, the condition in Theorem 2.2 is satisfied. So it follows from Theorem 2.2 that $h_{B,D}f = G_D\mu$ for some measure $\mu \in \mathscr{M}_D(B^r)$. This proves the second conclusion of the theorem.

Suppose $B_n \in \mathscr{B}$ for each n and $B_n \uparrow B$. Now $\tau_{B_n} \downarrow \tau_B$ by Proposition 2.3.8 so $\{\tau_B < T_D\} = \{\tau_{B_n} < T_D \text{ for } n \text{ sufficiently large}\}$. By the lower semicontinuity of f

$$\varliminf_n f(X(\tau_{B_n})) \ge f(X(\tau_B)) \qquad \text{on } \{\tau_B < T_D\}.$$

Therefore by Fatou's lemma

$$\varliminf_n h_{B_n,D}f = \varliminf_n E_{\cdot}(f(X(\tau_{B_n})); \tau_{B_n} < T_D)$$

$$\ge E_{\cdot}(f(X(\tau_B)); \tau_B < T_D) = h_{B,D}f.$$

Since $h_{B_n,D}f = h_{B_n,D}h_{B,D}f \le h_{B,D}f$ by Theorems 2.6.7 and 5.1.17 and $h_{B_n,D}f \uparrow$ by a similar argument, $h_{B_n,D}f \uparrow h_{B,D}f$.

In proving the last result it can be assumed that $B \subset D$. Let B_n be compact sets such that $B_n \uparrow B$. Then $h_{B_n,D}f = G_Dv_n$ for some measure $v_n \in \mathscr{M}_D(B_n^r) \subset \mathscr{M}_D(B^r)$ and $G_Dv_n \uparrow h_{B,D}f$. If $v \in \mathscr{M}_D(B^r)$ and $G_Dv \le f$ on D, then

$$G_Dv = G_D(vh_{B,D}) = h_{B,D}G_Dv \le h_{B,D}f \qquad \text{on } D.$$

Thus the last conclusion of the theorem is valid.

Theorem 3.2. *Let B be a compact nonpolar subset of D. Then there is a nonzero measure on D which is supported on B and whose Green potential is continuous on D.*

Proof. By Theorem 3.1, $P_{.}(\tau_B < T_D) = h_{B,D}1 = G_D v$ for some measure $v \in \mathcal{M}_D(B)$. Now $P_{.}(\tau_B < T_D) = 1$ on B^r, which is nonempty by Theorem 2.6.4, so v is not the zero measure. Choose ε, $0 < \varepsilon < v(B)$. By Theorem 1.10 there is a compact subset C of B such that $v(B \backslash C) \le \varepsilon$ and $G_D v|_C$ is continuous on D. Clearly $\mu = v|_C$ satisfies the conclusion of the Theorem.

Theorem 3.3. *Let B_n, $n \ge 1$, be subsets of $\mathbb{R}^d$ such that for each n every compact subset of B_n is polar. Then every compact subset of $\bigcup_n B_n$ is polar.*

Proof. Without loss of generality it can be assumed that $\bigcup_n B_n$ is bounded and hence contained in a Greenian open set D. Suppose the conclusion is false. Let B be a compact nonpolar subset of $\bigcup_n B_n$. It follows from Theorem 3.1, as in the proof of Theorem 3.2, that $P_{.}(\tau_B < T_D) = G_D\mu$ for some nonzero measure $\mu \in \mathcal{M}_D(B)$. By Theorem 2.8, μ does not change polar sets. Thus for each n, every compact subset of B_n has μ-measure zero and hence $\mu(B_n) = 0$. Consequently $\mu(\bigcup_n B_n) = 0$ and hence $\mu(B) = 0$, which contradicts the definition of μ. This completes the proof of the theorem.

A property is said to hold *quasi-everywhere* (q.e.) on B if there is a (Borel) set A such that the property holds on $B \backslash A$ and every compact subset of A is polar (see the end of Section 6.5 for an equivalent definition). Theorem 5.24 can be restated as follows: *if f is superharmonic on an open set, then f is finite quasi-everywhere on that set.* A property which holds q.e. on B also holds a.e. on B, but the converse is not true in general. If a property holds q.e. on B and if μ is a measure which is concentrated on B and does not charge polar sets, the property holds a.e. (μ). It follows from Theorem 3.2 that a property which holds q.e. on B_n for each n also holds q.e. on $\bigcup_n B_n$. It follows from the same theorem that if $f_1 \le f_2$ q.e. on B and $f_2 \le f_3$ q.e. on B, then $f_1 \le f_3$ q.e. on B.

The characterization of $h_{B,D}f$ in the next result is due to Brelot [6].

Theorem 3.4. *Let f be a nonnegative superharmonic function on D and let $B \in \mathcal{B}$. Then*

$$h_{B,D}f = \min[v : v \text{ is nonnegative and superharmonic on } D \text{ and } v \ge f \text{ q.e. on } B \cap D].$$

Proof. Let v_0 denote the infimum over the indicated collection. Now $h_{B,D}f$ is nonnegative and superharmonic on D and $h_{B,D}f = f$ on $B^r \cap D$, so $h_{B,D}f = f$ q.e. on $B \cap D$. Consequently $h_{B,D}f$ is in the indicated collection and hence $v_0 \le h_{B,D}f$. Let B_n be compact subsets of B such that $B_n \uparrow B$, and let v be in the indicated collection. Then $v \ge f$ q.e. on $B_n \cap D$ and hence $h_{B_n,D}f \le h_{B_n,D}v \le v$ on $D \backslash B_n^r$ by Theorem 2.6.5. Now $h_{B_n,D}f = f$ on B_n^r, so $h_{B_n,D}f \le v$ q.e. on D and hence a.e. on D. Therefore $h_{B_n,D}f \le v$ on D by Theorem 5.1.14. Since $h_{B_n,D}f \uparrow h_{B,D}f$ on D by Theorem 3.1, $h_{B,D}f \le v$ on D. Thus $h_{B,D}f \le v_0$ and therefore $h_{B,D}f = v_0$. This completes the proof of the theorem.

The next result is called the *domination principle* for Green potentials (see Brelot [8] and Theorem 2 on page 260 of Cartan [3]).

Theorem 3.5. *Let $B \in \mathscr{B}$, let μ be a measure on D which is concentrated on B^r and let f be a nonnegative superharmonic function on D such that $G_D\mu \le f$ q.e. on $B \cap D$. Then $G_D\mu \le f$ on D.*

Proof. Since $G_D\mu < \infty$ q.e. on $B \cap D$, $G_D\mu$ is superharmonic on D by Theorem 1.1. (Observe that if D_0 is a component of D and $\mu(D_0) > 0$, then $B \cap D_0$ is nonpolar.) Now $\mu h_{B,D} = \mu$, so by Theorem 3.4

$$f \ge h_{B,D}G_D\mu = G_D(\mu h_{B,D}) = G_D\mu \qquad \text{on } D$$

as desired.

The next result includes the maximum principle as a special case.

Theorem 3.6. *Let μ be a measure on D and f be a nonnegative superharmonic function on D. If $G_D\mu < \infty$ a.e. (μ) and $G_D\mu \le f$ a.e. (μ), then $G_D\mu \le f$ on D.*

Proof. Set

$$A = \{x \in D : G_D\mu(x) < \infty \text{ and } G_D\mu(x) \le f(x)\}.$$

Then μ is concentrated on A. Since μ is a Radon measure by Theorem 1.1, there exist compact subsets B_n of A such that $B_n \uparrow$ and $\mu(A \backslash \cup_n B_n) = 0$. Set $B = \cup_n B_n$. Then $B \in \mathscr{B}$, μ is concentrated on B, $G_D\mu < \infty$ on B and $G_D\mu \le f$ on B. Now μ doesn't charge polar sets by Theorem 2.8, so μ is concentrated on B^r by Theorem 2.6.3. Thus $G_D\mu \le f$ on D by the domination principle. (For an alternative proof of this result see Section 1 of Chapter 7 of Rao [1].)

The next result is called the *uniqueness principle* for Green potentials.

Theorem 3.7 *Let $B \in \mathscr{B}$, let $\mu \in \mathscr{M}_D(B^r)$ and let ν be a measure on D which is concentrated on B^r and satisfies $G_D\nu = G_D\mu$ q.e. on $B \cap D$. Then $\nu = \mu$.*

Proof. Observe first that $G_D\nu \le G_D\mu$ by the domination principle. Thus $G_D\nu$ is superharmonic by Theorem 1.1. By another application of the domination principle $G_D\nu = G_D\mu$ on D. Therefore $\nu = \mu$ by Theorem 1.3.

The characterization of $h_{B,D}f$ in the next result is also due to Brelot [6].

Theorem 3.8. *Let f be a nonnegative superharmonic function on D and let $B \in \mathscr{B}$. Set*

$$v_0 = \inf[v : v \text{ is nonnegative and superharmonic on } D \text{ and } v \ge f \text{ on } B \cap D].$$

Then $v_0 = h_{B,D}f$ on $D\backslash(B\backslash B^r)$ (and hence q.e. on D), $v_0 = f$ on $B \cap D$ and $\underline{v_0} = h_{B,D}f$ on D.

Proof. Suppose $d \ge 2$ (if $d = 1$, then $B\backslash B^r$ is empty and the proof simplifies considerably). Clearly $v_0 = f$ on $B \cap D$. By Theorem 3.4, $v_0 \ge h_{B,D}f$. Choose $x_0 \in D\backslash(B\backslash B^r)$. By Theorem 5.2.7 there is a nonnegative superharmonic function w on D such that $w = \infty$ on $B\backslash B^r$ and $w(x_0) < \infty$. Choose $\varepsilon > 0$. Since $h_{B,D}f = f$ on B^r, $h_{B,D}f + \varepsilon w$ is in the indicated class. Consequently $v_0 \le h_{B,D}f + \varepsilon w$. In particular $v_0(x_0) \le h_{B,D}f(x_0) + \varepsilon w(x_0)$. Since ε can be made arbitrarily small $v_0(x_0) \le h_{B,D}f(x_0)$. Thus $v_0 \le h_{B,D}f$ on $D\backslash(B\backslash B^r)$ and hence $v_0 = h_{B,D}f$ on $D\backslash(B\backslash B^r)$.

Since $h_{B,D}f$ is lower semicontinuous on D and $v_0 \ge h_{B,D}f$ on D, $\underline{v_0} \ge h_{B,D}f$ on D. On the other hand, since $B\backslash B^r$ is polar it is thin everywhere by Theorem 5.2.10. Thus $D\backslash(B\backslash B^r)$ is nowhere thin on D by Proposition 5.2.1. Consequently for $x \in D$

$$\underline{v_0}(x) \le \varliminf_{y \to x} v_0(y) \le \varliminf_{\substack{y \to x \\ y \in D\backslash(B\backslash B^r)}} h_{B,D}f(y) = h_{B,D}f(x).$$

Therefore $\underline{v_0} \le h_{B,D}f$ on D and hence $\underline{v_0} = h_{B,D}f$ on D. This completes the proof of the theorem.

The function v_0 occurring in the statement of Theorem 3.8 is called the *reduced function* (réduite) of f with respect to B on D. According to Theorem 3.8 the reduced function equals its lower regularization q.e. on D. For an example where equality need not occur everywhere on D, let $B \in \mathscr{B}$ be a nonempty polar subset of D. Then the reduced function of 1 with respect to B on D is I_B (by Theorem 5.2.7), while its lower regularization is identically zero on D.

The example of the above paragraph illustrates the fact the infimum of a collection of nonnegative superharmonic functions need not be lower semicontinuous. For a simpler example of this phenomenon suppose $d \ge 2$, let $y_0 \in D$ be fixed and set $f_n = G_D(\cdot, y_0)/n$ on D. Then each f_n is nonnegative and superharmonic on D and $f_n \downarrow f$ on D where $f(y_0) = \infty$ and $f = 0$ on $D \backslash \{y_0\}$. Again f equals its lower regularization $\underline{f} = 0$ q.e. on D.

According to the following theorem of Cartan [2] the lower regularization of the infimum of any collection of nonnegative superharmonic functions on D is superharmonic and equals the original infimum q.e. on D. Since this is really a local result the assumption that the open set D be Greenian is not required for its validity.

Theorem 3.9. *Let $\mathfrak{F}$ denote a collection of superharmonic functions on D such that $v_0 = \inf[v : v \in \mathfrak{F}]$ is locally bounded below on D. Then $\underline{v_0}$ is superharmonic on D and $\underline{v_0} = v_0$ q.e. on D.*

Proof. (Note that v_0 is not asserted to be a Borel function on D.) Without loss of generality it can be assumed that $v_0 \geq 0$ on D. Since the minimum of a finite number of superharmonic functions is superharmonic, the result is true if $\mathfrak{F}$ is a finite collection. Suppose next that $\mathfrak{F}$ is a countably infinite collection given by $\mathfrak{F} = \{v_n : n \geq 1\}$. Set $u_n = \min[v_1, \ldots, v_n]$. Then $u_n \downarrow v_0$.

Let D_0 be a relatively compact open subset of D and set $B = \overline{D_0}$. Set $f_n = h_{B,D}u_n$ for $n \geq 1$. Then $f_n \leq u_n \leq u_1 = v_1$, which is locally integrable on D, and $f_n = u_n$ on D_0. Also $f_n \downarrow f$ for some function f on D. Clearly $f = v_0$ and $\underline{f} = \underline{v_0}$ on D_0. By Theorem 3.1, $h_{B,D}f_n = G_D\mu_n$ for some measure $\mu_n \in \mathscr{M}_D(B)$. According to Theorem 2.4 there is a measure $\mu \in \mathscr{M}_D(B)$ such that μ_n converges vaguely to μ and $\underline{f} = G_D\mu$. In particular $\underline{f}$ is superharmonic on D_0.

It will now be shown that $\underline{f} = f$ q.e. on D_0. Suppose otherwise. Then for some $M \in (0, \infty)$ there is a compact nonpolar subset A of $\{x \in D_0 : \underline{f}(x) < f(x) \leq M\}$ (note that $v_1 < \infty$ q.e. on D by Theorem 5.2.4). By Theorem 3.2 there is a nonzero measure ν on D which is supported on A and such that $G_D\nu$ is continuous on D. Since $f_n = h_{B,D}f_n = G_D\mu_n$ on the support of μ,

$$\int f\,d\nu = \lim_n \int f_n\,d\nu = \lim_n \int G_D\mu_n\,d\nu = \lim_n \int G_D\nu\,d\mu_n = \int G_D\nu\,d\mu = \int \underline{f}\,d\nu.$$

Therefore $\int (f - \underline{f})\,d\nu = 0$, which is impossible since $\nu(A) > 0$ and $f > \underline{f}$ on A.

It has now been shown that $\underline{v_0}$ is superharmonic on D_0 and that $\underline{v_0} = v_0$ q.e. on D_0. Since D_0 is an arbitrarily relatively compact open subset of D, $\underline{v_0}$ is superharmonic on D and $\underline{v_0} = v_0$ q.e. on D. The theorem is therefore true if $\mathfrak{F}$ is countably infinite.

Suppose finally that $\mathfrak{F}$ is uncountably infinite. Let D_m be relatively compact open subsets of D such that $D_m \uparrow D$. Given a countable subcollection $\mathscr{G}$ of $\mathfrak{F}$, set $c_m(\mathscr{G}) = \int_{D_m} \inf[v(x) : v \in \mathscr{G}]\,dx$. Also set

$$c_m = \inf[c_m(\mathscr{G}) : \mathscr{G} \text{ is a countable subcollection of } \mathfrak{F}].$$

Let $\mathfrak{F}_{mn}$ be a countable subcollection of $\mathfrak{F}$ such that $c_m(\mathfrak{F}_{mn}) \downarrow c_m$ and set $\mathfrak{F}_0 = \bigcup_{m,n} \mathfrak{F}_{mn}$. Then $\mathfrak{F}_0$ is a countable subcollection of $\mathfrak{F}$ and $c_m(\mathfrak{F}_0) = c_m$ for all m. Set $u_0 = \inf[v : v \in \mathfrak{F}_0]$. Then $\underline{u_0}$ is superharmonic on D, $\underline{u_0} = u_0$ q.e. on D, and $\int_{D_m} u_0(x)\,dx = c_m(\mathfrak{F}_0) = c_m$ for $m \geq 1$. Choose $v \in \mathfrak{F}$ and $m \geq 1$. Then $\int_{D_m} (u_0 \wedge v)(x)\,dx = c_m = \int_{D_m} u_0(x)\,dx$ and hence $v \geq u_0$ a.e. on D_m. Thus $v \geq u_0 \geq \underline{u_0}$ a.e. on D. Since v and $\underline{u_0}$ are both superharmonic on D, it follows from Theorem 5.1.14 that $v \geq \underline{u_0}$ on D. Consequently $v_0 = \inf[v : v \in \mathfrak{F}] \geq \underline{u_0}$ on D and therefore $\underline{v_0} \geq \underline{u_0}$ on D. Since $\underline{v_0} \leq \underline{u_0}$ on D clearly holds, $\underline{v_0} = \underline{u_0}$ on D. Thus $\underline{v_0}$ is superharmonic on D. Observe that $\underline{u_0} = \underline{v_0} \leq v_0 \leq u_0$. Since $\underline{\mu_0} = \mu_0$. q.e. on D, $\underline{v_0} = v_0$ q.e. on D. This completes the proof of the theorem. (For another method of reducing the general case to the countable case see Section 3 of Chapter 7 of Rao [1].)

Theorem 3.10. *Let B_n be closed sets such that $B_n \downarrow B$ and let $\mu \in \mathcal{M}_D$ be such that $\mu(B \backslash B^r) = 0$ and $\mu h_{B_1,D}(D) < \infty$. Then $\mu h_{B_n,D}$ converges vaguely to $\mu h_{B,D}$ and $G_D(\mu h_{B_n,D}) \downarrow G_D(\mu h_{B,D})$ q.e. on D.*

Proof. Without loss of generality it can be assumed that μ is a finite measure on D (replace μ by $\mu h_{B_1,D}$ if necessary). Let φ be a nonnegative continuous function on D which vanishes outside a compact subset A of D and let M be an upper bound to φ. Observe that $h_{B_n,D}\varphi \leq M h_{B_n,D}(\cdot, A) \leq M P_{\cdot}(\tau_A < T_D)$. By Proposition 2.3.8, $P_{\cdot}(\tau_{B_n} \uparrow \tau_B) = 1$ on $D \backslash (B \backslash B^r)$. Thus

$$h_{B_n,D}\varphi = E_{\cdot}(\varphi(X(\tau_{B_n})); \tau_{B_n} < T_D) \to E_{\cdot}(\varphi(X(\tau_B); \tau_B < T_D) = h_{B,D}\varphi$$

on $D \backslash (B \backslash B^r)$ and hence a.e. (μ) on D. Consequently by the dominated convergence theorem

$$\int \varphi\, d(\mu h_{B_n,D}) = \int \mu(dy) h_{B_n,D}\varphi(y) \to \int \mu(dy) h_{B,D}\varphi(y) = \int \varphi\, d(\mu h_{B,D}).$$

Therefore $\mu h_{B_n,D}$ converges vaguely to $\mu h_{B,D}$. By Theorem 2.4, $G_D(\mu h_{B_n,D}) \downarrow f$ where $\underline{f} = G_D(\mu h_{B,D})$. By Theorem 3.9, $\underline{f} = f$ q.e. on D and hence $G_D(\mu h_{B_n,D}) \downarrow G_D(\mu h_{B,D})$ q.e. on D. This completes the proof of the theorem.

The next result implies that *if $\mu, \nu \in \mathcal{M}_D$, μ does not charge polar sets and $G_D\nu \leq G_D\mu$ on D, then ν does not charge polar sets.*

Theorem 3.11. *Let $\mu, \nu \in \mathcal{M}_D$ and let C be a compact polar subset of D such that $\mu(C) = 0$. Suppose there is an open subset U of D containing C such that $G_D\nu \leq G_D\mu + \alpha$ on U for some $\alpha \in \mathbb{R}$. Then $\nu(C) = 0$.*

Proof. Let B_n be compact subsets of U such that $C \subset \mathring{B}_n$ for each n and $B_n \downarrow C$. Now $P_{\cdot}(\tau_{B_n} < T_D) \downarrow 0$ on $D \backslash C$ by Proposition 5.2.6 and $G_D(\mu h_{B_n,D}) \downarrow 0$

q.e. on D by Theorem 3.10. Also

$$G_D\nu\Big|_C \le G_D(\nu h_{B_n,D}) = h_{B_n,D}G_D\nu \le G_D(\mu h_{B_n,D}) + \alpha P_{\cdot}(\tau_{B_n} < T_D),$$

so it follows by letting $n \to \infty$ that $G_D\nu|_C = 0$ a.e. on D. Thus $\nu|_C = 0$ by Theorem 1.3 and hence $\nu(C) = 0$ as desired.

If $\mu, \nu \in \mathscr{M}_D$, $G_D\nu \le G_D\mu$, and $G_D\mu < \infty$ a.e. (μ), then $G_D\mu < \infty$ a.e. (ν) and hence $G_D\nu < \infty$ a.e. (ν). It is left to the reader to derive this result as a consequence of Theorem 5.2.4, Theorem 2.8, and Theorem 3.11.

It is easily seen that the condition $\mu(B\backslash B^r) = 0$ in Theorem 3.10 and the next two results is required for their validity. This condition is automatically satisfied if μ is concentrated on $D\backslash B$.

Theorem 3.12. *Let B be a closed set. Let $\mu \in \mathscr{M}_D$ be such that $\mu(B\backslash B^r) = 0$ and let $\nu \in \mathscr{M}_D$. Then $\nu = \mu h_{B,D}$ if and only if $G_D\nu$ satisfies the following three conditions: $G_D\nu \le G_D\mu$ on D; $G_D\nu = G_D\mu$ q.e. on $B \cap D$; $G_D\nu$ is harmonic on $D\backslash B$.*

Proof. Suppose first that $\nu = \mu h_{B,D}$. Then $G_D\nu = h_{B,D}G_D\mu \le G_D\mu$. Also $G_D\nu = G_D\mu$ on B^r and hence q.e. on $B \cap D$. Now $\mu h_{B,D}$ is supported on B, so $G_D\nu$ is harmonic on $D\backslash B$ by Theorem 1.4. Suppose conversely that $G_D\nu$ satisfies the three indicated conditions. Then ν is supported on B by Theorem 1.4. Let C be a compact subset of $B\backslash B^r$. Then $\mu(C) = 0$ since $\mu(B\backslash B^r) = 0$. Thus $\nu(C) = 0$ by Theorem 3.11. Consequently $\nu(B\backslash B^r) = 0$ and hence ν is concentrated on B^r. Now $G_D\nu = G_D\mu = G_D(\mu h_{B,D})$ q.e. on B and $\mu h_{B,D}$ is concentrated on B^r, so $\nu = \mu h_{B,D}$ by the uniqueness principle.

Theorem 3.13. *Let B be a closed set and let $\mu \in \mathscr{M}_D$ be such that $\mu(B\backslash B^r) = 0$. Suppose $\nu_n \in \mathscr{M}_D$ and $G_D\nu_n = G_D\mu$ on $B \cap D$ for each n, $G_D\nu_1 \le G_D\mu$ on $D\backslash B$, and $G_D\nu_n \downarrow h$ on $D\backslash B$ where h is harmonic on $D\backslash B$. Then ν_n converges vaguely to $\mu h_{B,D}$ and $G_D\nu_n \downarrow h_{B,D}G_D\mu$ on $D\backslash B$.*

Proof. By Theorem 2.4 there is a measure $\nu \in \mathscr{M}_D$ such that ν_n converges vaguely to ν and $G_D\nu_n \downarrow f$ on D where $\underline{f} = G_D\nu$ on D. By Theorem 3.9, $f = G_D\nu$ q.e. on D. Now $G_D\nu \le G_D\mu$ on D, $G_D\nu = G_D\mu$ q.e. on B, and $G_D\nu$ equals the harmonic function h on $D\backslash B$. Thus $\nu = \mu h_{B,D}$ by Theorem 3.12. This completes the proof of the theorem.

Let B be closed and let $\{D_n\}$ be a sequence of relatively compact open subsets of $D\backslash B$ satisfying the following property: for each $x \in D\backslash B$ there is an $r > 0$ such that $B_r(x)$ is contained in infinitely many of the sets D_n. Set $B_n = D_n^c$ for $n \geq 1$. Let $\mu \in \mathscr{M}_D$ be such that $\mu(B \backslash B^r) = 0$. Define measures $\nu_n, n \geq 1$, by $\nu_1 = \mu h_{B_1,D}$ and $\nu_n = \nu_{n-1} h_{B_n,D}$ for $n \geq 2$. The process of going from ν_{n-1} to ν_n sweeps all the mass out of D_n, but mass can reappear in D_n at a later stage. Theorem 3.13 will now be used to show that ν_n *converges vaguely to* $\mu h_{B,D}$ and $G_D \nu_n \downarrow h_{B,D} G_D \mu$ on $D \backslash B$. Clearly $G_D \nu_n = G_D \mu$ on B for each n, $G_D \nu_1 \leq G_D \mu$ on $D \backslash B$, and $G_D \nu_n \downarrow h$ on $D \backslash B$ for some function h on $D \backslash B$. It remains to be shown that h is harmonic on $D \backslash B$. Choose $x \in D \backslash B$, $r > 0$ and positive integers $n_1 < n_2 < \cdots$ such that $D_{n_1}, D_{n_2}, \ldots$ each contain $B_r(x)$. Then $G_D \nu_{n_j}$ is harmonic on $\mathring{B}_r(x)$ and $G_D \nu_{n_j} \downarrow h$ on $\mathring{B}_r(x)$, so h is harmonic on $\mathring{B}_r(x)$ by Harnack's theorem (Theorem 4.3.6). Consequently h is harmonic on $D \backslash B$ as desired.

Poincare introduced the term "balayage" in connection with a setup of the type just described, which he used to solve the Dirichlet problem under the restriction on the boundary mentioned in Section 4.2 (see Sections 1 and 15 of Chapter XI of Kellogg [1]). Interest in balayage was rekindled by La Vallée Poussin [1].

4. Energy

The results of this section are due to Frostman [1] and especially Cartan [1], [2], [3]. The reader is encouraged to read these fundamental papers.

Let μ be a measure on D. The *energy* $I_D(\mu)$ of μ (relative to D) is defined by $I_D(\mu) = \int G_D \mu \, d\mu$.

Theorem 4.1. *Let μ be a measure on D having finite energy. Then $\mu \in \mathscr{M}_D$ and μ does not charge polar sets. If ν is a measure on D such that $G_D \nu \leq G_D \mu$ on D, then $I_D(\nu) \leq I_D(\mu)$.*

Proof. Let D_0 be a component of D such that $\mu(D_0) > 0$. Since $\int_{D_0} G_D \mu \, d\mu < \infty$, $G_D \mu$ must be finite somewhere on D_0. Thus $\mu \in \mathscr{M}_D$ by Theorem 1.1. Set $B = \{x \in D : G_D \mu(x) < \infty\}$, Then μ is concentrated on B and $G_D \mu < \infty$ on B. Thus by Theorem 2.8, μ does not charge polar sets. Let ν be a measure on D such that $G_D \nu \leq G_D \mu$ on D. Then

$$I_D(\nu) = \int G_D \nu \, d\nu \leq \int G_D \mu \, d\nu = \int G_D \nu \, d\mu \leq \int G_D \mu \, d\mu = I_D(\mu).$$

Theorem 4.2. *Let $\mu, \nu \in \mathscr{M}_D$ have finite energy. Then $(\int G_D \mu \, d\nu)^2 \leq I_D(\mu) I_D(\nu)$.*

Proof. Now $G_D(x, y) = \int_0^\infty Q_D(t, x, y)\,dt$ for $x, y \in D$, so by the semigroup property of the densities $Q_D(t, x, y)$, $t > 0$,

$$G_D(x, y) = \int_0^\infty \left(\int_D Q_D\left(\frac{t}{2}, x, z\right) Q_D\left(\frac{t}{2}, z, y\right) dz \right) dt \qquad \text{for } x, y \in D. \tag{1}$$

Thus by Schwarz's inequality and the symmetry of $Q_D(t/2, x, y)$ in x and y

$$\begin{aligned}
\int G_D\mu\,d\nu &= \int_0^\infty \left(\int_D Q_D^{t/2}\mu(z) Q_D^{t/2}\nu(z)\,dz \right) dt \\
&\leq \left[\int_0^\infty \left(\int_D (Q_D^{t/2}\mu(z))^2\,dz \right) dt \right]^{1/2} \left[\int_0^\infty \left(\int_D (Q_D^{t/2}\nu(z)^2\,dz) \right) dt \right]^{1/2} \\
&= (I_D(\mu) I_D(\nu))^{1/2}
\end{aligned}$$

as desired.

A set function of the form $\mu = \mu_1 - \mu_2$, where μ_1 and μ_2 are Radon measures on D, is called a *signed Radon measure* on D. Let μ be such a signed measure. Then $\mu(A)$ is well defined at least when A is a relatively compact subset of D. Let μ^+, μ^-, and $|\mu|$ denote, respectively, the postive part, negative part, and total variation of μ. Then μ^+, μ^-, and $|\mu|$ are Radon measures on D, $\mu = \mu^+ - \mu^-$ is the Jordan decomposition of μ, and $|\mu| = \mu^+ + \mu^-$ (see Rudin [1]). Also μ is said to be supported (concentrated) on B if $|\mu|$ is supported (concentrated) on B, and the support of μ is defined as the support of $|\mu|$.

Let $\mathscr{E}_D$ denote the vector space of signed Radon measures μ on D such that $|\mu|$ has finite energy and let $\mathscr{E}_D(B)$ denote the subspace of elements of $\mathscr{E}_D$ which are concentrated on B. Let $\mathscr{E}_D^+$ and $\mathscr{E}_D^+(B)$ denote, respectively, the (nonnegative) measures in $\mathscr{E}_D$ and $\mathscr{E}_D(B)$.

If $\mu \in \mathscr{E}_D$, then $G_D\mu = G_D\mu^+ - G_D\mu^-$ is well defined wherever $G_D|\mu| < \infty$ and hence is well defined q.e. on D. Let $\mu, \nu \in \mathscr{E}_D$. The *mutual energy* $(\mu, \nu) = \int G_D\mu\,d\nu$ of μ and ν is well defined and finite by Theorem 4.2 (for simplicity in notation the dependence of (,) on D is suppressed). Also (μ, ν) is symmetric in μ and ν and defines a bilinear form on $\mathscr{E}_D$. The next result is called the *energy principle* for Green potentials.

Theorem 4.3. *Let $\mu \in \mathscr{E}_D$. Then $(\mu, \mu) \geq 0$ with equality holding if and only if $\mu = 0$.*

Proof. By (1)

$$(\mu, \mu) = \int_0^\infty \left(\int_D Q_D^{t/2}\mu^+(z) - Q_D^{t/2}\mu^-(z))^2\,dz \right) dt \geq 0.$$

Suppose $(\mu, \mu) = 0$. Then $Q_D^t\mu^+(z) = Q_D^t\mu^-(z)$ for almost all $(t, z) \in (0, \infty) \times D$ and hence $G_D\mu^+ = G_D\mu^-$ a.e. on D. Therefore $\mu^+ = \mu^-$ by Theorem 1.3 and hence $\mu = \mu^+ - \mu^- = 0$. This completes the proof of the theorem.

For $\mu \in \mathscr{E}_D$ let the norm $\|\mu\|$ be the nonnegative square root of (μ, μ). The quantity $\|\mu\|^2 = (\mu, \mu)$ is called the *energy* of μ. If $\mu \in \mathscr{E}_D^+$ then $\|\mu\|^2 = I_D(\mu)$. Theorem 4.3 implies that $\|\mu\| = 0$ if and only if $\mu = 0$. Together with the usual elementary Hilbert space argument (see Rudin [1]) this yields the Schwarz inequality

$$|(\mu, \nu)| \leq \|\mu\| \, \|\nu\|, \qquad \mu, \nu \in \mathscr{E}_D,$$

with equality holding if and only if one of the elements μ, ν is a constant multiple of the other one. The Schwarz inequality in turn yields the triangle inequality

$$\|\mu + \nu\| \leq \|\mu\| + \|\nu\|, \qquad \mu, \nu \in \mathscr{E}_D.$$

The next result, due to Cartan [2], is an immediate consequence of Theorem 3.6. It can also be used to obtain that theorem, the details being left to the reader.

Theorem 4.4. *Let $\mu \in \mathscr{E}_D^+$ and let f be a nonnegative superharmonic function on D such that $G_D\mu \leq f$ a.e. (μ). Then $G_D\mu \leq f$ on D.*

Proof. By Theorems 2.5 and 4.1, $G_D\mu \wedge f$ is the Green potential of a measure $\nu \in \mathscr{E}_D^+$. Now $(\nu - \mu, \mu) = 0$ since $G_D\nu = G_D\mu$ a.e. (μ) and $(\nu - \mu, \nu) \leq 0$ since $G_D\nu \leq G_D\mu$ on D. Thus $\|\nu - \mu\|^2 = (\nu - \mu, \nu - \mu) \leq 0$ and hence $\nu = \mu$ by the energy principle. Therefore $G_D\mu = G_D\mu \wedge f$ and hence $G_D\mu \leq f$ on D.

If ν is a measure on D having compact support in D (i.e., supported on a compact subset of D) and a continuous Green potential, then ν has finite energy.

Theorem 4.5. *Let f_1 and f_2 be superharmonic functions on D such that $\int f_1 \, d\nu = \int f_2 \, d\nu$ for all measures $\nu \in \mathscr{E}_D^+$ having compact support in D. Then $f_1 = f_2$.*

Proof. Let A be a relatively compact subset of D. Define the measure ν on D by $\nu(dx) = I_A(x)\,dx$. Then ν has compact support in D and ν has a continuous Green potential by Theorem 4.6.6 so $\nu \in \mathscr{E}_D^+$. Thus $\int f_1 \, d\nu = \int f_2 \, d\nu$; that is, $\int_A f_1(x)\,dx = \int_A f_2(x)\,dx$. Consequently $f_1 = f_2$ a.e. and therefore $f_1 = f_2$ by Theorem 5.1.14.

Theorem 4.6. *Let $\{\mu_n\}$ be a sequence of elements of $\mathscr{E}_D^+$ such that $\|\mu_n\|$ is bounded in n. Then there is a strictly increasing sequence $\{n_j\}$ of positive integers and a measure $\mu \in \mathscr{E}_D^+$ such that μ_{n_j} converges vaguely to μ.*

Proof. Choose $x \in D$. Since $G_D(x, x) > 0$ and G_D is continuous on $D \times D$ in the extended sense there is an open neighborhood U of x and a $\delta > 0$ such that $G_D(y, z) \geq \delta$ for $y, z \in U$. Thus

$$\|\mu_n\|^2 \geq \int_U \int_U G_D(y, z)\mu_n(dy)\mu_n(dz) \geq \delta(\mu_n(U))^2$$

and hence $\mu_n(U)$ is bounded in n. Consequently $\mu_n(C)$ is bounded in n for every compact subset C of D. Therefore there is a strictly increasing sequence $\{n_j\}$ of positive integers and a Radon measure μ on D such that μ_n converges vaguely to μ. For each $x \in D$, $G_D(x, \cdot)$ is continuous in the extended sense on D, so $\underline{\lim}_n \int G_D(x, y)\mu_n(dy) \geq \int G_D(x, y)\mu(dy)$; i.e., $\underline{\lim}_n G_D\mu_n \geq G_D\mu$ on D. Set $M = \sup_n \|\mu_n\|$. Then $\int G_D\mu_m \, d\mu_n \leq M^2$ for $m, n \geq 1$ by Theorem 4.2. Thus

$$\int G_D\mu \, d\mu_n \leq \int \underline{\lim}_m G_D\mu_m \, d\mu_n \leq \underline{\lim}_m \int G_D\mu_m \, d\mu_n \leq M^2$$

and hence

$$\int G_D\mu \, d\mu \leq \int \underline{\lim}_n G_D\mu_n \, d\mu \leq \underline{\lim}_n \int G_D\mu \, d\mu_n \leq M^2,$$

so $\mu \in \mathscr{E}_D^+$. This completes the proof of the theorem.

Let $\mu_n, n \geq 1$, and μ be elements of $\mathscr{E}_D$. Then μ_n is said to *converge strongly* to μ if $\lim_n \|\mu_n - \mu\| = 0$ and μ_n is said to *converge weakly* to μ if $\|\mu_n\|$ is bounded in n and $\lim_n (\mu_n, \nu) = (\mu, \nu)$ for all $\nu \in \mathscr{E}_D$. Observe that strong convergence implies weak convergence.

Let $\{\mu_n\}$ be a sequence of elements of $\mathscr{E}_D$. It is called a *Cauchy sequence* if $\|\mu_n - \mu_m\| \to 0$ as $m, n \to \infty$ and a *weak Cauchy sequence* if $\|\mu_n\|$ is bounded in n and $(\mu_n - \mu_m, \nu) \to 0$ as $m, n \to \infty$ for all $\nu \in \mathscr{E}_D$. Clearly a Cauchy sequence is also a weak Cauchy sequence. If μ_n is strongly (weakly) convergent to μ, then $\{\mu_n\}$ is a Cauchy (weak Cauchy) sequence. *If $\{\mu_n\}$ is a Cauchy sequence and μ_n converges weakly to $\mu \in \mathscr{E}_D$, then μ_n converges strongly to μ.* For set $M = \sup_n \|\mu_n - \mu\|$. Then

$$\begin{aligned}\|\mu_n - \mu\|^2 &\leq |(\mu_m - \mu, \mu_n - \mu)| + |(\mu_n - \mu_m, \mu_n - \mu)| \\ &\leq |(\mu_m - \mu, \mu_n - \mu)| + M\|\mu_n - \mu_m\|.\end{aligned}$$

It follows by letting $m \to \infty$ that

$$\|\mu_n - \mu\|^2 \leq M \overline{\lim_m} \|\mu_n - \mu_m\|.$$

Consequently $\lim_n \|\mu_n - \mu\|^2 = 0$ and hence μ_n converges strongly to μ.

Theorem 4.7. *Suppose $\mu_n, n \geq 1$, and μ are in $\mathscr{E}_D^+$ and either $G_D\mu_n \uparrow G_D\mu$ on D or $G_D\mu_n \downarrow G_D\mu$ q.e. on D. Then μ_n converges strongly to μ.*

Proof. Suppose first that $G_D\mu_n \uparrow G_D\mu$ on D. Let $m \leq n$. Then

$$\begin{aligned}||\mu_n - \mu_m||^2 &= ||\mu_n||^2 + ||\mu_m||^2 - 2\int G_D\mu_n\,d\mu_m \\ &\leq ||\mu_n||^2 + ||\mu_m||^2 - 2\int G_D\mu_m\,d\mu_m = ||\mu_n||^2 - ||\mu_m||^2.\end{aligned}$$

Thus $||\mu_n||$ is increasing in n and, being bounded, it has a finite limit. Consequently $\{\mu_n\}$ is a Cauchy sequence. Similarly $\{\mu_n\}$ is a Cauchy sequence if $G_D\mu_n\downarrow$. Thus it suffices to show that μ_n converges weakly to μ or equivalently that $\int G_D\mu_n\,dv \to \int G_D\mu\,dv$ for all $v \in \mathscr{E}_D^+$. But this follows from the dominated convergence theorem since, by Theorem 4.1, v does not charge polar sets.

If $\mu \in \mathscr{E}_D$ and $B \in \mathscr{B}$, set $\mu h_{B,D} = \mu^+ h_{B,D} - \mu^- h_{B,D} \in \mathscr{E}_D$.

Theorem 4.8. *Let $\mu \in \mathscr{E}_D$. Suppose either that $B_n \in \mathscr{B}$ for each n and $B_n \uparrow B$ or that B_n is closed for each n, $|\mu|h_{B_1,D}(D) < \infty$ and $B_n \downarrow B$. Then $\mu h_{B_n,D}$ converges strongly to $\mu h_{B,D}$.*

Proof. Without loss of generality it can be assumed that $\mu \in \mathscr{E}_D^+$. Suppose first that $B_n \in \mathscr{B}$ for each n and $B_n \uparrow B$. Then $G_D(\mu h_{B_n,D}) \uparrow G_D(\mu h_{B,D})$ by Theorem 3.1, so $\mu h_{B_n,D}$ converges strongly to $\mu h_{B,D}$ by Theorem 4.7. Suppose next that B_n is closed for each n, $\mu h_{B_1,D}(D) < \infty$ and $B_n \downarrow B$. Then $G_D(\mu h_{B_n,D}) \downarrow G_D(\mu h_{B,D})$ q.e. on D by Theorem 3.10 (since μ has finite energy, $\mu(B \setminus B^r) = 0$ by Theorem 4.1), so $\mu h_{B_n,D}$ converges strongly to $\mu h_{B,D}$ by another application of Theorem 4.7.

A subset A_0 of a set $A \subset \mathscr{E}_D$ is said to be *dense* in A if for every $\mu \in A$ there is a subsequence $\{\mu_n\}$ of elements of A_0 such that μ_n converges strongly to μ.

Theorem 4.9. *The set of elements $v \in \mathscr{E}_D$ having compact support and such that $G_D v$ is continuous and has compact support is dense in $\mathscr{E}_D$.*

Proof. Let $v \in \mathscr{E}_D^+$. By Theorem 2.1 there exist measures $v_n \in \mathscr{E}_D^+$ such that $G_D v_n$ is continuous and $G_D v_n \uparrow G_D v$ on D. By Theorem 4.7, v_n converges strongly to v.

Let $v \in \mathscr{E}_D^+$ have a continuous Green potential. Let B_n be compact subsets of D such that $B_n^r = B_n$ for each n and $B_n \uparrow D$. Set $v_n = v h_{B_n,D}$. Then $G_D v_n = h_{B_n,D} G_D v \uparrow G_D v$ on D and hence v_n converges strongly to v by Theorem 4.7.

Now ν_n is supported on the compact set B_n and $G_D\nu_n = G_D\nu$ on B_n, so $G_D\nu_n$ is continuous on D by the continuity principle.

Let $\nu \in \mathscr{E}_D^+$ have compact support and a continuous Green potential. Let D_n be relatively compact regular open subsets of D such that D_1 contains the support of μ and $D_n \uparrow D$. Set $B_n = D_n^c$. Then $B_n^r \cap D = B_n \cap D$ for each n and $B_n \downarrow D^c$. Set $\mu_n = \nu h_{B_n,D} \in \mathscr{E}_D^+$. Then μ_n is supported on the compact subset ∂B_n of D and $G_D\mu_n = h_{B_n,D}G_D\nu$ on D. Now $G_D\mu_n = G_D\nu$ on ∂B_n and hence $G_D\mu_n$ is continuous on D by the continuity principle. By Theorem 3.10, $G_D\mu_n \downarrow 0$ q.e. on D. Thus μ_n converges strongly to zero by Theorem 4.7. Set $\nu_n = \nu - \mu_n$. Then ν_n converges strongly to ν. Also for each n, ν_n has compact support and continuous Green potential $G_D\nu_n = G_D\nu - G_D\mu_n = G_D\nu - h_{B_n,D}G_D\nu$, which vanishes outside the compact set $\bar{D}_n$.

The conclusion of the theorem follows from the above observations.

By the next result *if* μ_n, n ≥ 1, *and* μ *are in* $\mathscr{E}_D^+$ *and* μ_n *converges strongly to* μ, *then* μ_n *converges vaguely to* μ.

Theorem 4.10. *Let* μ_n, $n \geq 1$, *and* μ *be element of* $\mathscr{E}_D^+$ *such that* $\|\mu_n\|$ *is bounded in n. Then* μ_n *converges weakly to* μ *if and only if* μ_n *converges vaguely to* μ.

Proof. Suppose first that μ_n converges vaguely to μ. Then $\mu \in \mathscr{E}_D^+$ by Theorem 4.6. Let $\nu \in \mathscr{E}_D^+$ have a Green potential which is continuous and has compact support. Then

$$(\mu_n, \nu) = \int G_D\nu \, d\mu_n \to \int G_D\nu \, d\mu = (\mu, \nu).$$

By Theorem 4.9, $(\mu_n, \nu) \to (\mu, \nu)$ for all measures $\nu \in \mathscr{E}_D^+$ and hence μ_n converges weakly to μ.

Suppose conversely that μ_n converges weakly to μ. By Theorem 4.6 there is a strictly increasing sequence $\{n_j\}$ of positive integers and a measure $\gamma \in \mathscr{E}_D^+$ such that μ_n converges vaguely to γ. Then $(\gamma, \nu) = \lim_j (\mu_{n_j}, \nu) = (\mu, \nu)$ for all measures $\nu \in \mathscr{E}_D^+$ whose Green potential is continuous and has compact support. Thus $(\gamma, \nu) = (\mu, \nu)$ for all $\nu \in \mathscr{E}_D^+$ by Theorem 4.9 and hence $\gamma = \mu$ by the energy principle. It now follows that μ_n converges vaguely to μ. This completes the proof of the theorem.

It is now possible to strengthen part of Theorem 4.7.

Theorem 4.11. *Suppose* $\mu_n \in \mathscr{E}_D^+$ *for* $n \geq 1$, $\|\mu_n\|$ *is bounded in n and* $G_D\mu_n \uparrow$ *on D. Then there is a measure* $\mu \in \mathscr{E}_D^+$ *such that* $G_D\mu_n \uparrow G_D\mu$ *on D and* μ_n *converges strongly to* μ.

Proof. By Theorem 4.6 there is a strictly increasing sequence $\{n_j\}$ of positive integers and a measure $\mu \in \mathscr{E}_D^+$ such that μ_{n_j} converges vaguely to μ. By Theorem 4.10, μ_{n_j} converges weakly to μ. Thus $\int G_D\mu_{n_j}\,dv \to \int G_D\mu\,dv$ for all $v \in \mathscr{E}_D^+$. Therefore by the monotone convergence theorem $\int \lim_n G_D\mu_n\,dv = \int G_D\mu\,dv$ for all $v \in \mathscr{E}_D^+$. This shows that $\lim_n G_D\mu_n$ cannot be identically infinite on any component of D. (Note that if C is a compact subset of a component D_0 of D and C has positive Lebesgue measure, then $v(dx) = I_C(x)\,dx$ defines a measure having finite energy and assigning positive measure to D_0.) Thus $\lim_n G_D\mu_n$ is superharmonic on D by Proposition 5.1.1. Therefore $\lim_n G_D\mu_n = G_D\mu$ by Theorem 4.5. It now follows from Theorem 4.7 that μ_n converges strongly to μ. This completes the proof of the theorem.

A subset A of $\mathscr{E}_D$ is said to be *complete* if every Cauchy sequence of elements of A converges to an element of A; it is said to be *weakly complete* if every weak Cauchy sequence of elements of A converges weakly to an element of A. If A is weakly complete, it is complete. Cartan [2] showed that $\mathscr{E}_D^+$ is complete but that $\mathscr{E}_D$ is not complete (see Theorem 5.15 below).

Theorem 4.12. *The set $\mathscr{E}_D^+$ is weakly complete.*

Proof. Let $\{\mu_n\}$ be a weak Cauchy sequence of elements of $\mathscr{E}_D^+$. By definition, $\|\mu_n\|$ is bounded in n. Thus by Theorem 4.10, to show that μ_n converges weakly to a measure $\mu \in \mathscr{E}_D^+$ it suffices to show that μ_n converges vaguely to μ. By Theorem 4.6 it suffices to show that if $\{m_j\}$ and $\{n_j\}$ are strictly increasing sequences of positive integers such that $\mu_{m_j} \to \gamma_1 \in \mathscr{E}_D^+$ vaguely and $\mu_{n_j} \to \gamma_2 \in \mathscr{E}_D^+$ vaguely then $\gamma_1 = \gamma_2$. Choose $v \in \mathscr{E}_D$. Then $(\mu_{m_j}, v) \to (\gamma_1, v)$ and $(\mu_{n_j}, v) \to (\gamma_2, v)$ by Theorem 4.10. Now $(\mu_{m_j} - \mu_{n_j}, v) \to 0$ since $\{\mu_m\}$ is a weak Cauchy sequence. Thus $(\gamma_1, v) = (\gamma_2, v)$ for $v \in \mathscr{E}_D$. Consequently $\|\gamma_2 - \gamma_1\|^2 = 0$ and hence $\gamma_1 = \gamma_2$ by the energy principle. This completes the proof of the theorem.

Let $B \in \mathscr{B}$. Then $B \backslash B^r$ is polar by Theorem 2.6.3 and hence $\mathscr{E}_D(B) \subset \mathscr{E}_D(B^r)$ by Theorem 4.1. Let $\mu \in \mathscr{E}_D^+$ and set $\mu' = \mu h_{B,D}$. Then $\mu' \in \mathscr{E}_D^+(B^r)$ and $G_D\mu' = G_D\mu$ on B^r, so $\int G_D\mu\,d\mu' = \int G_D\mu'\,d\mu' = \|\mu'\|^2$. By the fundamental identity

$$\|\mu\|^2 = \int G_D\mu\,d\mu = \int g_{B \cup D^c}\mu\,d\mu + \int G_D\mu'\,d\mu' = \int g_{B \cup D^c}\mu\,d\mu + \|\mu'\|^2.$$

Consequently

$$\|\mu' - \mu\|^2 = \|\mu\|^2 - \|\mu'\|^2 = \int g_{B \cup D^c}\mu\,d\mu.$$

The characterization of balayage in the next result is due to Frostman [1].

Theorem 4.13. *Let $B \in \mathscr{B}$ and $\mu \in \mathscr{E}_D^+$ and set $\mu' = \mu h_{B,D}$. Then*

$$\min[\|\nu - \mu\| : \nu \in \mathscr{E}_D^+(B^r)] = \|\mu' - \mu\|$$

and the minimum occurs uniquely at $\nu = \mu'$. Also

$$\inf[\|\nu - \mu\| : \nu \in \mathscr{E}_D^+(B)] = \|\mu' - \mu\|;$$

if $\nu_n \in \mathscr{E}_D^+(B)$ and $\|\nu_n - \mu\| \to \|\mu' - \mu\|$, then ν_n converges strongly to μ'.

Proof. Now $\mu' \in \mathscr{E}_D^+(B^r)$. Also $G_D\mu' = G_D\mu$ on B^r, so $(\mu' - \mu, \nu) = 0$ for $\nu \in \mathscr{E}_D(B^r)$. Thus

$$\|\nu - \mu\|^2 = \|\nu - \mu'\|^2 + \|\mu' - \mu\|^2, \qquad \nu \in \mathscr{E}_D^+(B^r). \tag{2}$$

The first conclusion of the theorem now follows from the energy principle. Let B_n be compact subsets of B such that $B_n \uparrow B$ and set $\nu_n = \mu h_{B_n,D}$. Then $\nu_n \in \mathscr{E}_D^+(B)$. Also ν_n converges strongly to μ' by Theorem 4.8 and hence $\|\nu_n - \mu\| \to \|\mu' - \mu\|$. The second conclusion of the theorem now follows from (2) since $\mathscr{E}_D^+(B) \subset \mathscr{E}_D^+(B^r)$.

Let $B \in \mathscr{B}$ and $\mu \in \mathscr{M}_D$ and set $f = G_D\mu$. If μ has finite energy, then minimizing $\|\nu - \mu\|^2 = \|\mu\|^2 + \|\nu\|^2 - 2\int f\,d\nu$ as ν ranges over $\mathscr{E}_D^+(B^r)$ is equivalent to minimizing $\|\nu\|^2 - \int f\,d\nu$ as ν ranges over $\mathscr{E}_D^+(B^r)$. The latter problem makes sense even if μ has infinite energy.

Theorem 4.14. *Let $B \in \mathscr{B}$ and let f be a nonnegative superharmonic function on D such that $h_{B,D}f$ is the Green potential of a measure $\mu' \in \mathscr{E}_D^+$. Then*

$$\min\left[\|\nu\|^2 - 2\int f\,d\nu : \nu \in \mathscr{E}_D^+(B^r)\right] = -\|\mu'\|^2$$

and the minimum occurs uniquely at $\nu = \mu'$. Also

$$\inf\left[\|\nu\|^2 - 2\int f\,d\nu : \nu \in \mathscr{E}_D^+(B)\right] = -\|\mu'\|^2;$$

if $\nu_n \in \mathscr{E}_D^+(B)$ and $\|\nu_n\|^2 - 2\int f\,d\nu_n \to -\|\mu'\|^2$, then ν_n converges strongly to μ'.

Proof. Now $\mu' \in \mathscr{E}_D^+(B^r)$. Also $G_D\mu' = f$ on B^r so if $\nu \in \mathscr{E}_D^+(B^r)$, then $\int f\,d\nu = \int G_D\mu'\,d\nu$. Thus

$$\|\nu\|^2 - 2\int f\,d\nu = \|\nu - \mu'\|^2 - \|\mu'\|^2, \qquad \nu \in \mathscr{E}_D^+(B^r). \tag{3}$$

The first conclusion of the theorem now follows from the energy principle. The proof of the second conclusion uses (3) instead of (2) but is otherwise the same as the proof of the corresponding conclusion of Theorem 4.13.

Let $B \in \mathscr{B}$ be a polar subset of D. By Theorem 5.11 below there is a measure $\mu \in \mathscr{E}_D^+$ such that $G_D\mu = \infty$ on B. This result will be used in the proof of the following result, which is suggested by Theorems 3.4 and 3.8.

Theorem 4.15. *Let $B \in \mathscr{B}$ and let f be a nonnegative superharmonic function on D such that $h_{B,D}f$ is the Green potential of a measure $\mu' \in \mathscr{E}_D^+$. Then*

$$\min[\|\nu\|^2 : \nu \in \mathscr{E}_D^+ \text{ and } G_D\nu \geq f \text{ q.e. on } B \cap D] = \|\mu'\|^2$$

and the minimum occurs uniquely at $\nu = \mu'$. Also

$$\inf[\|\nu\|^2 : \nu \in \mathscr{E}_D^+ \text{ and } G_D\nu \geq f \text{ on } B \cap D] = \|\mu'\|^2;$$

if $\nu_n \in \mathscr{E}_D^+$, $G_D\nu_n \geq f$ on $B \cap D$ and $\|\nu_n\|^2 \to \|\mu'\|^2$, then ν_n converges strongly to μ'.

Proof. Observe that $\mu' \in \mathscr{E}_D^+(B^r)$ and $G_D\mu' = f$ on $B^r \cap D$ and hence q.e. on $B \cap D$. Suppose $\nu \in \mathscr{E}_D^+$ and $G_D\nu \geq f$ q.e. on $B \cap D$. Then $G_D\nu \geq h_{B,D}f = G_D\mu'$ on D by Theorem 3.4 and hence $(\nu - \mu', \mu') = \int (G_D\nu - G_D\mu')\, d\mu' \geq 0$. Thus

$$\|\nu\|^2 \geq \|\nu - \mu'\|^2 + \|\mu'\|^2,$$

so the first conclusion of the theorem follows from the energy principle.

By Theorem 5.11 below there is a measure $\mu_0 \in \mathscr{E}_D^+$ such that $G_D\mu_0 = \infty$ on $(B \backslash B^r) \cap D$. Choose $\varepsilon > 0$ and set $\nu = \mu' + \varepsilon\mu_0$. Then $\nu \in \mathscr{E}_D^+$, $G_D\nu \geq f$ on $B \cap D$ and $\|\nu\| \leq \|\mu'\| + \varepsilon\|\mu_0\|$ by the triangle inequality. Consequently

$$\inf[\|\nu\|^2 : \nu \in \mathscr{E}_D^+ \text{ and } G_D\nu \geq f \text{ on } B \cap D] \leq \|\mu'\|^2.$$

The second conclusion of the theorem now follows by the argument used to prove the first conclusion.

Theorem 4.16. *Let $B \in \mathscr{B}$ and let f be a nonnegative superharmonic function on D such that $h_{B,D}f$ is the Green potential of a measure $\mu' \in \mathscr{E}_D^+$. Then*

$$\max\left[\int f\, d\nu : \nu \in \mathscr{E}_D^+(B^r) \text{ and } G_D\nu \leq f\right] = \|\mu'\|^2$$

and the maximum occurs uniquely at $\nu = \mu'$. Also

$$\sup\left[\int f\, d\nu : \nu \in \mathscr{E}_D^+(B) \text{ and } G_D\nu \leq f\right] = \|\mu'\|^2;$$

if $\nu_n \in \mathscr{E}_D^+(B)$, $G_D\nu_n \leq f$ and $\int f\, d\nu_n \to \|\mu'\|^2$, then ν_n converges strongly to μ'.

Proof. If $\nu \in \mathscr{E}_D^+(B^r)$, then $\int f\,d\nu = \int G_D\mu'\,d\nu = (\nu, \mu')$. Observe that $\mu' \in \mathscr{E}_D^+(B^r)$, $G_D\mu' \leq f$ and $\int f\,d\mu' = \|\mu'\|^2$. Suppose $\nu \in \mathscr{E}_D^+(B^r)$ and $G_D\nu \leq f$. Then $(\mu' - \nu, \nu) = \int (f - G_D\nu)\,d\nu \geq 0$. Consequently

$$\|\mu'\|^2 = (\nu, \mu') + (\mu' - \nu, \mu' - \nu) + (\mu' - \nu, \nu) \geq (\nu, \mu') + \|\mu' - \nu\|^2.$$

The first conclusion of the theorem follows from these observations.

Let B_n be compact subsets of B such that $B_n \uparrow B$. Set $\nu_n = \mu' h_{B_n,D}$. Then $\nu_n \in \mathscr{E}_D^+(B)$ and $G_D\nu_n \leq G_D\mu' \leq f$. Also ν_n converges strongly to $\mu' h_{B,D} = \mu'$ by Theorem 4.8. Consequently

$$\int f\,d\nu_n = (\nu_n, \mu') \to (\mu', \mu') = \|\mu'\|^2.$$

The second conclusion now follows by the same argument used to prove the first conclusion.

The following characterization of $h_{B,D}f$ in terms of energy is due to Cartan [3].

Theorem 4.17. *Let f be a nonnegative superharmonic function on D and let $B \in \mathscr{B}$. Then $h_{B,D}f$ is the unique superharmonic function v on D such that $\int v\,d\mu = \int f\,d(\mu h_{B,D})$ holds for all measures $\mu \in \mathscr{E}_D^+$.*

Proof. Let $\mu \in \mathscr{E}_D^+$. Then

$$\int h_{B,D}f\,d\mu = \iint h_{B,D}(y, dz) f(z) \mu(dy) = \int f\,d(\mu h_{B,D}).$$

Suppose conversely that v is a superharmonic function on D such that $\int v\,d\mu = \int f\,d(\mu h_{B,D})$ for all measures $\mu \in \mathscr{E}_D^+$. Then $v = h_{B,D}f$ by Theorem 4.5. This completes the proof of the theorem.

5. Equilibrium Problem

Let $d \geq 3$ and let B be a compact nonpolar subset of $\mathbb{R}^d$. The third fundamental problem considered by Gauss in 1840, along with the Dirichlet problem and the balayage problem, is the equilibrium problem (sometimes called the Robin problem): find a measure μ which is supported on B and whose Green potential equals one on B. This problem also arises naturally in electrostatics. Let B be a (connected) conductor on which a charge of fixed total amount is placed. The charge will be redistributed so that in equilibrium there is no potential difference between any two points on the conductor; that is, the equilibrium charge distribution should have constant potential on the conductor. The constant value one in the definition of the equilibrium is a convenient normalizing constant. The equilibrium charge distribution should also minimize energy (see Theorem 5.16 below).

The discussion in Section 3 which showed that the balayage problem as originally formulated does not always have a solution also shows that the equilibrium problem as formulated above does not always have a solution. That discussion suggests the following generalized version of the equilibrium problem.

Let $B \in \mathscr{B}$. The *equilibrium problem* is to find a measure $\mu \in \mathscr{M}_D(B^r)$ such that $G_D\mu = 1$ on $B^r \cap D$. It follows from the uniqueness principle that there is at most one such measure. If such a measure exists it is called the *equilibrium measure* of B (relative to D), its total measure $C_D(B) = \mu_{B,D}(D)$ is called the *capacity* of B, and the Green potential $G_D\mu_{B,D}$ is called the *equilibrium potential* of B. The set B is called an *equilibrium set* if it has an equilibrium measure.

Theorem 5.1. *Let $B \in \mathscr{B}$. Then B is an equilibrium set if and only if $h_{B,D}1 = P_{.}(\tau_B < T_D)$ is a Green potential, in which case $G_D\mu_{B,D} = h_{B,D}1 = P_{.}(\tau_B < T_D)$ and $\mu_{B,D}$ is supported on ∂B and does not charge polar sets.*

Proof. It follows from the fundamental identity that if the equilibrium measure $\mu_{B,D}$ exists, its Green potential is given by $G_D\mu_{B,D} = h_{B,D}1 = P_{.}(\tau_B < T_D)$. Suppose conversely that $h_{B,D}1 = P_{.}(\tau_B < T_D)$ is the Green potential of a measure $\mu \in \mathscr{M}_D$. Then μ is concentrated on B^r by the argument preceding the statement of Theorem 3.1 and $G_D\mu = P_{.}(\tau_B < T_D) = 1$ on $B^r \cap D$. Thus μ is the equilibrium measure of B.

Suppose B has equilibrium measure $\mu_{B,D}$. Then $G_D\mu_{B,D} = P_{.}(\tau_B < T_D) < \infty$ so by Theorem 2.8, $\mu_{B,D}$ does not charge polar sets. Now $G_D\mu_{B,D}$ is harmonic on $\mathring{B} \cap D$, since it equals 1 on this set, so $\mu_{B,D}(\mathring{B} \cap D) = 0$ by Theorem 1.4. Thus μ is supported on ∂B, which completes the proof of the theorem.

Theorem 5.2. *Let B be a relatively closed equilibrium subset of D. Then ∂B is an equilibrium set, $\mu_{\partial B,D} = \mu_{B,D}$ and $C_D(\partial B) = C_D(B)$.*

Proof. Since $\mu_{B,D}$ is supported on ∂B and does not charge polar sets and $\partial B \backslash (\partial B)^r$ is polar, $\mu_{B,D}$ is concentrated on $(\partial B)^r$. Also $G_D\mu_{B,D} = 1$ on $B^r \cap D \supset (\partial B)^r \cap D$. The desired conclusion follows from these observations.

Theorem 5.3. *Let $B \in \mathscr{B}$ be a relatively compact subset of D. Then B is an equilibrium set and $C_D(B) < \infty$.*

Proof. By Theorem 3.1, $h_{B,D}1$ is a Green potential. Thus B is an equilibrium set by Theorem 5.1. Now $\mu_{B,D}$ is a Radon measure, so $C_D(B) = \mu_{B,D}(\bar{B}) < \infty$ since $\bar{B}$ is a compact subset of D.

Theorem 5.4. *Let $B \in \mathscr{B}$ be an equilibrium subset of D. Then B is polar if and only if $C_D(B) = 0$.*

Proof. Suppose B is polar. Then $G_D\mu_{B,D} = P_.(\tau_B < T_D) = 0$, so $\mu_{B,D} = 0$ and hence $C_D(B) = 0$. Suppose conversely that $C_D(B) = 0$. Then $\mu_{B,D} = 0$ and hence $P_.(\tau_B < T_D) = G_D\mu_{B,D} = 0$ on D. Consequently $B^r \cap D$ is empty. Let D_n be relatively compact open subsets of D such that $D_n \uparrow D$ and set $B_n = B \cap D_n$. For each n, B_n^r is empty and hence B_n is polar by Theorem 2.6.4. Thus $B = \bigcup_n B_n$ is polar.

Theorem 5.5. *Let $B \in \mathscr{B}$ be an equilibrium set and let $A \in \mathscr{B}$ be a subset of B. Then A is an equilibrium set,*

$$\mu_{A,D} = \mu_{B,D}h_{A,D},$$
$$G_D\mu_{A,D} = h_{A,D}G_D\mu_{B,D} \le G_D\mu_{B,D},$$

and

$$C_D(A) = \int P_.(\tau_A < T_D)\, d\mu_{B,D} \le C_D(B).$$

Proof. Set $\mu = \mu_{B,D}h_{A,D}$. It follows from the formulation and solution to the balayage problem in Section 3 that μ is concentrated on A^r and $G_D\mu = h_{A,D}G_D\mu_{B,D} = G_D\mu_{B,D} = 1$ on $A^r \cap D \subset B^r \cap D$. Thus A is an equilibrium set and $\mu_{A,D} = \mu$. The remaining conclusions now follow easily.

Note the common pattern in the first formula of the preceding theorem and the formulas for equilibrium measures corresponding to λ-potentials in Section 2.5, Newtonian potentials in Section 3.1, and logarithmic potentials in Section 3.4.

Theorem 5.6. *Let $A, B \in \mathscr{B}$ be equilibrium sets. Then $A \cup B$ is an equilibrium set and*

$$C_D(A \cup B) + C_D(A \cap B) \le C_D(A) + C_D(B).$$

Proof. Now $A \cap B$ is an equilibrium set by Theorem 5.5. Also

$$P_.(\tau_{A \cup B} < T_D) \le P_.(\tau_A < T_D) + P_.(\tau_B < T_D) = G_D(\mu_{A,D} + \mu_{B,D}),$$

so $P_\cdot(\tau_{A \cup B} < T_D)$ is a Green potential by Theorem 2.5 and hence $A \cup B$ is an equilibrium set. By Theorem 5.5

$$C_D(A \cup B) + C_D(A \cap B) = \int (P_\cdot(\tau_{A \cup B} < T_D) + P_\cdot(\tau_{A \cap B} < T_D))\, d\mu_{A \cup B, D}$$

and

$$C_D(A) + C_D(B) = \int (P_\cdot(\tau_A < T_D) + P_\cdot(\tau_B < T_D))\, d\mu_{A \cup B, D}.$$

Since

$$P_\cdot(\tau_{A \cup B} < T_D) \le P_\cdot(\tau_A < T_D, \tau_B < T_D) = P_\cdot(\tau_A < T_D) + P_\cdot(\tau_B < T_D) - P_\cdot(\tau_{A \cup B} < T_D),$$

the desired result holds.

Let $B \in \mathscr{B}$ be an equilibrium set. The energy of the equilibrium measure of B is given by

$$I_D(\mu_{B,D}) = \int G_D \mu_{B,D}\, d\mu_{B,D} = C_D(B).$$

Thus $\mu_{B,D} \in \mathscr{E}_D^+$ if and only if $C_D(B) < \infty$.

Theorem 5.7. *Suppose $B_n \in \mathscr{B}$ is an equilibrium set for each n, $B_n \uparrow B$ and $\lim_n C_D(B_n) < \infty$. Then B is an equilibrium set.*

Proof. It follows by applying Theorem 4.11 to the measures $\mu_n = \mu_{B_n,D}$ that there is a measure $\mu \in \mathscr{E}_D^+$ such that $h_{B_n,D}1 = G_D \mu_{B_n,D} \uparrow G_D \mu$. But $h_{B_n,D}1 \uparrow h_{B,D}1$ by Theorem 3.1, so $h_{B,D}1 = G_D\mu$. Thus B is an equilibrium set by Theorem 5.1.

If $B \in \mathscr{B}$ and B is not an equilibrium set, define $C_D(B) = \infty$. (By Theorem 5.7, this is the only definition of $C_D(B)$ which is compatible with the requirement that $C_D(A) \le C_D(B)$ if $A \in \mathscr{B}$ and $A \subset B$.)

Theorem 5.8. *Suppose $B_n \in \mathscr{B}$ for each n and $B_n \uparrow B$ where B is an equilibrium set. Then $\mu_{B_n,D}$ converges vaguely to $\mu_{B,D}$ and $\lim_n C_D(B_n) = C_D(B)$. If $C_D(B) < \infty$, then $\mu_{B_n,D}$ converges strongly to $\mu_{B,D}$.*

Proof. Now $G_D \mu_{B_n,D} = h_{B_n,D}1 \uparrow h_{B,D}1 = G_D \mu_{B,D}$ by Theorem 3.1, so $\mu_{B_n,D}$ converges vaguely to $\mu_{B,D}$ by Theorem 2.4. Consequently

$$\varliminf_n C_D(B_n) = \varliminf_n \mu_{B_n,D}(D) \ge \mu_{B,D}(D) = C_D(B).$$

Since $C_D(B_n) \le C_D(B)$ for each n by Theorem 5.5, $\lim_n C_D(B_n) = C_D(B)$. Suppose now that $C_D(B) < \infty$. Then $\mu_{B,D} \in \mathscr{E}_D^+$ and $\mu_{B_n,D} = \mu_{B,D} h_{B_n,D}$. Thus $\mu_{B_n,D}$ converges strongly to $\mu_{B,D}$ by Theorem 4.8.

Theorem 5.9. *Suppose B_n is closed for each n, $C_D(B_1) < \infty$ and $B_n \downarrow B$. Then $\mu_{B_n,D}$ converges strongly to $\mu_{B,D}$ and $\lim_n C_D(B_n) = C_D(B)$.*

Proof. Set $\mu = \mu_{B_1,D}$. Then $\mu \in \mathscr{E}_D^+$ and $\mu_{B_n,D} = \mu h_{B_n,D}$. By Theorem 4.8, $\mu_{B_n,D}$ converges strongly to $\mu h_{B,D} = \mu_{B,D}$. Thus $C_D(B_n) = \|\mu_{B_n,D}\|^2 \to \|\mu_{B,D}\|^2 = C_D(B)$ as $n \to \infty$. This completes the proof of the theorem.

Theorem 5.10. *Let $B_n \in \mathscr{B}$ for each n. Then $C_D(\bigcup_n B_n) \le \sum_n C_D(B_n)$.*

Proof. If $\sum_n C_D(B_n) = \infty$, the desired result holds trivially, so assume that $\sum_n C_D(B_n) < \infty$. Set $A_n = \bigcup_{1 \le m \le n} B_m$ and $A = \bigcup_n B_n$, so that $A_n \in \mathscr{B}$ for each n and $A_n \uparrow A$. Now $C_D(A_n) \le \sum_{1 \le m \le n} C_D(B_m)$ by Theorem 5.6 and hence $\lim_n C_D(A_n) \le \sum_n C_D(B_n) < \infty$. Thus A is an equilibrium set by Theorem 5.7. Consequently

$$C_D\left(\bigcup_n B_n\right) = C_D(A) = \lim_n C_D(A_n) \le \sum_n C_D(B_n)$$

by Theorem 5.8.

The result used to prove Theorem 4.15 above will now be obtained.

Theorem 5.11. *Let $B \in \mathscr{B}$ be a polar subset of D. Then there is a measure $\mu \in \mathscr{E}_D^+$ such that $G_D\mu = \infty$ on B.*

Proof. Suppose first that B is a compact polar subset of D. Now $C_D(B) = 0$ by Theorem 5.4, so it follows from Theorem 5.9 that for each $n \ge 1$ there is a compact subset B_n of D containing B in its interior and such that $C_D(B_n) \le n^{-4}$. Set $\mu = \sum_n \mu_{B_n,D}$ and note that $G_D\mu = \infty$ on B since $G_D\mu_{B_n,D} = 1$ on B for each n. Also set $\mu_n = \sum_{1 \le m \le n} \mu_{B_m,D}$. Then $G_D\mu_n \uparrow G_D\mu$ on D and

$$\|\mu_n\| \le \sum_1^n \|\mu_{B_m,D}\| = \sum_1^n (C_D(B_m))^{1/2} \le \sum_1^\infty m^{-2}.$$

Thus $\mu \in \mathscr{E}_D^+$ by Theorem 4.11.

Consider now the general case. Write $B = \bigcup_n B_n$ where each B_n is a compact polar subset of D. By what has already been shown there is a measure $\mu \in \mathscr{E}_D^+$ such that $\|\mu_n\| = n^{-2}$ and $G_D\mu_n = \infty$ on B_n. Set $\mu = \sum_n \mu_n$. Then $G_D\mu = \infty$ on B and $\mu \in \mathscr{E}_D^+$ by another application of Theorem 4.11. This completes the proof of the theorem.

Theorem 5.12. *Let $\mu, \nu \in \mathscr{M}_D$. If $G_D\mu \le G_D\nu$ on D, then $\mu(D) \le \nu(D)$. If $G_D\mu < G_D\nu$ on D and μ is supported on a compact subset of D, then $\mu(D) < \nu(D)$.*

Proof. Suppose $G_D\mu \le G_D\nu$ on D. Let D_n be relatively compact open subsets of D such that $D_n \uparrow D$ and set $B_n = \bar{D}_n$. Then

$$\mu(D_n) \le \int G_D\mu_{B_n,D}\, d\mu = \int G_D\mu\, d\mu_{B_n,D} \le \int G_D\nu\, d\mu_{B_n,D} = \int G_D\mu_{B_n,D}\, d\nu \le \nu(D)$$

and hence $\mu(D) \le \nu(D)$.

Suppose $G_D\mu < G_D\nu$ on D and μ is supported on a compact subset of D. Let B be a compact nonpolar subset of D containing the support of μ in its interior. Then

$$\mu(D) = \int G_D\mu_{B,D}\, d\mu = \int G_D\mu\, d\mu_{B,D} < \int G_D\nu\, d\mu_{B,D} = \int G_D\mu_{B,D}\, d\nu \le \nu(D).$$

This completes the proof of the theorem.

Suppose $d \ge 3$ and $D = \mathbb{R}^d$. Set $\mu = \sum_{n\ge 1} 2^{-n}\sigma_n$ and $\nu = \delta_0$. It follows from Proposition 3.1.7 that $g\sigma_n < g\nu$ on $\mathring{B}_n(0)$ and $g\sigma_n = g\nu$ on $(\mathring{B}_n(0))^c$; also $g\mu < \infty$ and hence $g\mu < g\nu$ on $\mathbb{R}^d$. Clearly $\mu(\mathbb{R}^d) = \nu(\mathbb{R}^d) = 1$. This example points out the need for the condition that μ have compact support in the statement of the second result of the previous theorem.

Let $B \in \mathscr{B}$ be an equilibrium set. Then $g_{D^c}\mu_{B,D} = G_D\mu_{B,D} = P_.(\tau_B < T_D)$ on D. Also $g_{D^c}\mu_{B,D} = 0 = P_.(\tau_B < T_D)$ on $(D^c)^r$. It will now be shown that the last equation holds everywhere on D^c.

Theorem 5.13. *Let $B \in \mathscr{B}$ be an equilibrium set. Then $g_{D^c}\mu_{B,D} = P_.(\tau_B < T_D)$ on $\mathbb{R}^d$.*

Proof. For $t > 0$ and $x \in \mathbb{R}^d$

$$\begin{aligned}\int \left(\int_t^\infty q_{D^c}(s, x, y)\, ds \right) \mu_{B,D}(dy) &= \int_D q_{D^c}(t, x, y) g_{D^c}\mu_{B,D}(y)\, dy \\ &= \int_D q_{D^c}(t, x, y) P_y(\tau_B < T_D)\, dy \\ &= P_x(T_D > t, t + \tau_B \circ \theta_t < T_D).\end{aligned}$$

The desired result now follows by letting $t \downarrow 0$.

Let A be a relatively closed subset of D and let B be a compact subset of $D\backslash A$. A signed measure $\nu \in \mathscr{E}_D$ is called the *condenser measure* corresponding to A, B if ν is concentrated on $A \cup B$, $G_D\nu = 1$ on B^r, and $G_D\nu = 0$ on $A^r \cap D$. The condenser measure, if it exists, is concentrated on $A^r \cup B^r$ (by Theorems

4.1 and 2.6.3) and is uniquely determined. To verify uniqueness observe that if μ and ν are both condenser measures corresponding to A, B, then $G_D(\nu - \mu) = 0$ on $(A^r \cup B^r) \cap D$, so $(\mu, \nu - \mu) = 0$ and $(\nu, \nu - \mu) = 0$; consequently $\|\nu - \mu\| = 0$ and hence $\mu = \nu$ by the energy principle. If A is polar the condenser measure corresponding to A, B is just the equilibrium measure of B.

The next result is called the *condenser theorem.* A similar result for Markov processes has been obtained by Chung and Getoor [1]. Such results arise in the theory of Dirichlet spaces developed by Beurling and Deny [1], [2] (see Landkof [1] and Rao [1]).

Theorem 5.14. *Let A be a relatively closed subset of D and let B be a compact subset of $D \backslash A$. Then A, B has the condenser measure $\nu = \mu_{B,D\backslash A} - \mu_{B,D\backslash A}h_{A,D}$. Moreover $|\nu|(D) < \infty$, ν^+ is concentrated on B^r, ν^- is concentrated on A^r, $0 \le G_D\nu \le 1$ on D and $G_D|\nu|$ is bounded on D.*

Proof. Set $\mu = \mu_{B,D\backslash A}$ and $\nu = \mu - \mu h_{A,D}$. Now μ, thought of as a measure on D, is a finite measure which is concentrated on B^r. Since the total measure of $\mu h_{A,D}$ is at most that of μ, $\mu h_{A,D}$ is a finite measure which is concentrated on A^r. Thus $\nu^+ = \mu$ is concentrated on B^r, $\nu^- = \mu h_{A,D}$ is concentrated on A^r and $|\nu|(D) < \infty$. By Theorem 5.13 and the fundamental identity

$$(1) \qquad G_D\mu = g_{A \cup D^c}\mu + h_{A,D}G_D\mu = P_{.}(\tau_B < T_{D\backslash A}) + h_{A,D}G_D\mu \qquad \text{on } D.$$

It follows from (4.5.1) that $h_{A,D}G_D\mu$ is bounded on D. Thus $G_D\mu$ is bounded on D by (1). Since $G_D(\mu h_{A,D}) = h_{A,D}G_D\mu$, $G_D(\mu h_{A,D})$ is bounded on D and hence $G_D|\nu|$ is bounded on D. Consequently $I_D(|\nu|) = \int G_D|\nu|\, d|\nu| < \infty$ and therefore $\nu \in \mathscr{E}_D$. It now follows from (1) that

$$G_D\nu = P_{.}(\tau_B < T_{D\backslash A}) \qquad \text{on } D.$$

Thus $G_D\nu = 1$ on B^r, $G_D\nu = 0$ on $A^r \cap D$ and $0 \le G_D\nu \le 1$ on D. This completes the proof of the theorem.

The condenser theorem will be used to prove the next result, which is due to Cartan [2].

Theorem 5.15. *The space $\mathscr{E}_D$ is not complete.*

Proof. Choose $x \in D$ and $r > 0$ such that $B_r(x) \subset D$. Let γ denote the equilibrium measure of $S_r(x)$. For $0 < s < r$ set $\gamma_s = \gamma h_{B_s(x),D}$. Then $\lim_{s \uparrow r} \|\gamma_s - \gamma\| = 0$ by Theorems 3.1 and 4.7. Thus there is a constant $c > 0$ and a strictly increasing sequence $\{r_n\}$ of numbers in $(0, r)$ such that $\gamma_{r_n}(D) \ge c$

for each n, $r_n \uparrow r$ and $\sum_n \|\gamma_{r_n} - \gamma\| < \infty$. Set $\mu_n = \sum_{1 \le m \le n} (\gamma_{r_m} - \gamma)$. Then $\{\mu_n\}$ is a Cauchy sequence of elements of $\mathscr{E}_D$. Thus to show that $\mathscr{E}_D$ is not complete it suffices to show that μ_n does not converge strongly to any element of $\mathscr{E}_D$. Suppose otherwise; that is, suppose there is an element $\mu \in \mathscr{E}_D$ such that μ_n converges strongly to μ. Given the positive integer m set $B_m = B_{r_m}(x)$ and $A_m = D \backslash \mathring{B}_{r_{m+1}}(x)$, and let v_m be the condenser measure corresponding to A_m, B_m. Then $v_m \in \mathscr{E}_D$, $G_D v_m = 1$ on B_m, $G_D v_m = 0$ on A_m and $0 \le G_D v_m \le 1$ on D. Consequently for $n \ge m$

$$mc \le \mu_n(B_m) = \int G_D v_m \, d\mu_n \to \int G_D v_m \, d\mu \le \mu^+(B_{m+1}).$$

Therefore $\mu^+(B_r(x)) = \infty$ and hence μ^+ is not a Radon measure. This contradicts the definition of $\mathscr{E}_D$, so the conclusion of the theorem is valid.

Deny [2] investigated the completion of $\mathscr{E}_D$.

Let $B \in \mathscr{B}$ be an equilibrium set. The various characterizations of balayage in Sections 3 and 4 yield corresponding characterizations of equilibrium measures, equilibrium potentials, and capacities.

By Theorem 3.1 the equilibrium potential is the largest (Green) potential of a measure on D which is concentrated on B^r and whose potential is bounded above by one on D. It is the supremum of the potentials of measures on D which are concentrated on B and whose potentials are bounded above by one on D. Suppose $C_D(B) < \infty$. By Theorem 4.16

$$C_D(B) = \max[v(D) : v \in \mathscr{M}_D(B^r) \text{ and } G_D v \le 1]$$

and the maximum occurs uniquely at $v = \mu_{B,D}$. Also

$$C_D(B) = \sup[v(D) : v \in \mathscr{M}_D(B) \text{ and } G_D v \le 1]; \tag{2}$$

if $v_n \in \mathscr{M}_D(B)$, $G_D v_n \le 1$ and $v_n(D) \to C_D(B)$, then v_n converges strongly to $\mu_{B,D}$. The characterization of capacity in (2) is due to La Vallée Poussin [1].

By Theorem 3.4 the equilibrium potential is the smallest nonnegative superharmonic function which is bounded below by one q.e. on B. By Theorem 3.8 it is the lower regularization of the infimum of the nonnegative superharmonic functions on D which are bounded below by one on B. Suppose $C_D(B) < \infty$. By Theorem 4.15

$$C_D(B) = \min[\|v\|^2 : v \in \mathscr{E}_D^+ \text{ and } G_D v \ge 1 \text{ q.e. on } B \cap D]$$

and the minimum occurs uniquely at $v = \mu_{B,D}$. Also

$$C_D(B) = \inf[\|v\|^2 : v \in \mathscr{E}_D^+ \text{ and } G_D v \ge 1 \text{ on } B \cap D];$$

if $v_n \in \mathscr{E}_D^+$, $G_D v_n \ge 1$ on $B \cap D$ and $\|v_n\|^2 \to C_D(B)$, then v_n converges strongly to $\mu_{B,D}$.

Suppose still that $C_D(B) < \infty$. By Theorem 4.14

$$-C_D(B) = \min[\|\nu\|^2 - 2\nu(D) : \nu \in \mathscr{E}_D^+(B^r)]$$

and the minimum occurs uniquely at $\nu = \mu_{B,D}$. Also

$$-C_D(B) = \inf[\|\nu\|^2 - 2\nu(D) : \nu \in \mathscr{E}_D^+(B)];$$

if $\nu_n \in \mathscr{E}_D^+(B)$ and $\|\nu_n\|^2 - 2\nu(D) \to -C_D(B)$, then ν_n converges strongly to $\mu_{B,D}$.

It follows from Theorem 4.1 that if $B \in \mathscr{B}$ is polar, then every probability measure on D which is concentrated on B has infinite energy. The next result, due to Frostman [1], characterizes the equilibrium measure of a compact nonpolar subset B of D as a multiple of that probability distribution concentrated on B which has minimum energy.

Theorem 5.16. *Suppose $B \in \mathscr{B}$ and $0 < C_D(B) < \infty$. Then*

$$\min[\|\nu\|^2 : \nu \in \mathscr{E}_D^+(B^r) \text{ and } \nu(D) = 1] = C_D(B)^{-1}$$

and the minimum occurs uniquely at $\nu = C_D(B)^{-1}\mu_{B,D}$. Also

$$\inf[\|\nu\|^2 : \nu \in \mathscr{E}_D^+(B) \text{ and } \nu(D) = 1] = C_D(B)^{-1};$$

if $\nu_n \in \mathscr{E}_D^+(B)$, $\nu_n(D) = 1$ and $\|\nu_n\|^2 \to C_D(B)^{-1}$, then ν_n converges strongly to $C_D(B)^{-1}\mu_{B,D}$.

Proof. Set $\mu = C_D(B)^{-1}\mu_{B,D}$. Then $\mu \in \mathscr{E}_D^+(B^r)$, $\mu(D) = 1$, and $\|\mu\|^2 = C_D(B)^{-1}$. Also $G_D\mu = C_D(B)^{-1}$ on B^r. Choose $\nu \in \mathscr{E}_D^+(B^r)$ such that $\nu(D) = 1$. Then $(\mu, \nu - \mu) = \int G_D\mu d(\nu - \mu) = 0$ and hence $\|\nu\|^2 = \|\mu\|^2 + \|\nu - \mu\|^2$. The first conclusion of the theorem now follows from the energy principle.

Let B_n be compact nonpolar sets such that $B_n \uparrow B$. Then $C_D(B_n) \uparrow C_D(B)$ by Theorem 5.8. Set $\mu_n = C_D(B_n)^{-1}\mu_{B_n,D}$. Then $\mu_n \in \mathscr{E}_D^+(B)$, $\mu_n(D) = 1$ and $\|\mu_n\|^2 = C_D(B_n)^{-1} \to C_D(B)^{-1}$. The second conclusion of the theorem now follows by the same argument used to prove the first conclusion.

Let $d \geq 3$ and $r > 0$ and let ν be a probability distribution on B_r. Then $\iint \|x - y\|^{2-d}\nu(dx)\nu(dy) \geq r^{2-d}$ with equality holding if and only if $\nu = \sigma_r$. This result follows from Theorem 5.16 and Equations (3.1.1) and (3.1.3).

The following characterization of the equilibrium measure is taken from La Vallée Poussin [1] and Section 2 of Chapter II of Landkof [1]. It is interesting to note that in this result the measure ν is not required a priori to be concentrated on B^r or even on $\overline{B}$.

Theorem 5.17. *Suppose $B \in \mathscr{B}$ and $C_D(B) < \infty$. Then $\mu_{B,D}$ is the unique measure $\nu \in \mathscr{M}_D$ such that $\nu(D) = C_D(B)$, $G_D\nu \le 1$ on D and $G_D\nu = 1$ q.e. on $B \cap D$.*

Proof. Now $\mu_{B,D}$ satisfies the indicated properties and $\int G_D\mu_{B,D}\,d\mu_{B,D} = C_D(B)$. Suppose that ν also satisfies the indicated properties. Then $\int G_D\nu\,d\nu \le C_D(B)$. Now $\mu_{B,D} \in \mathscr{M}_D(B^r)$ and $G_D\mu_{B,D} = 1$ q.e. on $B \cap D$. Thus by the domination principle

$$P_{.}(\tau_B < T_D) = G_D\mu_{B,D} \le G_D\nu \le 1 \qquad \text{on } D.$$

In particular $G_D\nu = 1$ on B^r and hence $\int G_D\nu\,d\mu_{B,D} = C_D(B)$. Therefore

$$\|\nu - \mu_{B,D}\|^2 = \int G_D\nu\,d\nu - 2\int G_D\nu\,d\mu_{B,D} + \int G_D\mu_{B,D}\,d\mu_{B,D} \le 0,$$

so $\nu = \mu_{B,D}$ by the energy principle. (Alternatively, this result follows from the energy principle and Theorem 5.8 without appeal to the domination principle.)

The next result involves the dependence of $C_D(B)$ on D.

Theorem 5.18. *Let D_0 be an open subset of D and let $B \in \mathscr{B}$ be a relatively compact subset of D_0. Then $C_D(B) \le C_{D_0}(B)$.*

Proof. Now $G_{D_0} \le G_D$ on $D_0 \times D_0$ and $\mu_{B,D}$ is concentrated on B^r. Thus $G_{D_0}\mu_{B,D} \le G_D\mu_{B,D} \le 1$ on D_0. Consequently

$$C_{D_0}(B) = \operatorname{Max}[\nu(D_0) : \nu \in \mathscr{M}_{D_0}(B^r) \text{ and } G_{D_0}\nu \le 1] \ge \mu_{B,D}(B) = C_D(B).$$

Theorem 5.19. *Let $B \in \mathscr{B}$ be a relatively compact subset of D and let D_n be open sets such that $D_n \uparrow D$. Then $\lim_n \mu_{B,D_n}(A) = \mu_{B,D}(A)$ for $A \subset D$.*

Proof. The proof depends on some preliminary results. Let A be a relatively compact subset of D. Then (i) $h_{B,D}(\cdot, A)$ is continuous on $D \backslash \bar{B}$; (ii) $h_{B,D_n}(\cdot, A) \to h_{B,D}(\cdot, A)$ uniformly on compact subsets of D; and (iii) $G_{D_n}(\cdot, A) \to G_D(\cdot, A)$ uniformly on compact subsets of D.

To prove (i) observe that $\int p(t, \cdot, y) h_{B,D}(y, A)\,dy$ is continuous on D for each $t > 0$ and converges to $h_B(\cdot, A)$ uniformly on compact subsets of $D \backslash \bar{B}$, so (i) holds. In proving (ii) it can be assumed that $\bar{B} \subset D_1$. Observe that

$$\begin{aligned} |h_{B,D_n}(\cdot, A) - h_{B,D}(\cdot, A)| &\le P_{.}(T_{D_n} < \tau_B < T_D) \\ &\le P_{.}(T_{D_n} < \infty, \tau_B \circ \theta_{T_{D_n}} < T_D - T_{D_n}) \qquad \text{on } D_n. \end{aligned}$$

Now $P_.(T_{D_n} \uparrow T_D) = 1$ on D by Proposition 2.3.8. By Propositions 2.2.10, 2.2.11, and 4.5.1 either D is recurrent or B is transient. Thus (by Proposition 2.2.9 if B is transient)

$$\lim_n P_.(T_{D_n} < \infty, \tau_B \circ \theta_{T_{D_n}} < T_D - T_{D_n}) = 0 \qquad \text{on } D;$$

by Theorem 4.3.9 the convergence is uniform on compact subsets of D, so (ii) holds. To prove (iii) observe that

$$G_D(\cdot, A) - G_{D_n}(\cdot, A) = E_. \int_{T_{D_n}}^{T_D} I_A(X(t))\, dt \qquad \text{on } D_n.$$

By the dominated convergence theorem the right-hand side of this equation converges to zero as $n \to \infty$; by Theorem 4.3.9 the convergence is uniform on compact subsets of D, so (iii) holds.

Suppose B is a compact subset of D and A is a relatively compact subset of D. Then

$$\int G_{D_n}(\cdot, A)\, d\mu_{B,D_n} = \int_A G_{D_n}\mu_{B,D_n}(y)\, dy = \int_A P_y(\tau_B < T_{D_n})\, dy.$$

Now $\int_A P_y(\tau_B < T_{D_n})\, dy \downarrow \int_A P_y(\tau_B < T_D)\, dy = \int_A G_D\mu_{B,D}(y)\, dy$ by Proposition 2.3.8. Thus by (iii) and Theorem 5.18

$$\text{(3)} \qquad \lim_n \int G_D(\cdot, A)\, d\mu_{B,D_n} = \int G_D(\cdot, A)\, d\mu_{B,D}.$$

It follows from Theorem 5.18 that there is a strictly increasing sequence $\{n_j\}$ of positive integers and a finite measure μ on D, necessarily supported on B, such that the equilibrium measure of B relative to D_{n_j} converges completely to μ. Since $G_D(\cdot, A)$ is continuous on D by Theorem 4.6.6, it follows from (3) that $\int_A G_D\mu(y)\, dy = \int_A G_D\mu_{B,D}(y)\, dy$. Consequently $G_D\mu = G_D\mu_{B,D}$ a.e. on D and hence $\mu = \mu_{B,D}$ by Theorem 1.3. This shows that μ_{B,D_n} converges completely to $\mu_{B,D}$.

Let C be a compact subset of D containing B in its interior. In proving the conclusion of the theorem it can be assumed that $A \subset B$. Now $\mu_{C,D}$ is supported on ∂C by Theorem 5.1, $h_{B,D}(\cdot, A)$ is continuous on ∂C by (i), $h_{B,D_n}(\cdot, A)$ converges uniformly to $h_{B,D}(\cdot, A)$ on C by (ii) and μ_{C,D_n} converges completely to $\mu_{C,D}$ by the result of the preceding paragraph. Thus by Theorem 5.5

$$\begin{aligned} \lim_n \mu_{B,D_n}(A) &= \lim_n \int \mu_{C,D_n}(dy) h_{B,D_n}(y, A) \\ &= \int \mu_{C,D}(dy) h_{B,D}(y, A) = \mu_{B,D}(A) \end{aligned}$$

as desired.

The relationship between $\mu_{B,D}$ and the equilibrium measure μ_B defined in Section 3.4 will now be obtained.

Theorem 5.20. *Let $d = 2$, let $B \in \mathscr{B}$ be bounded and nonpolar and let D_n be Greenian open subsets of $\mathbb{R}^2$ such that $D_n \uparrow \mathbb{R}^2$. Then* $\lim_n C_{D_n}(B)^{-1}\mu_{B,D_n}(A) = \mu_B(A)$ *for $A \subset \mathbb{R}^2$.*

Proof. Choose $r > 0$ such that $B \subset B_r$. By Theorem 5.5

$$\mu_{B,D_n}(A) = \int \mu_{B_r,D_n}(dy) h_{B,D_n}(y, A);$$

thus it follows from (3.4.12) that

$$\lim_{r\to\infty} \overline{\lim_{n}} \left|C_{D_n}(B_r)^{-1}\mu_{B,D_n}(A) - \mu_B(A)\right| = 0. \tag{4}$$

In particular (let $A = \bar{B}$)

$$\lim_{r\to\infty} \overline{\lim_{n}} \left|C_{D_n}(B_r)^{-1}C_{D_n}(B) - 1\right| = 0. \tag{5}$$

The desired result follows from (4) and (5).

Suppose $d = 1$, $D = (a, b) \neq \mathbb{R}$ and $B = [a_1, b_1] \subset D$. Now $P_.(\tau_B < T_D)$ is easily computed from Proposition 2.2.20 and G_D is determined in (3.5.8). It is left to the reader to use these results to show that

$$\mu_{B,D} = \frac{1}{2(a_1 - a)}\delta_{a_1} + \frac{1}{2(b - b_1)}\delta_{b_1}.$$

This example shows that the conclusion of Theorem 5.20 is not valid when $d = 1$.

Let $B \in \mathscr{B}$. The random variable $L_{B,D}$ defined by $L_{B,D} = \sup[t < T_D : X(t) \in B]$ on $\{\tau_B < T_D\}$ and $L_{B,D} = 0$ on $\{\tau_B \geq T_D\}$ is called the *last exit time* from B relative to D (since $\{L_{B,D} > t\} = \{t + \tau_B \circ \theta_t < T_D\}$ for $t \geq 0$, $L_{B,D}$ is $\mathfrak{F}_\infty$ measurable and hence a random variable). The set B is said to be *transient* relative to D (relatively transient) if $P_.(L_{B,D} < T_D) = 1$ on D. Observe that B is transient relative to $\mathbb{R}^d$ if and only if B is transient as defined in Chapter 2. A relatively compact subset of D is relatively transient.

Theorem 5.21. *Let $B \in \mathscr{B}$. Then $P_.(L_{B,D} = T_D)$ is the greatest harmonic minorant of $P_.(\tau_B < T_D)$ on D.*

Proof. Let D_n be relatively compact open subsets of D such that $D_n \uparrow D$. Then $P_.(T_{D_n} \uparrow T_D) = 1$ on D by Proposition 2.3.8. It follows from Theorem

5.1.23 that the greatest harmonic minorant h of $P_{.}(\tau_B < T_D)$ on D is given by

$$\begin{aligned} h = \lim_n \int H_{D_n}(\cdot, dz) P_z(\tau_B < T_D) &= \lim_n P_{.}(T_{D_n} + \tau_B \circ \theta_{T_{D_n}} < T_D) \\ &= \lim_n P_{.}(L_{B,D} > T_{D_n}) \\ &= P_{.}(L_{B,D} = T_D) \end{aligned}$$

as desired.

Theorem 5.22. *Let $B \in \mathscr{B}$. Then B is an equilibrium set if and only if B is relatively transient.*

Proof. By Theorem 2.6, $P_{.}(\tau_B < T_D)$ can be written uniquely in the form $P_{.}(\tau_B < T_D) = G_D\mu + h$ where $\mu \in \mathscr{M}_D$ and h is harmonic on D. Now h is the greatest harmonic minorant of $P_{.}(\tau_B < T_D)$. Thus B is an equilibrium set if and only if $P_{.}(\tau_B < T_D)$ has greatest harmonic minorant zero on D. The desired result now follows from Theorem 5.21.

Theorem 5.23. *Let $B \in \mathscr{B}$ be an equilibrium set. Then for $x \in D$*

$$P_x(L_{B,D} > 0, X(L_{B,D}) \in dy) = G_D(x, y)\mu_{B,D}(dy).$$

Proof. This result can be obtained as a straightforward modification of the proof of Theorem 3.2.1 (which was taken from that of Proposition 11.2 of Port and Stone [1]). The following alternative proof is taken from Section 1 of Chapter 7 of Rao [1].

Choose $A \subset D$. Now $P_{.}(L_{B,D} > 0, X(L_{B,D}) \in A)$ is excessive (on D) by the strong Markov property. It is bounded above by the Green potential $P_{.}(\tau_{B,D} < T_D)$, so by Theorem 2.5 it is the Green potential of a measure $\nu_A \in \mathscr{M}_D$. Similarly $P_{.}(L_{B,D} > 0, X(L_{B,D}) \in D\backslash A)$ is the Green potential of a measure $\nu_{D\backslash A} \in \mathscr{M}_D$. Now

$$G_D(\nu_A + \nu_{D\backslash A}) = P_{.}(\tau_B < T_D) = G_D\mu_{B,D},$$

so $\nu_A + \nu_{D\backslash A} = \mu_{B,D}$ by Theorem 1.3. In particular $\nu_A \leq \mu_{B,D}$. Suppose temporarily that A is a compact subset of D. It is easily seen that $P_{.}(L_{B,D} > 0, X(L_{B,D}) \in A)$ is harmonic on $D\backslash A$, so ν_A is supported on A by Theorem 1.4. Consequently $\nu_A \leq \mu_{B,D}|_A$ and hence

$$P_{.}(L_{B,D} > 0, X(L_{B,D}) \in A) \leq G_D\mu_{B,D}|_A.$$

This inequality therefore holds for all $A \subset D$. It applies to $D\backslash A$ as well; i.e.,

$$P_{.}(L_{B,D} > 0, X(L_{B,D}) \in D\backslash A) \leq G_D\mu_{B,D}|_{D\backslash A}.$$

Since the sum of the last two inequalities yields the equality

$$P_.(\tau_B < T_D) = G_D\mu_{B,D},$$

it follows that

$$P_.(L_{B,D} > 0, X(L_{B,D}) \in A) = G_D\mu_{B,D}|A,$$

which is equivalent to the desired result.

Kemeny, Snell, and Knapp [1] (see Chapter 8, Section 3 of their book) contains analogs of Theorems 5.21 and 5.22 for Markov chains. Chung [2] showed that Theorem 5.23 holds for a wide class of Markov processes.

Theorem 5.24. *Let $d = 2$, let $B \in \mathscr{B}$ be bounded and nonpolar, let D_n be Greenian open subsets of $\mathbb{R}^2$ such that $D_n \uparrow \mathbb{R}^2$, and let $x \in \mathbb{R}^2$ and $A \subset \mathbb{R}^2$. Then*

$$\lim_n P_x(L_{B,D_n} > 0, X(L_{B,D_n}) \in A) = \mu_B(A).$$

Proof. It suffices to prove the result for $x \notin \bar{B} \cup \{0\}$ and A bounded. It follows from the fundamental identity for logarithmic potentials and (3.4.2) that

$$\lim_n (G_{D_n}(x, y) - G_{D_n}(x, 0)) = k(x, y) - k(x, 0)$$

uniformly for $y \in \bar{B}$ and hence that

(6) $$\lim_n \frac{G_{D_n}(x, y)}{G_{D_n}(x, 0)} = 1 \qquad \text{uniformly for } y \in \bar{B}.$$

Now $\lim_n G_{D_n}\mu_{B,D_n}(x) = \lim_n P_x(\tau_B < T_{D_n}) = 1$, so by (6)

(7) $$\lim_n G_{D_n}(x, 0)C_{D_n}(B) = 1.$$

It follows from (6) and (7) that

(8) $$\lim_n C_{D_n}(B)G_{D_n}(x, y) = 1 \qquad \text{uniformly for } y \in \bar{B}.$$

The desired result now follows from (8) and Theorems 5.20 and 5.23.

The next result is due to Cartan [3] (see Section 1 of Chapter 5 of Landkof [1]).

Theorem 5.25. *Let $d \geq 3$, $B \in \mathscr{B}$, and $x \in B^c$. Let B^* denote the image of B under inversion relative to $S_1(x)$. Then x is irregular for B if and only if B^* is an equilibrium set (relative to Newtonian potentials).*

Proof. Without loss of generality it can be assumed that $x = 0$. By Theorem 5.22, B^* is an equilibrium set if and only if B^* is transient. Choose $\lambda \in (0, 1)$ and set $B_n = \{y \in B: \lambda^{n+1} < \|y\| \leq \lambda^n\}$. Then the image of B_n under inversion relative to $S_1(0)$ is $B_n^* = \{y \in B^*: \lambda^{-n} \leq \|y\| < \lambda^{-n-1}\}$. By Theorems 3.3.2 and 3.3.3 it must be shown that

$$(9) \qquad \sum_n \lambda^{n(2-d)} C(B_n) < \infty \text{ if and only if } \sum_n \lambda^{n(d-2)} C(B_n^*) < \infty.$$

Let μ_n and ν_n be probability measures on B_n and B_n^*, respectively, such that $\nu_n(A^*) = \mu_n(A)$ for $A \subset B_n$. Then $I(\nu_n) = \iint g(x^*, y^*)\mu_n(dx)\mu_n(dy)$, where $I(\mu) = I_{\mathbb{R}^d}(\mu)$. By (4.3.3) and the definition of the Newtonian potential kernel

$$g(x^*, y^*) = (\|x\|\,\|y\|)^{d-2} g(x, y), \qquad x, y \in \mathbb{R}^d \setminus \{0\}.$$

Consequently

$$\lambda^{2(n+1)(d-2)} I(\mu_n) \leq I(\nu_n) \leq \lambda^{2n(d-2)} I(\mu_n).$$

Theorem 5.16 now implies that

$$(10) \qquad \lambda^{2n(2-d)} C(B_n) \leq C(B_n^*) \leq \lambda^{2(n+1)(2-d)} C(B_n).$$

Since (9) follows from (10), the proof of the theorem is complete.

Let B be an arbitrary subset of D. The *inner capacity* $\underline{C}_D(B)$ is defined to be the supremum of the capacities $C_D(A)$ as A ranges over compact subsets of B. It follows from Theorems 5.7 and 5.8 that if $B \in \mathscr{B}$, then $\underline{C}_D(B) = C_D(B)$. The *outer capacity* $\bar{C}_D(B)$ is defined to be the infimum of the capacities $C_D(U)$ as U ranges over the open subsets of D containing B. Observe that $\underline{C}_D(B) \leq \bar{C}_D(B)$ by Theorem 5.5. The set B is said to be *capacitable* if $\underline{C}_D(B) = \bar{C}_D(B)$. An open subset of D is clearly capacitable.

Theorem 5.26. *Let $B \in \mathscr{B}$ be a subset of D. Then B is capacitable.*

Proof. If $C_D(B) = \infty$ the result is true. Suppose $C_D(B) < \infty$ and hence that B is an equilibrium set. The equilibrium measure $\mu_{B,D}$ of B has energy $\|\mu_{B,D}\|^2 = C_D(B)$ and Green potential $h_{B,D}1$. By Theorem 4.15

$$C_D(B) = \inf[\|\nu\|^2 : \nu \in \mathscr{E}_D^+ \text{ and } G_D\nu \geq 1 \text{ on } B].$$

Given ε, $0 < \varepsilon < 1$, choose $\nu \in \mathscr{E}_D^+$ such that $G_D\nu \geq 1$ on B and $\|\nu\|^2 \leq C_D(B) + \varepsilon$. Set $U = \{x \in D: G_D\nu > 1 - \varepsilon\}$. Then U is open, $U \supset B$, and $G_D((1 - \varepsilon)^{-1}\nu) > 1$ on U. Thus by Theorem 4.15

$$C_D(U) \leq \|(1 - \varepsilon)^{-1}\nu\|^2 = (1 - \varepsilon)^{-2}\|\nu\|^2 \leq (1 - \varepsilon)^{-2}(C_D(B) + \varepsilon).$$

Since ε can be made arbitrarily small, $\bar{C}_D(B) \leq C_D(B) = \underline{C}_D(B)$ and hence B is capacitable.

The capacitability theorem of Choquet [1] (see Brelot [8], Choquet [4], Helms [1], Meyer [1]) can be used to show that every Borel (more generally analytic) subset of D is capacitable. According to the definition of quasi-everywhere in Section 3, a property holds quasi-everywhere on $B \subset D$ if and only if there is a Borel set A having inner capacity zero such that the property holds on $B \backslash A$. By Choquet's theorem a property holds quasi-everywhere on B if and only if there is a Borel set A having outer capacity zero such that the property holds on $B \backslash A$. This is the usual definition of quasi-everywhere.

Hunt [2] used Choquet's theorem to establish the measurability and stopping time property of hitting times of Borel sets for a large class of Markov processes and to develop a potential theory for these processes. This theory is contained in Blumenthal and Getoor [1].

Let $P_.$ be appropriately completed and let B be a Borel set in $\mathbb{R}^d$. According to Exercise 10.24 of Chapter I of Blumenthal and Getoor's book there is an increasing sequence $\{B_n\}$ of compact subsets of B such that $P_.(\tau_{B_n} \downarrow \tau_B) = 1$ on $\mathbb{R}^d$. By Remark 11.3 of the same chapter for each probability measure μ on $\mathbb{R}^d$ such that $\mu(B \backslash B^r) = 0$ there is a decreasing sequence $\{U_n\}$ of open sets containing B such that $P_.(\tau_{U_n} \uparrow \tau_B) =$ a.e. (μ). By using these facts, the restriction that B belong to the collection $\mathscr{B}$ of F_σ sets can invariably be removed from the results of this book.

6. Application to Electrostatics

In this section the following problem is formulated precisely and solved. Consider an electrostatic system consisting of a body on which a fixed charge distribution is placed and a finite number of conductors on each of which the total charge is specified. Find the equilibrium charge distribution for the system.

Let $B_1, \ldots . B_n$ be disjoint compact nonpolar subsets of D and set $B = B_1 \cup \cdots \cup B_n$. Let $\nu_0 \in \mathscr{E}_D$ be such that $|\nu_0|(B) = 0$ and let $Q_1, \ldots, Q_n \in \mathbb{R}$. Let $\mathscr{N}$ denote the collection of elements $\nu \in \mathscr{E}_D$ such that $\nu|_{D \backslash B} = \nu_0$ and $\nu(B_j) = Q_j$ for $1 \leq j \leq n$. An element $\nu \in \mathscr{N}$ is called an *equilibrium charge distribution* if $G_D \nu$ is constant on B_j^r for $1 \leq j \leq n$.

Theorem 6.1. *There is a unique equilibrium charge distribution in $\mathscr{N}$, which uniquely minimizes $\|\nu\|^2$ as ν ranges over $\mathscr{N}$.*

Proof. Note first that if $\nu \in \mathscr{N}$, then ν does not charge any of the polar sets $B_1 \backslash B_1^r, \ldots, B_n \backslash B_n^r$.

Suppose ν' is an equilibrium charge distribution in $\mathcal{N}$ and let $\nu \in \mathcal{N}$. Then

$$(\nu', \nu - \nu') = \sum_1^n \int_{B_j} G_D\nu' \, d(\nu - \nu') = 0$$

and hence $\|\nu\|^2 = \|\nu' + \nu - \nu'\|^2 = \|\nu'\|^2 + \|\nu - \nu'\|^2$. Thus by the energy principle, $\|\nu'\|^2$ uniquely minimizes $\|\nu\|^2$ as ν ranges over $\mathcal{N}$. Suppose ν'' is also an equilibrium charge distribution in $\mathcal{N}$. Then ν'' minimizes $\|\nu\|^2$ as ν ranges over $\mathcal{N}$ and hence $\nu'' = \nu'$. Thus there is at most one equilibrium charge distribution in $\mathcal{N}$ and, if it exists, it uniquely minimizes $\|\nu\|^2$ as ν ranges over $\mathcal{N}$.

It remains to show that there is an equilibrium charge distribution in $\mathcal{N}$. To this end set $A_j = B\backslash B_j$ for $1 \le j \le n$. Then A_j and B_j are disjoint compact subsets of D. Let ν_j be the condenser measure corresponding to A_j, B_j, which was shown to exist in Theorem 5.14. Then $\nu_j \in \mathscr{E}_D$, ν_j^+ is concentrated on B_j, ν_j^- is concentrated on A_j, $G_D\nu_j = 1$ on B_j^r and $G_D\nu_j = 0$ on A_j^r. Set $\nu_0' = \nu_0 h_{B,D}$. Then $\nu_0' \in \mathscr{E}_D$, ν_0' is supported on B and $G_D\nu_0' = G_D\nu_0$ on B^r. Set $\nu = \nu_0 - \nu_0' + \sum_1^n \alpha_i\nu_i$, where $\alpha_1, \ldots, \alpha_n \in \mathbb{R}$. Then $\nu|_{D\backslash B} = \nu_0$ and $G_D\nu$ is constant on B_j^r for $1 \le j \le n$. It must be shown that $\alpha_1, \ldots, \alpha_n$ can be chosen so that $\nu(B_j) = Q_j$ for $1 \le j \le n$; i.e., so that

$$\sum_1^n \alpha_i\nu_i(B_j) = Q_j + \nu_0'(B_j), \qquad 1 \le j \le n.$$

To do so it is enough to show that if

$$\sum_1^n \alpha_i\nu_i(B_j) = 0, \qquad 1 \le j \le n, \tag{1}$$

then $\alpha_1 = \cdots = \alpha_n = 0$. Suppose (1) holds. Set $\mu = \sum_1^n \alpha_i\nu_i$. Then $\mu \in \mathscr{E}_D(B)$; also $\mu(B_j) = 0$ and $G_D\mu = \alpha_j$ on B_j^r for $1 \le j \le n$. Consequently

$$\|\mu\|^2 = \sum_1^n \int_{B_j} G_D\mu \, d\mu = 0$$

and hence $\mu = 0$ by the energy principle. Thus $G_D\mu = 0$ on D. Since $G_D\mu = \alpha_j$ on the nonempty set B_j^r, it follows that $\alpha_j = 0$ for $1 \le j \le n$. This completes the proof of the theorem.

7. Logarithmic Potential Theory

Consider in this section planar Brownian motion. Logarithmic potential theory will be thoroughly developed. Portions of this theory were obtained earlier in Section 3.4. For analytic approaches to logarithmic potential theory see La Vallée Poussin [1], Landkof [1], and Tsuji [1]. (The present approach is based in part on the theory of Green potentials. Alternatively, one could develop and use a corresponding theory of λ-potentials. With such modifications the results and methods of this section become applicable to a large class of recurrent symmetric infinitely divisible processes on $\mathbb{R}^d$, $d \leq 2$, namely those having, for each $t > 0$, an absolutely continuous distribution whose density $p(t, \cdot)$ is bounded and (without additional loss of generality) continuous. In this more general theory, which includes logarithmic potential theory and linear potential theory as special cases, the potential kernel k is given by

$$k = \int_0^1 p(t, \cdot)\, dt + \int_1^\infty (p(t, \cdot) - p(t, 0))\, dt,$$

the second integral here being a continuous nonpositive function on $\mathbb{R}^d$.)

Let $\mathscr{M}$ denote the collection of finite measures on $\mathbb{R}^2$ which have compact support. For $B \subset \mathbb{R}^2$ let $\mathscr{M}(B)$ denote the collection of measures $\mu \in \mathscr{M}$ which are concentrated on B. Recall from Section 3.4 that the logarithmic potential kernel k is given by $k(x) = \pi^{-1} \log(1/\|x\|)$ for $x \in \mathbb{R}^2$ and that k is integrable on compacts. Recall also that the logarithmic potential $k\mu$ of a measure $\mu \in \mathscr{M}$ is given by $k\mu = \int k(\cdot, y)\mu(dy)$ on $\mathbb{R}^2$, where $k(x, y) = k(y - x)$ for $x, y \in \mathbb{R}^2$, and that $k\mu$ is integrable on compacts.

Let $B \in \mathscr{B}$ be nonpolar. According to the fundamental identity for logarithmic potentials

$$k(x, y) = g_B(x, y) + \int h_B(x, dz)k(z, y) - W_B(x), \qquad x, y \in \mathbb{R}^2,$$

where W_B is nonnegative and bounded on compacts. Let D be a bounded open subset of $\mathbb{R}^2$. Then $W_{D^c} = 0$ on D by Proposition 3.4.7. Consequently

$$k(x, y) = G_D(x, y) + \int H_D(x, dz)k(z, y), \qquad x, y \in D. \tag{1}$$

Suppose $\mu \in \mathscr{M}(D)$. Then

$$k\mu = G_D\mu + H_D k\mu \qquad \text{on } D. \tag{2}$$

By Theorem 4.3.7 the function $H_D k\mu$ in (2) is harmonic on D. Equations (1) and (2) allow properties of logarithmic potentials to be deduced from those of Green potentials. For this purpose it is enough to let D be a sufficiently large open ball. For example, *if $B \in \mathscr{B}$ is bounded and polar, there is a measure $\mu \in \mathscr{M}$ such that $k\mu = \infty$ on B.* For let D be a bounded open set containing B in its interior. By Theorem 2.7 there is a measure $\mu \in \mathscr{M}_D$ such that $G_D\mu = \infty$ on B. It can be assumed that μ has compact support in D, in which case μ can be thought of as a measure in $\mathscr{M}$. Then $k\mu = \infty$ on B by (2).

Theorem 7.1. *Let $\mu \in \mathscr{M}$. Then $k\mu$ is superharmonic on $\mathbb{R}^2$.*

Proof. Let D be a bounded open subset of $\mathbb{R}^2$ such that $\mu \in \mathcal{M}(D)$. Then $G_D\mu$ is superharmonic on D by Theorem 1.1 and $H_Dk\mu$ is harmonic on D, so $k\mu$ is superharmonic on D by (2). Since D can be made arbitrarily large, $k\mu$ is superharmonic on $\mathbb{R}^2$.

Theorem 7.2. *Let $B \subset \mathbb{R}^2$. Suppose $\mu \in \mathcal{M}(B)$ and $k\mu < \infty$ on B. Then μ does not charge polar sets.*

Proof. Without loss of generality it can be assumed that B is bounded. Let D be a bounded open subset of $\mathbb{R}^2$ containing B. Then $G_D\mu < \infty$ on B by (2), so it follows from Theorem 2.8 that μ does not charge polar sets.

The next result is called the *maximum principle for logarithmic potentials.* For more general results see Theorems 7.13 and 7.14 below.

Theorem 7.3. *Let $B \subset \mathbb{R}^2$ and let $\mu \in \mathcal{M}(B)$. Then $\sup_{x \in \mathbb{R}^2} k\mu(x) = \sup_{x \in B} k\mu(x)$.*

Proof. Without loss of generality it can be assumed that $M = \sup_{x \in B} k\mu(x) < \infty$. Then by Theorem 7.2, μ does not charge polar sets. Choose $\varepsilon > 0$ and set $A = \{x \in \mathbb{R}^2 : k\mu(x) \leq M + \varepsilon\}$. Then A is a closed nonpolar subset of $\mathbb{R}^2$ and μ is concentrated on A^r. Thus by the fundamental identity for logarithmic potentials $k\mu \leq h_A k\mu \leq M + \varepsilon$. Since ε can be made arbitrarily small, $k\mu \leq M$ on $\mathbb{R}^2$ as desired.

The next result is called the *continuity principle for logarithmic potentials.*

Theorem 7.4. *Let B be compact and let $\mu \in \mathcal{M}(B)$ be such that the restriction of $k\mu$ to B is continuous on B. Then $k\mu$ is continuous on $\mathbb{R}^2$.*

Proof. Let D be a bounded open set containing B. By (2) the restriction of $G_D\mu$ to B is continuous on B. Thus $G_D\mu$ is continuous on D by the continuity theorem for Green potentials and hence $k\mu$ is continuous on D by another application of (2). Since D can be made arbitrarily large, $k\mu$ is continuous on $\mathbb{R}^2$.

Theorem 7.5. *Let B be a compact nonpolar set. Then there is a nonzero measure $\mu \in \mathcal{M}(B)$ such that $k\mu$ is continuous on $\mathbb{R}^2$.*

Proof. Let D be a bounded open set containing B. By Theorem 3.2 there is a nonzero measure $\mu \in \mathcal{M}(B)$ such that $G_D\mu$ is continuous on D. By (2),

$k\mu$ is continuous on D. Thus $k\mu$ is continuous on $\mathbb{R}^2$ by the continuity principle for logarithmic potentials.

Theorem 7.6. *Let B be a compact set and let $\mu \in \mathcal{M}(B)$ be such that $k\mu < \infty$ a.e. (μ). Then for every $\varepsilon > 0$ there is a compact subset C of B such that $\mu(B \backslash C) \le \varepsilon$ and $k\mu|_C$ is continuous on $\mathbb{R}^2$.*

Proof. Let D be a bounded open set containing B. Then $G_D\mu < \infty$ a.e. (μ) by (2). According to Theorem 1.10, for every $\varepsilon > 0$ there is a compact subset C of B such that $\mu(B \backslash C) \le \varepsilon$ and $G_D\mu|_C$ is continuous on D. Then $k\mu|_C$ is continuous on D by (2), so $k\mu|_C$ is continuous on $\mathbb{R}^2$ by the continuity principle for logarithmic potentials.

Theorem 7.7. *Let $\mu, \nu \in \mathcal{M}$ and let D_0 be an open subset of $\mathbb{R}^2$. Then $k\nu = k\mu + h$ on D_0 for some harmonic function h on D_0 if and only if $\nu|_{D_0} = \mu|_{D_0}$. In particular $k\mu$ is harmonic on D_0 if and only if $\mu(D_0) = 0$.*

Proof. Let D_1 be a bounded open subset of $\mathbb{R}^2$ containing D_0 and the supports of μ and ν. By (2), $k\mu = G_{D_1}\mu + H_{D_1}k\mu$ and $k\nu = G_{D_1}\nu + H_{D_1}k\nu$ on D_1. Thus $k\nu = k\mu + h$ on D_0 for some harmonic function h on D_0 if and only if $G_{D_1}\nu = G_{D_1}\mu + h_1$ on D_0 for some harmonic function h_1 on D_0. By Theorem 1.6 this is true if and only if $\nu|_{D_0} = \mu|_{D_0}$.

The next result due to Riesz [1], is called the *Riesz decomposition theorem for logarithmic potentials.*

Theorem 7.8. *Let U be an open subset of $\mathbb{R}^2$ and let f be superharmonic on U. Then there is a unique Radon measure μ on $\mathbb{R}^2$ which is concentrated on U such that, given any relatively compact open subset D_0 of U, $f = k\mu|_{D_0} + h_0$ on D_0, where h_0 is harmonic on D_0.*

Proof. Let D_0 and D_1 be relatively compact open subsets of U and let D be a relatively compact open subset of U containing $\bar{D}_0$ and $\bar{D}_1$. It follows from the Riesz decomposition theorem for Green potentials that for $i = 0, 1$ there is a measure $\mu_i \in \mathcal{M}(D_i)$ and a harmonic function h'_i on D_i such that $f = G_D\mu_i + h'_i$ on D_i. By (2) there is a harmonic function h_i on D_i such that $f = k\mu_i + h_i$ on D_i. By Theorem 7.7, $\mu_0|_{D_0 \cap D_1} = \mu_1|_{D_0 \cap D_1}$. The existence of a measure μ on $\mathbb{R}^2$ having the properties specified in the statement of the theorem follows easily from these observations. By Theorem 7.7, $\mu|_{D_0}$ is uniquely determined for every relatively compact subset D_0 of U. Thus μ is uniquely determined. This completes the proof of the theorem.

The next result is due to Bôcher [1].

Theorem 7.9. *Let $r > 0$ and let f be a harmonic function on $\mathring{B}_r \backslash \{0\}$ such that f/k is bounded above or below on $\mathring{B}_r \backslash \{0\}$. Then there is an $\alpha \in \mathbb{R}$ and a harmonic function h on $\mathring{B}_r$ such that $f = \alpha k + h$ on $\mathring{B}_r \backslash \{0\}$.*

Proof. The proof is an obvious modification of the corresponding result (Theorem 2.10) for Green potentials.

Theorem 7.10. *Let $r > 0$ and let f be a harmonic function on $\mathbb{R}^2 \backslash B_r$ such that f/k is bounded above or below on $\mathbb{R}^2 \backslash B_r$. Then there is an $\alpha \in \mathbb{R}$ such that $f - \alpha k$ has a finite limit at infinity.*

Proof. The proof is similar to that of Theorem 2.11. The details are left to the reader.

Let f be a harmonic function on $\mathbb{R}^2$ such that f/k is bounded above or below on $\{x \in \mathbb{R}^2 : \|x\| \geq 2\}$. It follows from the elementary complex variable argument used in the proof of Theorem 4.6.4 that f is necessarily constant on $\mathbb{R}^2$. Theorem 7.10 can be used to give an alternative proof of this result. According to this theorem, $f - \alpha k$ has a finite limit at infinity for some $\alpha \in \mathbb{R}$. Without loss of generality it can be assumed that $\alpha \leq 0$ (replace f by $-f$ if necessary). Then f is bounded below on $\{x \in \mathbb{R}^2 : \|x\| \geq 2\}$ and hence f is bounded below on $\mathbb{R}^2$. Thus f is constant by Theorem 4.3.4.

Theorem 7.10 will now be used to extend Theorem 4.6.8 in the planar case.

Theorem 7.11. *Suppose D^c is a compact nonpolar set, that ρ is bounded and continuous on D and vanishes outside a bounded subset of D, and that φ is essentially bounded and essentially continuous on ∂D. Let f be of the form*

$$(3) \qquad f = G_D\rho + H_D\varphi + \alpha W_{D^c} \qquad \text{on } D \text{ for some } \alpha \in \mathbb{R}.$$

Then f is a continuous solution to $\tilde{\Delta} f = -2\rho$ on D; f has boundary value $\varphi(b)$ at b for every $b \in \partial D \cap (D^c)^r$ at which φ is defined and continuous; f is bounded on bounded subsets of D; and f/k is bounded above or below on $\mathbb{R}^2 \backslash B_r$ for some $r > 0$. Conversely if f satisfies the properties of the last sentence, then f is of the form given by (3).

Proof. Let f be of the form given by (3). It follows from Proposition 4.5.2, Theorem 4.6.8, and the properties of W_{D^c} determined in Section 3.4 (W_{D^c} is nonnegative, upper semicontinuous and bounded on compacts and equals zero on $(D_c)^r$) that f satisfies the properties specified in the statement of the theorem. Suppose conversely that f satisfies these properties. Then $f - G_D\rho$ is harmonic on D and $(f - G_D\rho)/k$ is bounded above or below on $\mathbb{R}^2 \backslash B_r$ for some $r > 0$. Thus by Theorem 7.10 there is an $\alpha \in \mathbb{R}$ such that $f - G_D\rho + \alpha k$ has a finite limit at infinity. Theorem 3.4.12 now implies that $f - G_D\rho - \alpha W_{D^c}$ has a finite limit at infinity and hence is a bounded solution to the generalized Dirichlet problem for φ. Consequently $f - G_D\rho - \alpha W_{D^c} = H_D\varphi$ on D by Theorem 4.2.10 (D^c is recurrent by Proposition 2.2.10) and hence f is of the form given by (3).

Theorem 7.12. *Let $\mu \in \mathscr{M}$ and suppose that $B_n \in \mathscr{B}$ for each n and $B_n \uparrow B$ where B is bounded. Then* $\lim_n h_{B_n} k\mu = h_B k\mu$ on $\mathbb{R}^2$.

Proof. It can be assumed that B_n is nonpolar for each n. Since $k\mu$ is bounded below on $\bar{B}$ and lower semicontinuous, it follows from Proposition 2.3.8 and Fatou's lemma that $\underline{\lim}_n h_{B_n} k\mu \geq h_B k\mu$. Thus it suffices to show that

$$\overline{\lim_n}\, h_{B_n} k\mu \leq h_B k\mu. \tag{4}$$

To this end observe first that by Equations (2.4.2) and (2.4.3) and Proposition 2.4.4

$$E_x(p(t - \tau_{B_n}, X(\tau_{B_n}), y); \tau_{B_n} < t) \leq E_x(p(t - \tau_B, X(\tau_B), y); \tau_B < t)$$

for $n \geq 1$, $x, y \in \mathbb{R}^2$, and $t > 0$. Consequently for $\lambda > 0$, $h_{B_n}^\lambda g^\lambda \leq h_B^\lambda g^\lambda$ or equivalently

$$h_{B_n}^\lambda k^\lambda - W_{B_n}^\lambda \leq h_B^\lambda k^\lambda - W_B^\lambda.$$

It follows by letting $\lambda \downarrow 0$ that

$$h_{B_n} k - W_{B_n} \leq h_B k - W_B$$

and hence that

$$h_{B_n} k\mu - \mu(\mathbb{R}^2) W_{B_n} \leq h_B k\mu - \mu(\mathbb{R}^2) W_B.$$

Now $W_{B_n} \downarrow W_B$ by Theorem 3.4.15, so (4) holds as desired.

The next result, called the *complete maximum principle for logarithmic potentials*, is the analog of the domination principle for Green potentials.

Theorem 7.13. *Let $B \in \mathscr{B}$ be nonpolar and let $\mu \in \mathscr{M}(B^r)$. Suppose that $k\mu \leq k\nu + \alpha$ q.e. on B, where $\nu \in \mathscr{M}$, $\nu(\mathbb{R}^2) \leq \mu(\mathbb{R}^2)$, and $\alpha \in \mathbb{R}$. Then $k\mu \leq k\nu + \alpha$ on $\mathbb{R}^2$.*

Proof. Without loss of generality it can be assumed that B is bounded. Let B_n be compact nonpolar sets such that $B_n \uparrow B$. Then

$$(5) \qquad h_{B_n} k\mu \le h_{B_n} k\nu + \alpha$$

on B_n^c by Theorem 2.6.5 and (5) holds q.e. on B_n^r. Thus (5) holds a.e. on $\mathbb{R}^2$. It now follows from Theorem 7.12 that

$$h_B k\mu \le h_B k\nu + \alpha \qquad \text{a.e. on } \mathbb{R}^2.$$

Consequently by the fundamental identity for logarithmic potentials

$$k\mu = h_B k\mu - \mu(\mathbb{R}^2) W_B \le g_B \nu + h_B k\nu - \nu(\mathbb{R}^2) W_B + \alpha = k\nu + \alpha$$

a.e. on $\mathbb{R}^2$. Therefore $k\mu \le k\nu + \alpha$ on $\mathbb{R}^2$ by Theorem 5.1.14.

The next result includes the maximum principle for logarithmic potentials as a special case.

Theorem 7.14. *Let $\mu, \nu \in \mathscr{M}$ and $\alpha \in \mathbb{R}$ be such that $k\mu < \infty$ a.e. (μ), $k\mu \le k\nu + \alpha$ a.e. (μ) and $\nu(\mathbb{R}^2) \le \mu(\mathbb{R}^2)$. Then $k\mu \le k\nu + \alpha$ on $\mathbb{R}^2$.*

Proof. Set

$$A = \{x \in \mathbb{R}^2 : k\mu(x) < \infty \text{ and } k\mu(x) \le k\nu(x) + \alpha\}.$$

Since μ is a finite measure concentrated on A, there exist compact subsets B_n of A such that $B_n \uparrow$ and $\mu(B_n^c) \downarrow 0$. Set $B = \bigcup_n B_n$. Then $B \in \mathscr{B}$, $\mu \in \mathscr{M}(B)$, $k\mu < \infty$ on B and $k\mu \le k\nu + \alpha$ on B. Now μ does not charge polar sets by Theorem 7.2, so $\mu \in \mathscr{M}(B^r)$ by Theorem 2.6.3. Therefore $k\mu \le k\nu + \alpha$ on $\mathbb{R}^2$ by the complete maximum principle for logarithmic potentials.

The next result, called the *uniqueness principle for logarithmic potentials*, is stronger than Theorem 3.4.3.

Theorem 7.15. *Let $B \in \mathscr{B}$ and let $\mu, \nu \in \mathscr{M}(B^r)$ be such that $\mu(\mathbb{R}^2) = \nu(\mathbb{R}^2)$ and $k\mu = k\nu + \alpha$ q.e. on B for some $\alpha \in \mathbb{R}$. Then $\mu = \nu$.*

Proof. Without loss of generality it can be assumed that B is nonpolar. It follows from the complete maximum principle for logarithmic potentials that $k\mu = k\nu + \alpha$ on $\mathbb{R}^2$. Thus $\mu = \nu$ by Theorem 7.7 or Theorem 3.4.1.

Let $\mu, \nu \in \mathscr{M}$. According to the next result, *if μ does not charge polar sets and if $k\nu \le k\mu + \alpha$ on $\mathbb{R}^2$ for some $\alpha \in \mathbb{R}$, then ν does not charge polar sets.*

Theorem 7.16. *Let $\mu, \nu \in \mathscr{M}$ and let C be a compact polar set such that $\mu(C) = 0$. Suppose there is an open set U containing C such that $k\nu \leq k\mu + \alpha$ on U for some $\alpha \in \mathbb{R}$. Then $\nu(C) = 0$.*

Proof. Without loss of generality it can be assumed that ν is supported on C. Let D be a bounded open set containing C and the support of μ. Since $k\nu$ is bounded above on $D \backslash U$ and $k\mu$ is bounded below on D there is an $\alpha_1 \in \mathbb{R}$ such that $k\nu \leq k\mu + \alpha_1$ on D. Now $k\mu$ and $k\nu$ are bounded on ∂D. Thus there is an $\alpha_2 \in \mathbb{R}$ such that $H_D k\mu \leq \alpha_2$ and $H_D k\nu \geq -\alpha_2$ on D. By (2)

$$k\mu = G_D\mu + H_D k\mu \leq G_D\mu + \alpha_2 \qquad \text{on } D$$

and

$$k\nu = G_D\nu + H_D k\nu \geq G_D\nu - \alpha_2 \qquad \text{on } D.$$

Therefore

$$G_D\nu \leq k\nu + \alpha_2 \leq k\mu + \alpha_1 + \alpha_2 \leq G_D\mu + \alpha_1 + 2\alpha_2 \qquad \text{on } D$$

and hence $\nu(C) = 0$ by Theorem 3.11.

Suppose $B \in \mathscr{B}$ is bounded and nonpolar and let $\mu \in \mathscr{M}$. The *balayage problem for logarithmic potentials* is to find a measure $\nu \in \mathscr{M}(B^r)$ such that $\nu(\mathbb{R}^2) = \mu(\mathbb{R}^2)$ and $k\nu = k\mu + \alpha$ on B^r for some $\alpha \in \mathbb{R}$. It follows from the uniqueness principle for logarithmic potentials that there is at most one solution to this balayage problem. To see that there is a solution recall from (3.4.10) that

$$(6) \quad \int h_B(x, dz)k(z, y) = \int h_B(y, dz)k(z, x) + W_B(x) - W_B(y), \qquad x, y \in \mathbb{R}^2.$$

Set $\mu h_B = \int \mu(dy) h_B(y, \cdot)$. Then μh_B is concentrated on $B^r \cap (C \cup \partial B)$, where C denotes the support of μ, so $\mu h_B \in \mathscr{M}(B^r)$. Since B is recurrent by Proposition 2.2.10, $\mu h_B(\mathbb{R}^2) = \mu(\mathbb{R}^2)$. By (6)

$$(7) \qquad k(\mu h_B) = h_B k\mu + \int W_B \, d\mu - \mu(\mathbb{R}^2) W_B \qquad \text{on } \mathbb{R}^2.$$

Now $W_B = 0$ on B^r by Theorem 3.4.2, so

$$(8) \qquad k(\mu h_B) = k\mu + \int W_B \, d\mu \qquad \text{on } B^r.$$

Thus μh_B is the unique solution to the balayage problem. By Proposition 3.4.7 the constant $\int W_B \, d\mu$ is zero if B is compact and the support of μ is disjoint from the unbounded component of B^c. It follows from the fundamental identity for logarithmic potentials that

$$(9) \qquad k\mu = g_B\mu + k(\mu h_B) - \int W_B \, d\mu \qquad \text{on } \mathbb{R}^2$$

and hence that

$$(10) \qquad k(\mu h_B) \le k\mu + \int W_B \, d\mu \qquad \text{on } \mathbb{R}^2.$$

Theorem 7.17. *Let B be a compact nonpolar set, let $\mu \in \mathscr{M}$ be such that $\mu(B \backslash B^r) = 0$, and let $v \in \mathscr{M}$ satisfy $v(\mathbb{R}^2) = \mu(\mathbb{R}^2)$. Then $v = \mu h_B$ if and only if kv satisfies the following three conditions: $kv \le k\mu + \alpha$ on $\mathbb{R}^2$ for some $\alpha \in \mathbb{R}$; $kv = k\mu + \alpha$ q.e. on B for some $\alpha \in \mathbb{R}$; and kv is harmonic on B^c.*

Proof. Suppose first that $v = \mu h_B$. Then the first two conditions hold with $\alpha = \int W_B \, d\mu$ and the third condition follows from Theorem 7.7. Suppose conversely that kv satisfies the three indicated conditions. Then v is supported on B by Theorem 7.7. By Theorem 7.16, v assigns measure zero to every compact subset of $B \backslash B^r$ and hence $v(B \backslash B^r) = 0$. Thus v is concentrated on B^r and hence $v \in \mathscr{M}(B^r)$. Now $kv = k\mu + \alpha = k(\mu h_B) - \int W_B \, d\mu + \alpha$ q.e. on B, so $v = \mu h_B$ by the uniqueness principle for logarithmic potentials.

Let $B \in \mathscr{B}$ be bounded and nonpolar. The *equilibrium problem for logarithmic potentials* is to find a probability measure $\mu \in \mathscr{M}(B^r)$ such that $k\mu$ is constant on B^r. This problem was shown in Section 3.4 to have a unique solution μ_B, the equilibrium measure of B. The constant value $R(B)$ of μ_B on B^r, called the Robin constant of B, is necessarily finite. Recall that $k\mu_B = R(B) - W_B \le R(B)$ on $\mathbb{R}^2$. If $B \in \mathscr{B}$ is bounded and polar its Robin constant was defined as $R(B) = \infty$.

Theorem 7.18. *Let $A, B \in \mathscr{B}$ be bounded nonpolar sets such that $A \subset B$. Then $\mu_A = \mu_B h_A$ and $R(A) = R(B) + \int W_A \, d\mu_B$.*

Proof. By (7)

$$k(\mu_B h_A) = R(B) + \int W_A \, d\mu_B - W_A \qquad \text{on } \mathbb{R}^2.$$

Thus $\mu_B h_A$ is the equilibrium measure of A and $R(B) + \int W_A \, d\mu_B$ is the Robin constant of A.

Let $\mu \in \mathscr{M}$. The *energy* $I(\mu)$ of μ is defined by $I(\mu) = \int k\mu \, d\mu$. The condition $k\mu < \infty$ a.e. (μ) in Theorem 7.14 is implied by the condition that μ have finite energy. Let $B \in \mathscr{B}$. If B is bounded and nonpolar its equilibrium measure has energy $I(\mu_B) = R(B)$. It follows from Theorem 2.6.3 and the next result that if $\mu \in \mathscr{M}(B)$ and μ has finite energy, then $\mu \in \mathscr{M}(B^r)$.

Theorem 7.19. *Let $\mu \in \mathscr{M}$ have finite energy. Then μ does not charge polar sets. If $v \in \mathscr{M}$ also has finite energy, then $\int k\mu \, dv < \infty$.*

Proof. Let D be a bounded open set containing the support of μ. Now $k\mu = G_D\mu + H_D k\mu$ on D and $H_D k\mu$ is bounded on D. Consequently $I_D(\mu) = \int G_D\mu \, d\mu < \infty$ and hence μ does not charge polar sets by Theorem 4.1. Suppose $\nu \in \mathscr{M}$ has finite energy and let D also contain the support of ν. Then $I_D(\nu) < \infty$ and hence $\int G_D\mu \, d\nu < \infty$ by Theorem 4.2. Since $H_D k\mu$ is bounded on the support of ν, $\int k\mu \, d\nu < \infty$. This completes the proof of the theorem.

Let $\mathscr{E}$ denote the vector space of signed Radon measures μ on $\mathbb{R}^2$ such that $|\mu| \in \mathscr{M}$ and $|\mu|$ has finite energy. If $\mu \in \mathscr{E}$ then $k\mu = k\mu^+ - k\mu^-$ is well defined whenever $k|\mu| < \infty$ and hence is well defined q.e. on $\mathbb{R}^2$. Let $\mathscr{E}(B)$ denote the subspace of elements $\mu \in \mathscr{E}$ such that $|\mu|$ is concentrated on B. Let $\mathscr{E}^+$ and $\mathscr{E}^+(B)$ denote, respectively, the (nonnegative) measures in $\mathscr{E}$ and $\mathscr{E}(B)$.

Let C be a fixed compact nonpolar set and let $c > 0$. For $\mu, \nu \in \mathscr{E}(C)$

$$\begin{aligned}(\mu, \nu) &= \int k(\mu - \mu(\mathbb{R}^2)\mu_C) \, d(\nu - \nu(\mathbb{R}^2)\mu_C) + c\mu(\mathbb{R}^2)\nu(\mathbb{R}^2) \\ &= \int k\mu \, d\nu - (R(C) - c)\mu(\mathbb{R}^2)\nu(\mathbb{R}^2)\end{aligned}$$

is well defined and finite by Theorem 7.19. If either $\mu(\mathbb{R}^2) = 0$ or $\nu(\mathbb{R}^2) = 0$, then $(\mu, \nu) = \int k\mu \, d\nu$ independently of C and c. In particular if $\mu(\mathbb{R}^2) = 0$, then $(\mu, \mu) = \int k\mu \, d\mu$. Note that (μ, ν) is symmetric in μ and ν and defines a bilinear form on $\mathscr{E}(C)$. It follows from the next result, called the *energy principle for logarithmic potentials*, that (μ, ν) determines an inner product on $\mathscr{E}(C)$.

Theorem 7.20. *Let $\mu \in \mathscr{E}(C)$. Then $(\mu, \mu) \geq 0$ with equality holding if and only if $\mu = 0$.*

Proof. Set $\nu = \mu(\mathbb{R}^2)\mu_C$. Recall the definition of k^λ in Section 3.4. It follows from the monotonicity properties of k^λ in $\lambda > 0$ described in that section that $\lim_{\lambda \to 0} \int k^\lambda \mu_1 \, d\nu_1 = \int k\mu_1 \, d\nu_1$ for $\mu_1, \nu_1 \in \mathscr{M}$. Consequently

$$\lim_{\lambda \to 0} \int k^\lambda(\mu - \nu) \, d(\mu - \nu) = \int k(\mu - \nu) \, d(\mu - \nu).$$

In other words

$$(\mu, \mu) = \lim_{\lambda \to 0} \int k^\lambda(\mu - \nu) \, d(\mu - \nu) + c(\mu(\mathbb{R}^2))^2. \tag{11}$$

Let $\lambda > 0$ and recall that $k^\lambda = g^\lambda - g^\lambda(u)$ where $\|u\| = 1$. Since $\nu(\mathbb{R}^2) = \mu(\mathbb{R}^2)$,

$$\int k^\lambda(\mu - \nu) \, d(\mu - \nu) = \int g^\lambda(\mu - \nu) \, d(\mu - \nu). \tag{12}$$

Observe that $\int g^\lambda |\mu|\, d|\mu| < \infty$ and $\int g^\lambda |\nu|\, d|\nu| < \infty$. Now

$$g^\lambda(x, y) = \int_0^\infty e^{-\lambda t} p(t, x, y)\, dt = \int_0^\infty e^{-\lambda t}\left(\int p\left(\frac{t}{2}, x, z\right) p\left(\frac{t}{2}, z, y\right) dz\right) dt$$

and hence

$$\int g^\lambda(\mu - \nu)\, d(\mu - \nu) = \int_0^\infty e^{-\lambda t}\left(\int (p^{t/2}(\mu - \nu)(z))^2\, dz\right) dt, \tag{13}$$

where $p^{t/2}(\mu - \nu) = p^{t/2}\mu - p^{t/2}\nu$. It follows from (11)–(13) that

$$(\mu, \mu) = \int_0^\infty \left(\int (p^{t/2}(\mu - \nu)(z))^2\, dz\right) dt + c(\mu(\mathbb{R}^2))^2. \tag{14}$$

Consequently $(\mu, \mu) \geq 0$. Suppose $(\mu, \mu) = 0$. By (14), $\mu(\mathbb{R}^2) = 0$, so $\nu = 0$ and hence by another application of (14)

$$p^t \mu(z) = 0 \qquad \text{a.e. on } (0, \infty) \times \mathbb{R}^2.$$

Therefore for $t > 0$

$$\int_0^t p^s \mu\, ds = 0 \qquad \text{a.e. on } \mathbb{R}^2.$$

It now follows as in the proof of Proposition 3.1.1 that $\nu = 0$. This completes the proof of the theorem.

Suppose C is a closed ball of unit radius and $\mu \in \mathscr{E}(\mathring{C})$. *Then* $\int k\mu\, d\mu \geq 0$ *with equality holding if and only if* $\mu = 0$. This result follows from Theorem 7.20 and Proposition 3.4.11, the details being left to the reader.

The norm $\|\mu\|$ for $\mu \in \mathscr{E}(C)$, the Schwarz and triangle inequalities and the notions of strong convergence, weak convergence, Cauchy sequence, weak Cauchy sequence, denseness, completeness, and weak completeness are given exactly as in Section 4 with $\mathscr{E}_D$ replaced by $\mathscr{E}(C)$. If $\mu(\mathbb{R}^2) = 0$, then $\|\mu\|^2 = \int k\mu\, d\mu$ independently of C.

Theorem 7.21. *The set of elements* $\nu \in \mathscr{E}(C)$ *such that* $k\nu$ *is continuous on* $\mathbb{R}^2$ *is dense in* $\mathscr{E}(C)$.

Proof. Choose $\mu \in \mathscr{E}^+(C)$ and $\varepsilon > 0$. Since $\iint |k(x, y)| \mu(dx)\mu(dy) < \infty$, there is a δ, $0 < \delta < \varepsilon$, such that if $A \subset \mathbb{R}^2 \times \mathbb{R}^2$ and

$$\iint_A \mu(dx)\mu(dy) \leq \delta \tag{15}$$

then

$$\iint_A |k(x, y)|\mu(dx)\mu(dy) \le \varepsilon. \tag{16}$$

By Theorem 7.6 there is a compact subset B of C such that $\mu(C\backslash B) \le \delta^{1/2}$ and kv is continuous on $\mathbb{R}^2$, where $v = \mu|_B$. Also (15) and hence (16) hold with $A = (C\backslash B) \times (C\backslash B)$. Consequently

$$\begin{aligned}\|\mu - v\|^2 &= \int_{B\backslash C} \int_{B\ C} k(x, y)\mu(dx)\mu(dy) - (R(C) - c)(\mu(B\backslash C))^2 \\ &\le \varepsilon(1 + c + |R(C)|).\end{aligned}$$

Thus the indicated set of elements is dense in $\mathscr{E}^+(C)$. It is therefore dense in $\mathscr{E}(C)$.

Let μ_n, $n \ge 1$, and μ be elements of $\mathscr{E}^+(C)$. Observe that μ_n converges completely to μ if and only if μ_n converges vaguely to μ. By the next result *if μ_n converges strongly to μ, then μ_n converges completely to μ.*

***Theorem* 7.22.** *Let μ_n, $n \ge 1$, be elements of $\mathscr{E}^+(C)$ such that $\|\mu_n\|$ is bounded in n. Then there is a strictly increasing sequence $\{n_j\}$ of positive integers and an element $\mu \in \mathscr{E}^+(C)$ such that μ_{n_j} converges completely to μ. Also μ_n converges weakly to μ if and only if μ_n converges completely to μ.*

Proof. Suppose first that μ_n converges completely to some finite measure μ. Then $\mu_n(C)$ is bounded in n. Now $\|\mu_n\|^2 = \int k\mu_n \, d\mu_n - (R(C) - c)(\mu_n(C))^2$ and hence $\int k\mu_n \, d\mu_n$ is bounded in n. Since $k(x, y)$ is bounded below and continuous in the extended sense on $C \times C$

$$\int k\mu \, d\mu \le \underline{\lim}_n \int k\mu_n \, d\mu_n < \infty$$

and hence $\mu \in \mathscr{E}^+(C)$. Choose $v \in \mathscr{E}(C)$ such that kv is continuous on $\mathbb{R}^2$. Then

$$\begin{aligned}(\mu_n, v) &= \int kv \, d\mu_n - (R(C) - c)\mu_n(C)v(C) \\ &\to \int kv \, d\mu - (R(C) - c)\mu(C)v(C) = (\mu, v).\end{aligned}$$

Thus μ_n converges weakly to μ by Theorem 7.21.

To prove the first conclusion of the theorem it is enough to show that $\mu_n(C)$ is bounded in n. Assume the contrary. Then there exist probability measures v_n, $n \ge 1$, and v all supported on C such that $v_n \to v$ completely and $\|v_n\| \to 0$. By the first paragraph of this proof, $v \in \mathscr{E}^+(C)$ and v_n converges weakly to v. Consequently by Schwarz's inequality

$$\|v\|^2 = (v, v) = \lim_n (v_n, v) = 0$$

and hence $v = 0$ by the energy principle for logarithmic potentials. But this contradicts the fact that $v(C) = 1$.

Suppose finally that μ_n converges weakly to $\mu \in \mathscr{E}^+(C)$. Now $\mu_n(C)$ is bounded in n as was just shown. Let $\{n_j\}$ be a strictly increasing sequence of positive integers such that μ_{n_j} converges completely to some element μ_0 of $\mathscr{E}^+(C)$. Then μ_{n_j} converges weakly to μ_0 and hence $\mu_0 = \mu$. Consequently μ_n converges completely to μ. This completes the proof of the theorem.

Theorem 7.23. *The set $\mathscr{E}^+(C)$ is weakly complete.*

Proof. Let $\{\mu_n\}$ be a weak Cauchy sequence of elements of $\mathscr{E}^+(C)$. By definition, $\|\mu_n\|$ is bounded in n. Thus by Theorem 7.22 (or from the proof of that theorem) $\mu_n(C)$ is bounded in n. According to that theorem, to show that μ_n converges weakly to a measure $\mu \in \mathscr{E}^+(C)$, it is enough to show that μ_n converges completely to μ. It suffices to show that if $\{m_j\}$ and $\{n_j\}$ are strictly increasing sequences of positive integers such that $\mu_{m_j} \to \gamma_1 \in \mathscr{E}^+(C)$ completely and $\mu_{n_j} \to \gamma_2 \in \mathscr{E}^+(C)$ completely, then $\gamma_1 = \gamma_2$. Choose $v \in \mathscr{E}(C)$. Then $(\mu_{m_j}, v) \to (\gamma_1, v)$ and $(\mu_{n_j}, v) \to (\gamma_2, v)$ by Theorem 7.22. Now $(\mu_{m_j} - \mu_{n_j}, v) \to 0$ since $\{\mu_n\}$ is a weak Cauchy sequence. Thus $(\gamma_1, v) = (\gamma_2, v)$ for $v \in \mathscr{E}(C)$. Consequently $\|\gamma_2 - \gamma_1\|^2 = 0$ and hence $\gamma_1 = \gamma_2$ by the energy principle for logarithmic potentials. This completes the proof of the theorem.

Let $B \in \mathscr{B}$ be a nonpolar subset of C and let $\mu \in \mathscr{E}^+(C)$. By (10)

$$\int k(\mu h_B)\, d(\mu h_B) \le \int \left(k\mu + \int W_B\, d\mu\right) d(\mu h_B) = \int k(\mu h_B)\, d\mu + \mu(\mathbb{R}^2) \int W_B\, d\mu,$$

so by another application of (10)

$$\int k(\mu h_B)\, d(\mu h_B) \le \int k\mu\, d\mu + 2\mu(\mathbb{R}^2) \int W_B\, d\mu.$$

Thus $\mu h_B \in \mathscr{E}^+(C)$. Suppose now that $\mu \in \mathscr{E}(C)$. Set $\mu h_B = \int \mu(dy) h_B(y, \cdot) = \mu^+ h_B - \mu^- h_B$. Then $\mu h_B \in \mathscr{E}(C)$. It follows from (9) that

$$(\mu - \mu h_B, \mu h_B) = \int \left(g_B \mu - \int W_B\, d\mu\right) d(\mu h_B) = \int g_B(\mu h_B)\, d\mu - \mu(\mathbb{R}^2) \int W_B\, d\mu;$$

now $g_B(\mu h_B) = 0$, so

(17) $$(\mu - \mu h_B, \mu h_B) = -\mu(\mathbb{R}^2) \int W_B\, d\mu.$$

Similarly it follows from (9) that

$$(\mu - \mu h_B, \mu) = \int g_B\mu\, d\mu - \mu(\mathbb{R}^2)\int W_B\, d\mu. \tag{18}$$

By (17) and (18)

$$\|\mu - \mu h_B\|^2 = \int g_B\mu\, d\mu \tag{19}$$

and

$$\|\mu h_B\|^2 = \|\mu\|^2 + 2\mu(\mathbb{R}^2)\int W_B\, d\mu - \int g_B\mu\, d\mu \le \|\mu\|^2 + 2\mu(\mathbb{R}^2)\int W_B\, d\mu. \tag{20}$$

Theorem 7.24. *Let $\mu \in \mathscr{E}$. Suppose either that $B_n \in \mathscr{B}$ for $n \ge 1$ and $B_n \uparrow B$ where B is bounded and nonpolar or that B_n is compact for $n \ge 1$ and $B_n \downarrow B$ where B is nonpolar. Then μh_{B_n} converges strongly to μh_B.*

Proof. Without loss of generality it can be assumed that $\mu \in \mathscr{E}^+(C)$ and that B_n is a nonpolar subset of C for each n. Suppose first that $B_n \uparrow B$. By (19)

$$\|\mu h_B - \mu h_{B_n}\|^2 = \int g_{B_n}(\mu h_B)\, d(\mu h_B).$$

It follows from Proposition 2.4.5 and the dominated convergence theorem that

$$\lim_n \int g_{B_n}(\mu h_B)\, d(\mu h_B) = \int g_B(\mu h_B)\, d(\mu h_B) = 0$$

and hence that $\lim_n \|\mu h_B - \mu h_{B_n}\|^2 = 0$ as desired.

Suppose now that B_n is compact for each n and $B_n \downarrow B$. Set $\mu_n = \mu h_{B_n}$. Then

$$\lim_n \int W_B\, d\mu_n = 0. \tag{21}$$

To verify (21) observe first that

$$\int W_B\, d\mu_n = \int W_B(z) \int \mu(dy) h_{B_n}(y, dz) = \int E_{.} W_B(X(\tau_{B_n}))\, d\mu.$$

Now W_B is bounded on compacts and upper semicontinuous and $W_B = 0$ on B^r. Since $X(\tau_B) \in B^r$ a.s. $(P_.)$ it follows from Proposition 2.3.8 that $\lim_n E_. W_B(X(\tau_{B_n})) = 0$ on $B^r \cup B^c$. Since μ has finite energy and $B \backslash B^r$ is polar, μ is concentrated on $B^r \cup B^c$. Thus by the dominated convergence theorem

$$\lim_n \int W_B\, d\mu_n = \lim_n \int E_. W_B(X(\tau_{B_n}))\, d\mu = 0$$

as desired.

Choose $m > n$. Then $\mu_n h_{B_m} = \mu_m$. Since $W_{B_m} \le W_B$ and $\mu_n(\mathbb{R}^2) = \mu(\mathbb{R}^2)$, it follows from (19) and the equality in (20) that

$$(22) \qquad \|\mu_n - \mu_m\|^2 + \|\mu_m\|^2 \le \|\mu_n\|^2 + 2\mu(\mathbb{R}^2)\int W_B\, d\mu_n.$$

It follows easily from (21) and (22) that

$$(23) \qquad \lim_n \|\mu_n\|^2 \text{ exists and is finite.}$$

Equations (21)–(23) together imply that $\{\mu_n\}$ is a Cauchy sequence. Thus to prove that μ_n converges strongly to μh_B it suffices to prove that μ_n converges weakly to μh_B. By Theorem 7.22 it suffices to prove that μ_n converges completely to μh_B.

Let φ be a bounded continuous function on $\mathbb{R}^2$. Then

$$\int \varphi\, d\mu_n = \int \varphi\, d\left(\int \mu(dy) h_{B_n}(y, \cdot)\right) = \int E_{\cdot}\varphi(X(\tau_{B_n}))\, d\mu$$

and similarly

$$\int \varphi\, d(\mu h_B) = \int E_{\cdot}\varphi(X(\tau_B))\, d\mu.$$

It now follows from Proposition 2.3.8 (as in the proof of (21)) that $\lim_n \int \varphi\, d\mu_n = \int \varphi\, d\mu$. Thus μ_n converges completely to μh_B, which completes the proof of the theorem.

For some applications of Theorem 7.24 let B_n, $n \ge 1$, and B be as in the statement of that theorem and let μ be the equilibrium measure of a compact set containing the B_n's and B. Then $\mu h_{B_n} = \mu_{B_n}$ and $\mu h_B = \mu_B$, so μ_{B_n} converges strongly to μ_B. Consequently μ_{B_n} converges completely to μ_B (this fact also follows easily from Theorem 7.18 and Proposition 2.3.8); also $\|\mu_{B_n}\|^2 \to \|\mu_B\|^2$ or equivalently $R(B_n) \to R(B)$. The last result was obtained in Theorem 3.4.15.

Let A and B be disjoint nonpolar compact sets. A signed measure $\nu \in \mathscr{E}$ is called the *condenser measure* corresponding to A, B if ν is supported on $A \cup B$, $\nu(\mathbb{R}^2) = 0$, $k\nu = \alpha$ on A^r and $k\nu = \alpha + 1$ on B^r for some $\alpha \in \mathbb{R}$. The condenser measure, if it exists, is concentrated on $(A \cup B)^r = A^r \cup B^r$ and is uniquely determined. The uniqueness follows easily from the energy principle. The next result is called the *condenser theorem for logarithmic potentials.*

Theorem 7.25. *Let A and B be disjoint nonpolar compact sets. Then A, B has the condenser measure $\nu = \mu_{B,A^c} - \mu_{B,A^c} h_A$. Moreover ν^+ is supported on B, ν^- is supported on A, $k|\nu| < \infty$ on $\mathbb{R}^2$, $k\nu = \alpha$ on A^r, $k\nu = \alpha + 1$ on B^r, and $\alpha \le k\nu \le \alpha + 1$ on $\mathbb{R}^2$, where $\alpha = -\int W_A\, d\nu^+ \in [-1, 0]$.*

Proof. By the fundamental identity for logarithmic potentials

$$k\mu_{B,A^c} \le G_{A^c}\mu_{B,A^c} + H_{A^c}k\mu_{B,A^c} \le 1 + H_{A^c}k\mu_{B,A^c} \qquad \text{on } A^c.$$

Thus $k\mu_{B,A^c}$ is bounded above on A^c and hence finite on $\mathbb{R}^2$. Also μ_{B,A^c} has finite energy, so $\mu_{B,A^c} \in \mathscr{E}^+$. Consequently $\mu_{B,A^c}h_A \in \mathscr{E}^+$. Set $v = \mu_{B,A^c} - \mu_{B,A^c}h_A$. Then $v \in \mathscr{E}$, v is supported on $A \cup B$, $v(\mathbb{R}^2) = 0$, v^+ is supported on B and v^- is supported on A. By Theorem 5.13

$$g_A\mu_{B,A^c} = P_{\cdot}(\tau_B < \tau_A) \qquad \text{on } \mathbb{R}^2.$$

It now follows from (9) that

$$k\mu_{B,A^c} = P_{\cdot}(\tau_B < \tau_A) + k(\mu_{B,A^c}h_A) - \int W_A \, d(\mu_{B,A^c}).$$

Thus $k(\mu_{B,A^c}h_A)$ is finite on $\mathbb{R}^2$. Therefore $k|v|$ is finite on $\mathbb{R}^2$ and

$$kv = P_{\cdot}(\tau_B < \tau_A) - \int W_A \, dv^+ \qquad \text{on } \mathbb{R}^2.$$

Set $\alpha = -\int W_A \, dv^+$. Then $kv = \alpha$ on A^r, $kv = \alpha + 1$ on B^r and $\alpha \le kv \le \alpha + 1$ on $\mathbb{R}^2$. Since $v(\mathbb{R}^2) = 0$ it follows from (3.4.2) that kv has limit zero at infinity. Consequently $\alpha \le 0 \le \alpha + 1$ and hence $\alpha \in [-1,0]$. This completes the proof of the theorem.

It follows from Proposition 3.4.7 that the constant α in Theorem 7.25 equals zero if B is disjoint from the unbounded component of A^c.

The application to electrostatics discussed in Section 6 has an obvious analog for logarithmic potentials. The details are left to the reader.

Let $0 < r < q$. It follows from Proposition 3.4.9 that the condenser measure for S_q, S_r is given by

$$v = \frac{\pi}{\log q/r}(\sigma_r - \sigma_q).$$

Its logarithmic potential kv is given by $kv(x) = 1$ for $\|x\| \le r$, $kv(x) = 0$ for $\|x\| \ge q$, and

$$kv(x) = \frac{\log q/\|x\|}{\log q/r} \qquad \text{for } r < \|x\| < q.$$

The proof of the next result is a simplification of the proof of Theorem 5.15, the details being left to the reader.

Theorem 7.26. *The space $\mathscr{E}(C)$ is not complete.*

A characterization of balayage in terms of energy will now be obtained.

Theorem 7.27. *Let $B \in \mathscr{B}$ be bounded and nonpolar, let $\mu \in \mathscr{E}^+$ and set $\mu' = \mu h_B$. Then*

$$\min[\|\nu - \mu\| : \nu \in \mathscr{E}^+(B^r) \text{ and } \nu(\mathbb{R}^2) = \mu(\mathbb{R}^2)] = \|\mu' - \mu\|$$

and the minimum occurs uniquely at $\nu = \mu'$. Also

$$\inf[\|\nu - \mu\| : \nu \in \mathscr{E}^+(B) \text{ and } \nu(\mathbb{R}^2) = \mu(\mathbb{R}^2)] = \|\mu' - \mu\|;$$

if $\nu_n \in \mathscr{E}^+(B)$, $\nu_n(\mathbb{R}^2) = \mu(\mathbb{R}^2)$, and $\|\nu_n - \mu\| \to \|\mu' - \mu\|$, then ν_n converges strongly to μ'.

Proof. Now $\mu' \in \mathscr{E}^+(B^r)$, $\mu'(\mathbb{R}^2) = \mu(\mathbb{R}^2)$ and $k\mu' = k\mu + \int W_B \, d\mu$ on B^r by (8), so $(\mu' - \mu, \nu - \mu') = 0$ for $\nu \in \mathscr{E}^+(B^r)$ with $\nu(\mathbb{R}^2) = \mu(\mathbb{R}^2)$. Consequently

(24) $\|\nu - \mu\|^2 = \|\nu - \mu'\|^2 + \|\mu' - \mu\|^2$ for $\nu \in \mathscr{E}^+(B^r)$ with $\nu(\mathbb{R}^2) = \mu(\mathbb{R}^2)$.

The first conclusion of the theorem now follows from the energy principle for logarithmic potentials.

Observe that $\mathscr{E}^+(B) \subset \mathscr{E}^+(B^r)$. Let B_n be compact nonpolar subsets of B such that $B_n \uparrow B$ and set $\nu_n = \mu h_{B_n}$. Then $\nu_n \in \mathscr{E}^+(B)$, $\nu_n(\mathbb{R}^2) = \mu(\mathbb{R}^2)$, and ν_n converges strongly to μ' by Theorem 7.24. Thus $\|\nu_n - \mu\| \to \|\mu' - \mu\|$. The second conclusion of the theorem now follows from (24).

A characterization due to Frostman [1] of the equilibrium measure and Robin constant in terms of energy will now be obtained.

Theorem 7.28. *Let $B \in \mathscr{B}$ be a bounded nonpolar set. Then*

$$\min[I(\nu) : \nu \in \mathscr{E}^+(B^r) \text{ and } \nu(\mathbb{R}^2) = 1] = R(B)$$

and the minimum occurs uniquely at $\nu = \mu_B$. Also

$$\inf[I(\nu) : \nu \in \mathscr{E}^+(B) \text{ and } \nu(\mathbb{R}^2) = 1] = R(B);$$

if $\nu_n \in \mathscr{E}^+(B)$, $\nu_n(\mathbb{R}^2) = 1$ and $I(\nu_n) \to R(B)$, then ν_n converges strongly to μ_B.

Proof. This result can be obtained directly. Alternatively let μ be the equilibrium measure of a compact set C containing B. Then $\mu \in \mathscr{E}^+$ and $\mu h_B = \mu_B$. Choose $\nu \in \mathscr{E}^+(B^r)$ such that $\nu(\mathbb{R}^2) = 1 = \mu(\mathbb{R}^2)$. Then $\int k\nu \, d\mu = \int k\mu \, d\nu = R(C)$ and hence

$$\|\nu - \mu\|^2 = \int k(\nu - \mu) \, d(\nu - \mu) = \int (k\nu - R(C)) \, d(\nu - \mu) = I(\nu) - R(C).$$

In particular

$$\|\mu_B - \mu\|^2 = I(\mu_B) - R(C) = R(B) - R(C).$$

The theorem now follows from Theorem 7.27.

Let $r > 0$ *and let* μ *be a probability distribution supported on* B_r. *Then*

$$\iint \log(1/\|y - x\|)\mu(dx)\mu(dy) \geq \log(1/r)$$

with equality holding if and only if $\mu = \sigma_r$. This result follows immediately from Theorem 7.28 and Proposition 3.4.11.

Theorem 7.29. *Let* $B \in \mathscr{B}$ *be relatively compact and nonpolar. Then*

$$\min[\sup_x kv(x) : v \in \mathscr{M}(B^r) \text{ and } v(\mathbb{R}^2) = 1] = R(B)$$

and the minimum occurs uniquely at $v = \mu_B$. *Also*

$$\inf[\sup_x kv(x) : v \in \mathscr{M}(B) \text{ and } v(\mathbb{R}^2) = 1] = R(B);$$

if $v_n \in \mathscr{M}(B)$, $v_n(\mathbb{R}^2) = 1$, and $\sup_x v_n(x) \to R(B)$, *then* v_n *converges strongly to* μ_B.

Proof. Now $\mu_B \in \mathscr{M}(B^r)$ and $\mu_B(\mathbb{R}^2) = 1$. Since $W_B \geq 0$ on $\mathbb{R}^2$ and $W_B = 0$ on the nonempty set B^r, $\sup_x k\mu_B(x) = \sup_x (R(B) - W_B(x)) = R(B)$. Let v be a probability measure in $\mathscr{M}(B^r)$. Now $I(v) \leq \sup_x kv(x)$; so it follows from Theorem 7.28 that $\sup_x kv(x) \geq R(B)$, with equality holding if and only if $v = \mu_B$. This completes the proof of the first conclusion.

Let $\{B_n\}$ be an increasing sequence of compact nonpolar subsets of $\mathbb{R}^2$ such that $B_n \uparrow B$. Then $R(B_n) \downarrow R(B)$ by Theorem 3.4.15. Now $\mu_{B_n} \in \mathscr{M}(B)$, $\mu_{B_n}(\mathbb{R}^2) = 1$ and

$$\sup_x k\mu_{B_n}(x) = R(B_n) \downarrow R(B).$$

Let $\{v_n\}$ be a sequence of probability measures in $\mathscr{M}(B)$. Then $M_n = \sup_x k\mu_n(x) \geq I(\mu_n)$. Thus by Theorem 7.28, $\underline{\lim}_n M_n \geq R(B)$ and if $\lim_n M_n = R(B)$, then v_n converges strongly to μ_B.

The next result, due to La Vallée Poussin [2], is a direct application of Theorem 7.29. The details are left to the reader.

Theorem 7.30. *Let* $B \in \mathscr{B}$ *be a relatively compact nonpolar set such that* $R(B) > 0$. *Then*

$$\max[v(\mathbb{R}^2) : v \in \mathscr{M}(B^r) \text{ and } kv \leq 1] = R(B)^{-1}$$

and the maximum occurs uniquely at $v = \mu_B$. *Also*

$$\sup[v(\mathbb{R}^2) : v \in \mathscr{M}(B) \text{ and } kv \leq 1] = R(B)^{-1};$$

if $v_n \in \mathscr{M}(B)$, $kv_n \leq 1$ *and* $v_n(\mathbb{R}^2) \to R(B)^{-1}$, *then* v_n *converges strongly to* $R(B)^{-1}\mu_B$.

The following characterization of the equilibrium measure is also due to La Vallée Poussin [2]. It is interesting that in this result the measure ν is not required a priori to be concentrated on B^r or even on $\bar{B}$.

Theorem 7.31. *Let $B \in \mathscr{B}$ be relatively compact and nonpolar. Then μ_B is the unique probability measure $\nu \in \mathscr{M}$ such that $k\nu = R(B)$ q.e. on B and $k\nu \leq R(B)$ on $\mathbb{R}^2$.*

Proof. Now μ_B satisfies the indicated properties and $\int k\mu_B \, d\mu_B = R(B)$. Suppose that ν also satisfies the indicated properties. Then $\int k\nu \, d\nu \leq R(B)$. Now $\mu_B \in \mathscr{M}(B^r)$ and $k\mu_B = R(B)$ on B^r, so $k\mu_B = k\nu$ q.e. on B. Thus by the complete maximum principle for logarithmic potentials

$$R(B) - W_B = k\mu_B \leq k\nu \leq R(B) \qquad \text{on } \mathbb{R}^2.$$

In particular $k\nu = R(B)$ on B^r and hence $\int k\nu \, d\mu_B = R(B)$. Therefore

$$\|\nu - \mu_B\|^2 = \int k\nu \, d\nu - 2\int k\nu \, d\mu_B + \int k\mu_B \, d\mu_B = \int k\nu \, d\nu - R(B) \leq 0,$$

so $\nu = \mu_B$ by the energy principle for logarithmic potentials. (Alternatively, this result follows from the energy principle for logarithmic potentials and the remarks following the proof of Theorem 7.24 without appeal to the complete maximum principle for logarithmic potentials.)

A map T from a set B into $\mathbb{R}^2$ is called a *contraction mapping* of B if $\|Ty - Tx\| \leq \|y - x\|$ for $x, y \in B$. Such a mapping is clearly continuous. Let TB denote the image of B under a contraction mapping. If B is compact, then TB is compact and if $B \in \mathscr{B}$, then $TB \in \mathscr{B}$.

Theorem 7.32. *Let $B \in \mathscr{B}$ be bounded and let T be a contraction mapping of B. Then $R(TB) \geq R(B)$.*

Proof. Let ν be a probability measure which is concentrated on TB. Then there is a probability measure μ which is concentrated on B and such that $\mu(T^{-1}A) = \nu(A)$ for $A \subset TB$. The measure μ may not be uniquely determined if T is not one-to-one on B. One approach to proving existence of the measure μ is to find a measurable transformation U from TB to B such that $T(Uy) = y$ for $y \in TB$. It is easier to proceed by a different route.

Suppose first that B and hence TB are compact. Let ν_n, $n \geq 1$, be probability measures on TB each of which is supported on a finite subset of TB and which converge completely to ν as $n \to \infty$. For each n there is clearly a probability measure μ_n on B such that $\mu_n(T^{-1}A) = \nu_n(A)$ for $A \subset TB$. There is a strictly increasing sequence $\{n_j\}$ of positive integers and a probability measure μ on B such that μ_{n_j} converges completely to μ. Let φ be a bounded continuous function on TB. Then $\varphi \circ T$ is a bounded continuous function on B. Also $\int \varphi \circ T\, d\mu_{n_j} = \int \varphi\, d\nu_{n_j}$ for each j, so $\int \varphi \circ T\, d\mu = \int \varphi\, d\nu$. This implies that

$$\mu(T^{-1}A) = \int I_A \circ T\, d\mu = \int I_A\, d\nu = \nu(A), \qquad A \subset TB.$$

Consider now the general case. Let B_m be compact subsets of B such that $B_m \uparrow B$ and let ν_m denote the restriction of ν to $TB_m \backslash TB_{m-1}$ (where $B_0 = \varnothing$). Then there is a measure μ_m supported on B_m such that $\mu_m(T^{-1}A) = \nu_m(A)$ for $A \subset TB$. The measure $\mu = \sum_m \mu_m$ satisfies the desired property.

Observe that

$$k(x, y) \le k(Tx, Ty), \qquad x, y \in B.$$

Thus

$$I(\mu) = \iint k(x, y)\mu(dx)\mu(dy) \le \iint k(Tx, Ty)\mu(dx)\mu(dy) = \iint k(x, y)\nu(dx)\nu(dy)$$
$$= I(\nu).$$

Therefore $R(TB) \ge R(B)$ by Theorem 7.28.

The next result provides a useful approximation of the logarithmic potential kernel in terms of Green potential kernels.

Theorem 7.33. *Let D_n be nonempty bounded open subsets of $\mathbb{R}^2$ such that $D_n \uparrow \mathbb{R}^2$. Then*

$$k(x, y) = G_{D_n}(x, y) + \int H_{D_n}(0, dz)k(z, 0) + L_n(x, y),$$

where $L_n(x, y) \to 0$ uniformly for x, y in compacts.

Proof. By (1)

$$k(x, y) = G_{D_n}(x, y) + \int H_{D_n}(x, dz)k(z, y), \qquad x, y \in D_n. \tag{25}$$

It follows from (3.4.2) that

$$\lim_n \int H_{D_n}(x, dz)(k(z, y) - k(z, 0)) = 0 \quad \text{uniformly for } x, y \text{ in compacts.} \tag{26}$$

By (6) and Proposition 3.4.7

$$\int H_{D_n}(x, dz)k(z, 0) = \int H_{D_n}(0, dz)k(z, x), \qquad x \in D_n. \tag{27}$$

By (3.4.2)

$$(28)\quad \lim_n \int H_{D_n}(0, dz)(k(z,x) - k(z,0)) = 0 \qquad \text{uniformly for } x \text{ in compacts.}$$

The desired result follows from (25)–(28).

Next an approximation of the Robin constant in terms of capacity with respect to Green potentials will be obtained. Recall that C is a compact nonpolar set.

Theorem 7.34. *For every $\varepsilon > 0$ there is a bounded open set D containing C and an $M \in \mathbb{R}$ such that if $B \in \mathscr{B}$ is a subset of C, then*

$$(29)\qquad C_D(B)^{-1} + M - \varepsilon \le R(B) \le C_D(B)^{-1} + M + \varepsilon.$$

Proof. Choose $\varepsilon > 0$. By Theorem 7.33 there is a bounded open set D containing C and an $M \in \mathbb{R}$ such that

$$(30)\qquad G_D(x,y) + M - \varepsilon \le k(x,y) \le G_D(x,y) + M + \varepsilon, \qquad x, y \in C.$$

Let $B \in \mathscr{B}$ be a subset of C. If B is polar, then $C_D(B)^{-1} = R(B) = \infty$ so (29) holds. Suppose B is nonpolar. It follows from Theorem 5.16 that $I_D(\mu_B) \ge C_D(B)^{-1}$. Thus by (30)

$$C_D(B)^{-1} + M - \varepsilon \le I_D(\mu_B) + M - \varepsilon \le I(\mu_B) = R(B),$$

so the first inequality in (29) holds. Set $\mu = C_D(B)^{-1}\mu_{B,D}$. Then $I_D(\mu) = C_D(B)^{-1}$. Also μ is a probability measure so by Theorem 7.28, $R(B) \le I(\mu)$. Thus by (30)

$$R(B) \le I(\mu) \le I_D(\mu) + M + \varepsilon = C_D(B)^{-1} + M + \varepsilon$$

and hence (29) holds. This completes the proof of the theorem.

The planar version of *Wiener's test* (Wiener [3] in slightly different form) will now be obtained.

Theorem 7.35. *Let $B \in \mathscr{B}$, $x \in \mathbb{R}^2$, and $\lambda \in (0,1)$. Set $B_n = \{y \in B : \lambda^{n+1} < \|y - x\| \le \lambda^n\}$ for $n \ge 1$. Then $x \in B^r$ if and only if $\sum_n nR(B_n)^{-1} = \infty$.*

Proof. Observe that B_n is contained in the closed ball $B_\lambda(x)$ and hence that $R(B_n) \ge \pi^{-1}\log(1/\lambda) > 0$ by Theorem 3.4.12. Let D be a bounded open set containing $B_\lambda(x)$. It follows by imitating the proof of Theorem 3.3.2 that $x \in B^r$ if and only if $\sum_n nC_D(B_n) = \infty$. The desired result now follows from Theorem 7.34.

Wiener's test can be used to construct an example of a closed set B such that B is the closure of $\mathring{B}$ and $0 \in B \backslash B^r$. Choose $\lambda \in (0, 1)$, choose $b_n \in \mathbb{R}^2$ for $n \geq 1$ such that $\|b_n\| = (\lambda^{n+1} + \lambda^n)/2$, and choose $r_n > 0$ such that $r_n < (\lambda^n - \lambda^{n+1})/2$. Set $B = \bigcup_n B_{r_n}(b_n)$. Then $\{y \in B : \lambda^{n+1} < \|y\| \leq \lambda^n\} = B_{r_n}(b_n)$, which has Robin constant $\pi^{-1}\log(1/r_n)$. It now follows from Theorem 7.35 that the origin is regular for B if and only if $\sum_n n/\log(1/r_n) = \infty$. Thus if $r_n = c\lambda^{n^2}$ for some $c > 0$, the origin is regular for B; but if $r_n = c\lambda^{n^3}$ for some $c > 0$, the origin is irregular for B.

Wiener's test will now be used to give an alternative proof of Theorem 2.7.2 Let $B \in \mathscr{B}$ be a nonempty set such that no component of B reduces to a single point and let $x \in B$. Then $x \in B^r$. For let $y \neq x$ be in the same component of B as x and let T be the contraction mapping of $\mathbb{R}^2$ which maps z to $x + (\|z - x\|, 0)$. Then TB contains the line segment from x to $x + (\|y - x\|, 0)$. Choose $\lambda \in (0, 1)$ and set $B_n = \{y \in B : \lambda^{n+1} < \|y - x\| \leq \lambda^n\}$ for $n \geq 1$. Then for n sufficiently large TB_n contains the line segment from $x + (\lambda^{n+1}, 0)$ to $x + (\lambda^n, 0)$. By Theorems 3.4.14 and 7.32

$$\begin{aligned} R(B_n) \leq R(TB_n) &= c + \pi^{-1}\log(1/(\lambda^n - \lambda^{n+1})) \\ &= c - \pi^{-1}\log(1 - \lambda) + n\pi^{-1}\log(1/\lambda), \end{aligned}$$

where c is the Robin constant for a line segment of unit length. Since the uniform distribution on such a line segment has finite energy, the line segment is nonpolar and hence c is finite. Consequently $\sum_n nR(B_n)^{-1} = \infty$ and hence x is regular for B by Wiener's test. (For a slightly different proof, which doesn't appeal to Theorem 7.32, see page 258 of Itô and McKean [2].)

Let A be a bounded (Borel) set. Set

$$\underline{R}(A) = \inf[R(B) : B \text{ is a compact subset of } A]$$

and

$$\overline{R}(A) = \sup[R(U) : U \text{ is a bounded open set containing } A].$$

Recall from Section 3.4 that the (logarithmic) capacity $C(B)$ of a bounded set $B \in \mathscr{B}$ is defined by $C(B) = \exp(-R(B))$. The *inner capacity* $\underline{C}(A)$ and *outer capacity* $\overline{C}(A)$ of A are defined by

$$\underline{C}(A) = \sup[C(B) : B \text{ is a compact subset of } A]$$

and

$$\overline{C}(A) = \inf[C(U) : U \text{ is a bounded open set containing } A].$$

Observe that $\underline{C}(A) = \exp(-\underline{R}(A))$ and $\overline{C}(A) = \exp(-\overline{R}(A))$. The set A is said to be *capacitable* if $\underline{C}(A) = \overline{C}(A)$ or equivalently if $\underline{R}(A) = \overline{R}(A)$. A bounded open set is clearly capacitable. The capacitability theorem of

Choquet will now be used to show that every bounded (Borel) set is capacitable, a result due to Choquet [2].

Theorem 7.36. *Let A be bounded. Then A is capacitable.*

Proof. Since $\underline{R}(A) \geq \overline{R}(A)$ by Theorem 3.4.14, it suffices to show that $\underline{R}(A) \leq \overline{R}(A)$. It can be assumed that $\bar{A}$ is contained in the interior of the compact set C. Choose $\varepsilon > 0$ and let D and M be as in Theorem 7.34. Then $\underline{C}_D(A) = \overline{C}_D(A)$ by Choquet's theorem (if $A \in \mathscr{B}$ this result follows from Theorem 5.26). Let B be a compact subset of A and let U be an open subset of C containing A. Then

$$R(B) \leq C_D(B)^{-1} + M + \varepsilon$$

and

$$R(U) \geq C_D(U)^{-1} + M - \varepsilon.$$

Consequently

$$\underline{R}(A) \leq \underline{C}_D(A)^{-1} + M + \varepsilon$$

and

$$\overline{R}(A) \geq \overline{C}_D(A)^{-1} + M - \varepsilon = \underline{C}_D(A)^{-1} + M - \varepsilon.$$

Thus $\underline{R}(A) \leq \overline{R}(A) + 2\varepsilon$. Since ε can be made arbitrarily small $\underline{R}(A) \leq \overline{R}(A)$ as desired. This completes the proof of the theorem.

References

AHLFORS, L. V.

[1] *Complex Analysis*, 2nd ed., McGraw Hill, New York, 1966 (1st ed., 1953).

BACHELIER, L.

[1] Théorie de la spéculation, *Ann. Sci. École. Norm. Sup.* **17**, 21–86 (1900) (English translation in *The Random Character of Stock Market Prices* (revised ed.), edited by Paul H. Cootner, M. I. T. Press, Cambridge, Massachusetts, 1964.

BEURLING, A., and J. DENY

[1] Espaces de Dirichlet I. Le cas élémentaire, *Acta Math.* **99**, 203–224 (1958).

[2] Dirichlet spaces, *Proc. Nat. Acad. Sci. USA* **45**, 208–215 (1959).

BLUMENTHAL, R. M.

[1] An extended Markov property, *Trans. Amer. Math. Soc.* **85**, 52–72 (1957).

BLUMENTHAL, R. M., and R. K. GETOOR

[1] *Markov Processes and Potential Theory*, Academic Press, New York, 1968.

BÔCHER, M.

[1] Singular points of functions which satisfy partial differential equations of the elliptic type, *Bull. Amer. Math. Soc.* **9**, 455–465 (1903).

BRELOT, M.

[1] Familles de Perron et problème de Dirichlet, *Acta Litt. Sci. Szeged* **9**, 133–153 (1939).

[2] Points irréguliers et transformations continues en théorie du potentiel, *J. Math. Pure Appl.* **19**, 319–337 (1940).

[3] Sur la théorie autonome des fonctions sous-harmoniques, *Bull. Sci. Math.* **65**, 72–98 (1941).

[4] Sur les ensembles effilés *Bull. Sci. Math.* **68**, 12–36 (1944).

[5] Sur le rôle du point à l'infini dans la théorie des functions harmoniques, *Ann. Sci. École Norm. Sup.* **61**, 301–332 (1944).

[6] Minorantes sous-harmoniques, extrémales et capacités, *J. Math. Pures Appl.* **24**, 1–32 (1945).

[7] La théorie moderne du potential, *Ann. Inst. Fourier* **4**, 113–140 (1952).

[8] *Eléments de la théorie classique du potential*, 4th ed., Centre de Documentation Universitaire, Paris, 1969 (1st ed., 1959).
[9] Les étapes et les aspects multiples de la théorie du potentiel, *Enseignement Math.* **18**, 1–36 (1972).

CARTAN, H.
[1] Sur less fondements de la théorie du potentiel, *Bull. Soc. Math. France* **69**, 71–96 (1941).
[2] Théorie du potentiel newtonien: énergie, capacité, suites de potentials, *Bull. Soc. Math. France* **73**, 74–106 (1945).
[3] Théorie générale du balayage en potential newtonien, *Ann. Univ. Grenoble Math. Phys.* **22**, 221–280 (1946).

CHACON, R. V.
[1] Potential processes, *Trans. Amer. Math. Soc.* **226**, 39–58 (1977).

CHOQUET, G.
[1] Theory of capacities, *Ann. Inst. Fourier* **5**, 131–295 (1953–1954).
[2] Capacibilité en potentiel logarithmique, *Acad. Roy. Bull. Cl. Sci.* **44**, 321–326 (1958).
[3] Sur les G_δ de capacité nulle, *Ann. Inst. Fourier* **9**, 75–83 (1959).
[4] *Lectures on Analysis* (3 vol.), vol. 1, Benjamin, New York, 1969.

CHUNG, K. L.
[1] *A Course in Probability Theory*, 2nd ed., Academic Press, New York 1974 (1st ed., 1968).
[2] Probabilistic approach in potential theory to the equilibrium problem, *Ann. Inst. Fourier* **23**, (3) 313–322 (1973).

CHUNG, K. L., and R. K. GETOOR
[1] The condenser problem, *Ann. Probability* **5**, 82–86 (1977).

COURANT, R., K. FRIEDRICHS, and H. LEWY
[1] Über die partiellen Differenzengleichungen der mathematischen Physik, *Math. Ann.* **100**, 32–74 (1928) (English translation in *IBM J. Res. Develop.* **11**, 215–234 (1967)).

DENY, J.
[1] Sur les infinis d'un potential, *C. R. Acad. Sci. Paris* **224**, 524–525 (1947).
[2] Les potentiels d'énergie finie, *Acta Math.* **82**, 107–183 (1950).

DOOB, J. L.
[1] Semimartingales and subharmonic functions. *Trans. Amer. Math. Soc.* **77**, 86–121 (1954).

DYNKIN, E. B.
[1] *Markov Processes* (2 vol.), Springer-Verlag, Berlin, 1965 (English translation from 1959 Russian edition).

DYNKIN, E. B., and A. A. YUSHEKEVICH
[1] Strong Markov property, *Theory Probability and Math. Statist.* **1**, 134–139 (1956) (English Translation from *Teor. Verojatnost. Mat. Statist.* **1**, 149–155 (1956)).

EINSTEIN, A.
[1] *Investigations on the Theory of the Brownian Movement* edited with notes by R. Fürth, translated by A. D. Cooper, Dover, New York, 1956.

EVANS, G. C.
[1] On potentials of positive mass I, *Trans. Amer. Math. Soc.* **37**, 226–253 (1935).
[2] Potentials and positively infinite singularities of harmonic functions, *Monatsh. Math.* **43**, 419–424 (1936).

FELLER, W.
[1] *An Introduction to Probability Theorey and its Applications* (2 vol.), vol. 2, 2nd ed., Wiley, New York, 1971.

FROSTMAN, O.
[1] Potentiel d'équilibre et capacité des ensembles avec quelques applications à la théorie des functions, *Medd. Lunds. Univ. Math. Sem.* **3**, 1–118 (1935).

HELMS, L. L.
[1] *Introduction to Potential Theory*, Wiley (Interscience), New York, 1969.
HUNT, G. A.
[1] Some theorems concerning Brownian motion, *Trans. Amer. Math. Soc.* **81**, 294–319 (1956).
[2] Markov processess and potentials, *Illinois J. Math.* **1**, 44–93, 316–369 (1957); **2**, 151–213 (1958).
ITÔ, K., and H. P. McKEAN, JR.
[1] Potentials and the random walk, *Illinois J. Math.* **4**, 119–132 (1960).
[2] *Diffusion Processes and Their Sample Paths*, Springer-Verlag, Berlin, 1965.
KAC, M.
[1] On some connections between probability theory and differential and integral equations, *Proc. Second Berkeley Symp. Math. Statist. Probability*, pp. 189–215, University of California Press, Berkeley, California, 1951.
KAKUTANI, S.
[1] Two-dimensional Brownian motion and harmonic functions, *Proc. Imp. Acad. Tokyo* **20**, 706–714 (1944).
KELLOGG, O. D.
[1] *Foundations of Potential Theory*, Springer, Berlin, 1929 (reprinted by Dover, New York, 1955).
KEMENY, J. G., J. L. SNELL, and A. W. KNAPP
[1] *Denumerable Markov Chains*, 2nd ed., Springer-Verlag, New York, 1976 (1st ed., 1966).
LAMPERTI, J.
[1] Wiener's test and Markov chains, *J. Math. Anal. Appl.* **6**, 58–66 (1963).
LANDKOF, N. S.
[1] *Foundations of Modern Potential Theory*, Springer-Verlag, Berlin, 1972 (English translation from 1966 Russian edition).
LA VALLÉE POUSSIN, CH.-J. DE
[1] L'extension de la méthode du balayage de Poincaré et problème de Dirichlet, *Ann. Inst. H. Poincaré* **2**, 169–232 (1932).
[2] *Le Potentiel logarithmique: Balayage et Representation conforme*, Libraire Univ., Louvain, and Gauthier-Villars, Paris, 1949.
MARIA, A. J.
[1] The potential of a positive mass and the weight function of Wiener, *Proc. Nat. Acad, Sci. USA* **20**, 485–489 (1934).
MARTIN, R. S.
[1] Minimal positive harmonic functions, *Trans. Amer. Math. Soc.* **49**, 137–172 (1941).
MEYER, P. A.
[1] *Probability and Potentials*, Ginn (Blaisdell), Waltham, Massachusetts, 1966.
MONNA, A. F.
[1] *Dirichlet's Principle: A Mathematical Comedy of Errors and its Influence on the Development of Analysis*. Oostoek, Scheltema, and Holkema, Utrecht, 1975.
NELSON, E.
[1] *Dynamical Theories of Brownian Motion*, Princeton Univ. Press, Princeton, New Jersey, 1967.
PERRON, O.
[1] Eine neue Behandlung der ersten Randwertaufgabe für $\Delta u = 0$, *Math. Z.* **18**, 42–54 (1923).
PORT, S. C.
[1] Hitting times for transient stable processes, *Pacific J. Math.* **21**, 161–165 (1967).

PORT, S. C., and C. J. STONE

[1] Infinitely divisible processes and their potential theory, *Ann. Inst. Fourier* **21**, (2) 157–275 and (4) 179–265 (1971).

RAO, M.

[1] *Brownian Motion and Classical Potential Theorey*, Lecture Notes Series, No. 47, Aarhus Universitet, 1977.

REVUZ, D.

[1] *Markov Chains*, North–Holland Publ., Amsterdam, 1975.

RIESZ, F.

[1] Sur les fonctions subharmoniques et leur rapport à la théorie du potentiel, *Acta Math.* **48**, 329–343 (1926); **54**, 321–360 (1930).

RIESZ, F., and B. SZ.-NAGY

[1] *Functional Analysis*, Ungar, New York, 1955 (English translation from 2nd ed. of *Leçons d'analyse fonctionelle*, 1953).

RUDIN, W.

[1] *Real and Complex Analysis*, 2nd ed., McGraw-Hill, New York, 1974 (1st ed., 1966).

SPITZER, F. L.

[1] *Principles of Random Walk*, 2nd ed., Springer-Verlag, Berlin and New York, 1976 (1st ed., 1964).

[2] Electrostatic capacity, heat flow, and Brownian motion, *Z. Wahrscheinlichkeitstheor. Verw. Gebiete* **3**, 110–121 (1964).

TSUJI, M.

[1] *Potential Theory in Modern Function Theory*, Maruzen, Tokyo, 1959.

VASILESCO, F.

[1] Sur la continuité du potentiel à travers des masses et la démonstration d'un lemme de Kellogg, *C. R. Acad. Sci. Paris* **200**, 1173–1174 (1935).

WERMER, J.

[1] *Potential Theory*, Springer-Verlag, Berlin and New York, 1974.

WIENER, N.

[1] Differential space, *J. Math. Phys.* **2**, 131–174 (1923).

[2] Certain notions in potential theory, *J. Math. Phys.* **3**, 24–51 (1924).

[3] The Dirichlet problem, *J. Math. Phys.* **3**, 127–146 (1924).

[4] Note on a paper of O. Perron, *J. Math. Phys.* **4**, 21–32 (1925).

Index

Probability and Mathematical Statistics

A Series of Monographs and Textbooks

Thomas Ferguson. Mathematical Statistics: A Decision Theoretic Approach. 1967

Howard Tucker. A Graduate Course in Probability. 1967

K. R. Parthasarathy. Probability Measures on Metric Spaces. 1967

P. Révész. The Laws of Large Numbers. 1968

H. P. McKean, Jr. Stochastic Integrals. 1969

B. V. Gnedenko, Yu. K. Belyayev, and A. D. Solovyev. Mathematical Methods of Reliability Theory. 1969

Demetrios A. Kappos. Probability Algebras and Stochastic Spaces. 1969

Ivan N. Pesin. Classical and Modern Integration Theories. 1970

S. Vajda. Probabilistic Programming. 1972

Sheldon M. Ross. Introduction to Probability Models. 1972

Robert B. Ash. Real Analysis and Probability. 1972

V. V. Fedorov. Theory of Optimal Experiments. 1972

K. V. Mardia. Statistics of Directional Data. 1972

H. Dym and H. P. McKean. Fourier Series and Integrals. 1972

Tatsuo Kawata. Fourier Analysis in Probability Theory. 1972

Fritz Oberhettinger. Fourier Transforms of Distributions and Their Inverses: A Collection of Tables. 1973

Paul Erdös and Joel Spencer. Probabilistic Methods in Combinatorics. 1973

K. Sarkadi and I. Vincze. Mathematical Methods of Statistical Quality Control. 1973

Michael R. Anderberg. Cluster Analysis for Applications. 1973

W. Hengartner and R. Theodorescu. Concentration Functions. 1973

Kai Lai Chung. A Course in Probability Theory, Second Edition. 1974

L. H. Koopmans. The Spectral Analysis of Time Series. 1974

L. E. Maistrov. Probability Theory: A Historical Sketch. 1974

William F. Stout. Almost Sure Convergence. 1974

E. J. McShane. Stochastic Calculus and Stochastic Models. 1974

Robert B. Ash and Melvin F. Gardner. Topics in Stochastic Processes. 1975

Avner Friedman, Stochastic Differential Equations and Applications, Volume 1, 1975; Volume 2. 1975

Roger Cuppens. Decomposition of Multivariate Probabilities. 1975

Eugene Lukacs. Stochastic Convergence, Second Edition. 1975

H. Dym and H. P. McKean. Gaussian Processes, Function Theory, and the Inverse Spectral Problem. 1976

N. C. Giri. Multivariate Statistical Inference. 1977

Lloyd Fisher and John McDonald. Fixed Effects Analysis of Variance. 1978

Sidney C. Port and Charles J. Stone. Brownian Motion and Classical Potential Theory. 1978

A
B
C 8
D 9
E 0
F 1
G 2
H 3
I 4
J 5